Intelligent Systems Reference Library

Volume 261

The aim of this series is to publish a Reference Library, including novel advances and developments in all aspects of Intelligent Systems in an easily accessible and well structured form. The series includes reference works, handbooks, compendia, textbooks, well-structured monographs, dictionaries, and encyclopedias. It contains well integrated knowledge and current information in the field of Intelligent Systems. The series covers the theory, applications, and design methods of Intelligent Systems. Virtually all disciplines such as engineering, computer science, avionics, business, e-commerce, environment, healthcare, physics and life science are included. The list of topics spans all the areas of modern intelligent systems such as: Ambient intelligence, Computational intelligence, Social intelligence, Computational neuroscience, Artificial life, Virtual society, Cognitive systems, DNA and immunity-based systems, e-Learning and teaching, Human-centred computing and Machine ethics, Intelligent control, Intelligent data analysis, Knowledge-based paradigms, Knowledge management, Intelligent agents, Intelligent decision making, Intelligent network security, Interactive entertainment, Learning paradigms, Recommender systems, Robotics and Mechatronics including human-machine teaming, Self-organizing and adaptive systems, Soft computing including Neural systems, Fuzzy systems, Evolutionary computing and the Fusion of these paradigms, Perception and Vision, Web intelligence and Multimedia.

Indexed by SCOPUS, DBLP, zbMATH, SCImago.

All books published in the series are submitted for consideration in Web of Science.

Peter Ilic · Imogen Casebourne · Rupert Wegerif
Editors

Artificial Intelligence in Education: The Intersection of Technology and Pedagogy

Springer

Editors
Peter Ilic
School of Computer Science
and Engineering
Center for Language Research
University of Aizu
Aizuwakamatsu, Fukushima, Japan

Imogen Casebourne
Digital Education Futures Initiative (DEFI)
University of Cambridge
Cambridge, UK

Rupert Wegerif
Faculty of Education
University of Cambridge
Cambridge, UK

ISSN 1868-4394 ISSN 1868-4408 (electronic)
Intelligent Systems Reference Library
ISBN 978-3-031-71231-9 ISBN 978-3-031-71232-6 (eBook)
https://doi.org/10.1007/978-3-031-71232-6

This Springer imprint is published by the registered company Springer Nature Switzerland AG
The registered company address is: Gewerbestrasse 11, 6330 Cham, Switzerland

Contents

Chapter 1
Introduction: Constructive Dialogue Between Technology and Pedagogy

Peter Ilic, Imogen Casebourne, and Rupert Wegerif

Abstract The rapidly evolving field of artificial intelligence (AI) in education necessitates a convergence of technological innovation and pedagogical wisdom. This book represents a crucial step towards fostering a constructive dialogue between these often-siloed perspectives, essential for developing educational approaches that truly enhance human potential. This introductory chapter sets the stage for this interdisciplinary discourse, highlighting the urgent need for a synergy between the voice of technology and the voice of pedagogy in the AI era. What distinguishes this volume is its unique assembly of contributors who embody both educational expertise and technological acumen. Most authors bring dual perspectives as both educators and engineers or AI experts, providing a natural bridge between these disciplines. This integration of viewpoints offers readers a holistic understanding of AI's role in education, moving beyond simplistic applications to explore transformative possibilities. The chapter provides an overview of pressing issues in AI and education, followed by summaries of the book's contributions. These span theoretical frameworks, design principles, and case studies of AI implementation in various educational contexts. From exploring ecological learning spaces and human-centred AI design to examining AI's potential in language learning and mathematics education, each chapter represents the intersection of multiple fields. Collectively, they offer a comprehensive outlook on AI's capacity to enhance learning outcomes, address community needs, and reshape educational paradigms. By uniting the perspectives of those who understand both the intricacies of pedagogy and the capabilities of AI, this book aims to chart a path towards educational practices that are both technologically advanced

P. Ilic (✉)
Center for Language Research, University of Aizu, Aizu, Japan
e-mail: pilic@u-aizu.ac.jp

I. Casebourne
DEFI, Hughes Hall, University of Cambridge, Cambridge, UK
e-mail: ic407@hughes.cam.ac.uk

R. Wegerif
Faculty of Education, University of Cambridge, Cambridge, UK
e-mail: rw583@cam.ac.uk

P. Ilic et al. (eds.), *Artificial Intelligence in Education: The Intersection of Technology and Pedagogy*, Intelligent Systems Reference Library 261,
https://doi.org/10.1007/978-3-031-71232-6_1

and pedagogically sound. It serves as a foundation for future research and practice in the thoughtful integration of AI in education.

Keywords Artificial intelligence in education · Generative AI · Human-centred design · Adaptive learning · Formative assessment · Educational technology integration · Educational ecosystems

Artificial intelligence (AI) in education is not new, but in the past 18 months, there has been global consternation about the potential of generative AI to unsettle and upend traditional approaches to learning and teaching. Between OpenAI's release of ChatGPT 3.5 in November 2022 and the publication of this book, multiple AI tutors have been released, and in 2024, OpenAI announced Open Edu, a version tailored specifically for education, Google reported on their Illuminate experiment, which aims to enable people to learn in dialogue with an academic paper and released a paper about the development of LearnLM (Jurenka et al., 2024). In addition to OpenAI's ChatGPT, Microsoft's Copilot, Elon Musk's Grok, Anthropic's Claude, and Meta's Llama have all been considered as potential educational tools, while a Large language model (LLM) trained specifically for education was announced by Merlyn Mind (Mind, 2024), and there have been AI offerings from existing educational platforms, such as the Khan Academy's Khanmigo https://www.khanmigo.ai/.

The ability for individuals to create custom generative pre-trained transformers (GPTs) has resulted in academics creating interactive dialogic GPTs based on their books or papers (such as Digital Don or TeachSmart). At the same time, both Apple and Microsoft have released hardware that is optimised for AI processing. The remarkable speed with which generative AI has made inroads into education has caused alarm—with concerns rightly raised about the potential of generative AI to be biased (as a result of its training data), unreliable, and incomplete. Meanwhile, legislators are starting to release guidelines specifically aimed at AI and education, such as the UNESCO Beijing Consensus.

Alongside alarm about issues of academic integrity, there has been excitement about the possibility of personal tutors and new forms of assessment. However, realising even a small part of AI's potential to transform education in positive ways will require constructive dialogue between the very different perspectives of engineering and of pedagogy. This book offers an introduction to that necessary dialogue: it is a tentative step towards forging an interdisciplinary collaboration of the kind essential for redefining education in the light of new technology, especially generative AI.

We think that this new technology does indeed offer the potential to transform learning for the better, and it is also likely to change the world of work, which may, in turn, impact aspects of our current curriculums. By bringing together experienced educators who are also developers and experts in technologies such as generative Pre-trained Transformer AI, this book presents a multidisciplinary perspective on leveraging technology to meet the diverse needs of education.

One of the central challenges we have set out to address is the historical lack of communication and knowledge-sharing between educators, developers, and engineers with a deep understanding of AI. Too often, these two communities have operated in silos, with educators focused on pedagogical approaches and learning theories while engineers concentrated on developing technological solutions, with insufficient common understanding and dialogue between the groups. This volume aims to bridge that divide, fostering dialogues that deepen the understanding of the respective challenges and opportunities faced by each discipline. The authors, from across the globe, bring a background in and an understanding of both technological and pedagogical issues to bear on questions of AI in education. Together, they challenge the notion of an unrelenting push for increasingly technology-driven education, instead advocating for a measured approach that leverages the strengths of both human educators and technological advancements.

Despite the surge in technologically-centred educational literature, a disconnect persists between the creation of technological competencies and the pedagogical processes underpinning their development. This book is about research that strikes a balance between educational methodologies and engineering innovations, united by the common objective of optimising technology's service to education. In addition to providing a series of case studies that illustrate the diverse ways in which AI is currently being integrated into (and thereby potentially altering) educational practices, the authors offer design frameworks for further development that draw on a deep understanding of the technical aspects and possibilities of AI.

We hope that this book can serve as a guide to help navigate the dynamic confluence of pedagogy and technology, serving as a call to embrace a future where the amalgamation of engineering innovation and pedagogical acumen is not merely an aspiration but a realised ethos, reshaping the nature of education.

1.1 The Role of Technology in Educational Theory

While books and articles about educational technology don't always engage with educational theory, when they do, they often cite theories like constructivism that have no input from theories of technology (Wegerif & Major, 2023). This is unfortunate because education as we practice it today around the world is crucially intertwined with technology. The very first schools, schools with practices and a curriculum that are quite recognisable as similar to schools today, were developed in ancient Sumeria to train operators called scribes for the then-new technology of cuneiform writing with a stylus on clay tablets. The current mass schooling system is similarly focused on literacy and numeracy, in other words, training operators of communications technology. Of course, there is more to education than that, but the literacy element, learning how to use established communications technology, has a big impact that is often unacknowledged in theories of education.

Educational theory has traditionally drawn from a range of disciplines, including psychology, philosophy, history, and sociology, to answer the key questions faced by education: 'What should we teach, how should we teach it, and, more ultimately, why should we teach it?'. Historically, however, one critical aspect has been largely overlooked—the fundamental role of technology in shaping educational content and pedagogy. Though deeply embedded within the educational fabric, technology's profound impact and agency in transforming the foundations of teaching and learning have been underestimated.

To develop a theory of education truly reflective of our technological era, we must recognise technology not merely as an ancillary tool but as a central, transformative force within educational discourse itself. This reconceptualisation aligns with philosopher Gilbert Simondon's notion of "transindividuation" (Simondon & Du, 1958)—the process by which collectives form shared identities and agencies intertwined with the technologies they employ. Simondon illustrates how education, as an inherently social process, harnesses technology to draw individual learners into broader communities.

Simondon influenced Bruno Latour (2007), Gilles Deleuze (1987), Bernard Steigler (1998), and other theorists to argue that technology has agency and a kind of voice that needs to be included in any design thinking about society and education. Claiming that technology has agency and a voice does not mean that it operates independently of human engineers. It is rather about recognising that the design process is not one where human engineers think up a design and then implement it with technology. In practice, new designs are almost invariably a product of a dialogue between human and machine. There is a collaborative engagement whereby engineers enter a dialogue with technology, interlacing their foresight with the machine's intrinsic potential to co-create an ever-evolving technical reality. This dialogue with technology is what has been lacking in education. Providing a space for more such dialogue is our motive in creating this book that brings the perspective of education and of engineering together.

Our experience with mass literacy over the last two hundred years or so suggests that education serves a role in promoting technologies that change humans and human societies from the inside. Literate people are different from non-literates in clear and measurable ways (Dehaene, 2010). The advent of effective generative AI might herald an equally profound transformation in education and in what it means to be human. In this context, then, it is essential that we are aware of what we are doing when we design technology for education and that we take into account all the consequences that designs have. In education, learning experiences emerge from the dynamic interplay between learners and technological systems. Acknowledging the central role that educational technology plays as a vital component of our collective learning process, we position technology not as a supplementary aid but as an essential partner in pedagogy itself. It follows that the perspective of the technology itself, mediated by engineers who know and understand that perspective, is a vital one for the development of education.

1.2 Chapters Included in the Book

This book is divided into two sections. In the opening section, 'Frameworks and design principles,' educators introduce pedagogical frameworks and design approaches aimed at integrating AI into learning and assessment. The contributions from various authors focus on the theoretical, social, and practical aspects of employing AI in education and assessment.

A recurrent theme across this section is the emphasis on human-centred AI. The authors advocate for designs that prioritise human values, needs, and ethical considerations, ensuring that AI serves as a supportive tool rather than a replacement for human interaction and decision-making. This theme is evident in discussions of Human-in-the-Loop (HITL) practices and the emphasis on ethically aligned design principles.

The integration of AI into education is viewed not simply as a technological shift but as a complex process that requires input from multiple stakeholders in order to navigate ethical, pedagogical, and practical issues. This is particularly emphasised in calls for cybernetic frameworks and design-based research methodologies. Beyond simply supplementing or scaling existing practices, the authors explore ways in which AI might fundamentally transform educational paradigms. They argue that the implementation of AI in learning and in assessment involves rethinking how educational content is designed, delivered, assessed, and managed.

1.3 Sect. 1: Frameworks and Design Principles

In the opening chapter, Ilic and Sato-Ilic consider the implications of integrating AI into ecological learning spaces. They start by presenting the historical context of generative AI, tracing its evolution from early theoretical foundations to contemporary advanced models. This lays the groundwork for understanding how AI technologies have progressed and become integral to modern educational environments.

They introduce the concept of ecological learning spaces, drawing on Bronfenbrenner's ecological systems theory (Bronfenbrenner, 1979), of learning within a complex system of relationships influenced by multiple environmental levels. This framework highlights the importance of a holistic consideration of the social, cultural, and technological dimensions that shape learning experiences. Ilic and Sato-Ilic argue that such spaces, characterised by adaptability, personalisation, and interconnectivity, are essential for fostering holistic educational development.

They go on to consider how AI might enhance ecological learning spaces, focusing on dynamic adaptability, emphasising the ways in which AI technologies might create responsive learning environments that adjust to the needs and actions of learners.

This adaptability is crucial in a rapidly changing world, where the skills and knowledge required can shift quickly. AI contributes to adaptability by enabling learning environments to evolve in response to student needs.

Ilic and Sato-Ilic also explore the principle of interconnectedness in ecological learning spaces. They discuss how AI can link various elements, such as technology, the physical environment, social interactions, and learning materials, creating a comprehensive and integrated educational experience. This interconnected approach ensures that learning is not viewed in isolation but as part of a broader system, facilitating collaboration and idea exchange across different disciplines and geographical locations.

Guralnick's chapter opens with a reflection on his early educational experiences at a non-traditional school, which emphasised project-based learning and individual exploration over conventional assessments. This formative experience shaped his perspective on the transformative potential of AI in education. Guralnick recounts a memorable project from his childhood where students created scrapbooks on U.S. state governments, fostering a deep, practical understanding through hands-on activities and collaborative efforts rather than rote memorisation. Such approaches are of increasing interest to educational institutions wishing to foster and teach future skills, and Guralnick goes on to consider how AI might foster active learning.

Guralnick focuses on the ways in which AI, in combination with other emerging technologies such as Virtual Reality (VR) and Augmented Reality (AR), can enhance workplace training by underpinning highly immersive, *learning-by-doing* simulations where learning is experiential, intrinsic, and to an extent tacit. He discusses the theoretical importance of *learning-by-doing* in vocational training and considers the ways in which AI can scaffold these experiences. He reviews early attempts to use AI, including intelligent tutoring systems that delivered adaptive feedback and skills simulations (including those he designed and developed), which provided realistic practice environments. Drawing on this background, the chapter outlines a framework for designing immersive simulations underpinned by AI while critiquing the current use of AI to reproduce textbooks or classroom experiences. He advocates for using AI to create simulations and interactive scenarios that mimic real-world challenges, allowing learners to engage in problem-solving and critical-thinking activities. By providing immediate feedback, he argues that AI systems have the potential to provide compelling simulations where learners can explore and practice skills.

Guralnick goes on to describe current tools and potential and planned AI-based learning experiences that create full immersion through VR, with AI coaches guiding exploration based on individual interests and needs. Guralnick also considers the benefits and challenges of augmented workplace performance support, where AI can act as an ever-present coaching aide, providing real-time guidance and facilitating collaboration. Throughout, his emphasis is on designing human-centred, holistic experiences that deeply engage learners and performers and provide an intrinsic appetite for learning—a departure from simply replicating classroom lecture models through technology.

Watson and Shi consider the integration of generative AI into educational systems more broadly. They begin by defining generative AI and explaining its underlying mechanisms, distinguishing it from other forms of artificial intelligence. The authors explore how these AI models generate content, highlighting their potential for educational applications beyond mere content creation. They emphasise the importance of understanding generative AI's capabilities, particularly its semantic transduction ability, which allows it to transform and translate text while maintaining contextual meaning. They point to the need to develop generative AI literacy among both educators and learners. They argue that this should involve not only technical knowledge but also ethical awareness and the ability to critically evaluate AI outputs. The goal should be to enable users to harness generative AI's capabilities for innovative and effective teaching and learning while mitigating potential risks.

To achieve this, they propose a human-centred approach to integrating generative AI in education. They advocate for direct interaction between educators, learners, and AI tools to develop generative AI literacy, enabling them to employ these tools responsibly and creatively. They go on to outline three broad categories of generative AI integration in education:

- **Autonomous Generative AI Tools (GAITs):** These tools, such as ChatGPT, enable users to interact directly with AI models. They offer high levels of personalisation and flexibility but require significant user proficiency and understanding.
- **Integrated Generative AI Applications (IGAAs):** Examples include Microsoft Copilot, which enhances existing software applications by embedding AI functionalities. This approach improves productivity and accessibility but offers limited direct interaction with AI models.
- **Generative AI-based Automatic Decision-Making Systems (GADMS):** These systems, like intelligent tutoring systems, use AI for data analytics and decision-making to personalise learning experiences. While efficient and scalable, they lack human-centricity and flexibility.

To address the complexities of integrating generative AI into education, the authors propose the adoption of a Design-Based Research (DBR) (Reimann, 2010) methodology. DBR involves iterative cycles of design, implementation, and evaluation, allowing for contextual and participatory development of AI integration strategies. It is an approach that can enable the emergence of new educational practices that are systemically coherent and sustainable. The authors argue that DBR can serve as a catalyst for systemic change by bridging the gap between research and practice, legitimising the adoption of specific pedagogical approaches, and fostering a more adaptive and effective educational system.

Yuan, Hoel and Powell also call for DBR. They start by providing a historical overview of AI in education, tracing its roots from early teaching machines to advanced adaptive learning systems. They point out that these technologies have previously promised to revolutionise education by enabling personalised learning at scale but argue that existing Adaptive Intelligent Tutoring Systems (ITS) often uncritically reinforce traditional content delivery models, thereby limiting pedagogical innovation and flexibility.

They go on to propose an analytic framework based on the cybernetic principle of variety to evaluate and design AI-driven educational change. This framework aims to address issues of curriculum, pedagogy, and assessment at the classroom, institutional, and governmental level. They discuss the Ultraversity project at Anglia Ruskin University (Powell et al., 2008) as an illustration of how cybernetic analysis and systems-thinking can inform the design of innovative technology-supported educational programmes that involve authentic assessment and address complex educational challenges.

The authors emphasise the need for collaborative efforts among educators, policymakers, and technologists to navigate the uncertainties and maximise the benefits of AI in education. By drawing on cybernetics and complexity theory, they advance a framework intended to help navigate operational challenges, data privacy and use, issues of academic integrity, and other ethical and social implications. This, they argue, will be essential to avoid the risk of simply scaling up poor pedagogical practices.

In the final chapter of this section, Von Davier and Burstein offer a framework for AI in assessment that encompasses test development and measurement as well as security. They point to the importance of a commitment to human-centred AI and of prioritising human values, needs, goals, and decision-making power in system design. Their approach emphasises embedding social and moral norms into AI assessment systems to ensure they serve and respect human users. They introduce three critical drivers for AI-powered assessments:

- **Validity Argument**: constructing validity arguments that account for the use of AI in assessments. Traditional validity arguments, as discussed by Kane (Kane, 1992), involve a chain of inferences that support the interpretation and use of test scores. In the context of AI, these arguments must address the capabilities and limitations of AI technologies in producing valid, fair, and reliable test scores.
- **Human-in-the-Loop (HITL)** (Ye et al., 2023) **AI Practices**: HITL AI integrates human intelligence with AI systems, enhancing their performance and ensuring that human oversight is maintained. This practice is crucial for tasks such as automatic item generation, where AI-generated content is reviewed by human experts, and for maintaining fairness and transparency in the assessment process. HITL AI practices ensure that AI systems can learn from human feedback, thereby continuously improving their accuracy and reliability.
- **Responsible AI (RAI) Standards**: addressing ethical considerations, including fairness, explainability, privacy, and accountability. RAI standards ensure that AI systems are developed and used in ways that are transparent, fair, and respectful of user privacy.

They discuss the implementation of these principles in the Duolingo English Test (DET), a digital-first, computer-adaptive English language assessment. The DET's assessment ecosystem is composed of integrated frameworks for test design, measurement, and security, with HITL AI practices applied throughout. For example, human experts review AI-generated test content and label data for building automated scoring systems, ensuring that the AI's decisions are valid and fair. Key considerations for human-centred AI assessment include fairness, transparency, explainability, privacy and security, customisation, and collaboration. They argue that by combining human expertise with AI capabilities, it is possible to develop assessment tools that are both innovative and ethical, ultimately enhancing the educational experience for all stakeholders.

1.4 Sect. 2: Implementation, Experimentation, and Case Studies

The chapters in this section report on findings from research into the implementation of AI in education and assessment and studies of generative AI for educational purposes.

As AI offers the potential of simultaneous translation, it raises significant questions for the future of teaching and learning foreign languages, and two of the chapters in this section address this, reporting on approaches and techniques for teaching English using AI. They suggest ways in which AI might move language teaching beyond a focus on the mechanics to more sophisticated language use.

Other chapters in this section consider AI and the teaching of maths, with Maclaren focusing on learning through games and Bimpeh on formative assessment. In both chapters, the potential of AI to offer personalised and adaptive learning experiences is a significant theme. Bimpeh's study of AI-powered formative assessment and McLaren's decades of research on AI learning games both illustrate ways in which AI might tailor educational content to individual learners' needs, enhancing engagement and learning outcomes.

Casebourne and Wegerif note that the release of ChatGPT to the general public in November 2022 (Ye et al., 2023) led to headlines suggesting that it could pass a variety of exams (Ali et al., 2023; Varanasi, 2023), in addition to reports of children using it to do their homework (Casbourne, 2023). They ask what the ability of generative AI (LLM) to write essays and pass exams means for current models of education, concluding that it suggests a need to move to a more dialogic understanding of education.

This chapter offers a background to the history and approaches to AI, distinguishing symbolic AI from neural networks. It discusses debates around AI, education, and intelligence, including the Turing test and its focus on outputs versus arguments that process and intention also matter. Wegerif and Casebourne present education as an ongoing process of dialogue and reflection, grounding this in the initial insights offered by Socrates and tracing it through to modern concepts of dialogic space. They propose that generative AI offers the potential for a new form of collective intelligence that could support student participation in cultural dialogues but caution that issues of bias must be addressed and that current LLM do not fully replicate all aspects of debate with peers.

The authors go on to share details of experiments conducted with ChatGPT and other LLMs at the Digital Education Futures Initiative (DEFI) that aimed to test its capabilities. These involved using AI to create worked examples for statistics teachers, to role-play philosophers conducting dialogues, to generate counterarguments, and to participate in literature reviews. They reflect on the promise and limitations revealed through these trials. They conclude by arguing that LLM AI holds significant potential benefits for dialogic education if challenges relating to bias and representation can be successfully navigated.

Bimpeh argues that the incorporation of AI into formative assessment has the potential to overcome some of the limitations of traditional testing methods, which often fail to capture the intricacies of individual learning paths. In contrast, AI-powered systems have the potential to offer personalised and timely feedback, potentially enhancing the overall learning experience and outcomes.

He discusses the design, implementation, and evaluation of an AI-powered formative assessment system focused on secondary school maths. The platform used a closed-loop system to collect data from learners to assess their progress, recommend learning activities, and provide personalised feedback. The platform integrated learning objectives, cognitive mapping, and detailed competency measures, metacognition, and time management to tailor the assessment.

The primary objectives of his study were to:

- Assess whether AI-powered formative assessment could differentiate students of varying ability levels.
- Evaluate the consistency and reliability of the feedback provided by the system.
- Establish the concurrent validity of the system with existing mathematics competency tests.
- Provide empirical evidence to support the system's diagnostic accuracy.

He explains that the methodological framework drew on both quantitative and qualitative analyses. Data was collected from a diverse sample of secondary school students in England, ensuring broad applicability. Students' item response data from the diagnostic tests, along with their interactions with the platform, was collected and analysed. Additionally, feedback from teachers and students on each of the key mathematical concepts was gathered. Structural equation modelling and Rasch-based analysis were used to validate the constructs and reliability of the assessment system.

Preliminary results indicate that the system both engages students and differentiates students of varying ability levels, providing consistent and dependable information about each student's knowledge and skills. It also had strong concurrent validity with existing Key Stage 3 and 4 mathematics competency tests. Finally, empirical evidence supported the accuracy of inferences based on the diagnostic/formative test. Thus, the study is broadly positive for the future of adaptive formative assessment. A recent futures study (Abu Sitta et al.) concluded that we may well expect to see more of these types of assessment, and it will be important that these draw on evidence for effectiveness.

McLaren and colleagues also consider the potential of AI to support intrinsic learning in combination with other approaches to aid immersion but in the context of mathematics for elementary and middle school children. They report on a series of design experiments undertaken with an AI learning game, Decimal Point, which they have used as a research platform to explore a series of questions related to educational gaming over the course of six studies.

After setting the scene by introducing the potential and challenges of designing games to support learning, McLaren introduces Decimal Point, an educational game consisting of 24 mini-games and 48 problems. Each of the mini-games is essentially a self-contained intelligent tutor built with the Cognitive Tutor Authoring Tools (CTAT) platform. However, unlike more straightforward ITS, Decimal Point incorporates a variety of gaming features, so for example, while the original tutoring system, Decimal Tutor, presents a straightforward maths problem and offers a series of options, the same problem in Decimal Point is presented as a challenge where students must capture ghosts. In the first of the six studies McLaren reports, researchers compared these two approaches and found that students learned more by playing the game (with a relatively high effect size) and appeared to enjoy the learning process more. As with the previous two chapters, we are presented here with an argument about the benefits of using AI in support of experiential and challenge-based learning, as opposed to simply using it to underpin an adaptive approach to the presentation of learning that is more similar to traditional textbooks or e-learning.

Two other studies involved comparing variations of low- and high-agency versions of the game (yielding little difference in results) and adding a third high-agency option (students dropped out of the high-agency versions more quickly, indicating higher learning efficiency). The research team also compared versions of the game designed to focus more on either learning or enjoyment (but again found little difference). A further interesting experiment compared a version of the game that offered hints with one that did not. Interestingly, a learning curve analysis indicated that *No-Hint* students initially did worse than *Hint* students but eventually caught up with their *Hint* counterparts, doing better on a delayed post-test. The authors speculate that *No-Hint* students may have worked harder to construct their knowledge and thus learned more. This interesting result aligns with work in the vocational field, for example, by Billett, which finds apprentices learn actively when immersed in experiences where they may be given relatively few hints and little guidance and are forced to pay active attention to construct learning models. Similarly, the researchers found that students in a self-explanation group learned more than students in a prompted self-explanation group. Experiments with LLM found that these were more suited to giving conceptual feedback than procedural feedback for self-explanations.

A final pair of studies focused on affective experience and the potential of game-play to promote mindfulness (mindful flowlike experiences is something that Guralnick suggests vocational simulations should aspire to), although it is possible that the mindfulness reporting mechanism employed in these studies may have taken children out of the mindfulness state. An interesting finding across the series of studies was that girls performed better than expected in the games—they achieved higher learning gains than boys in comparison with using the intelligent tutoring system. The researchers point to further research directions to build on their results.

Across Africa, students face significant educational challenges, including limited access to essential resources such as computers, internet connectivity, reliable electricity, and a shortage of qualified teachers. Boateng's chapter discusses a range of problems such as lack of access to computers/internet, lack of regulatory support, heterogeneity in educational systems across Africa, lack of digitised materials, bias in AI systems which have not been developed in an African context, lack of support for African languages and potential inaccuracies with easily accessible generative AI models.

It also highlights opportunities, such as the proliferation of smartphones to deliver AI in Education (AIED) tools, free-to-use generative AI that can act as a personalised learning assistant, crowdsourcing ideas through initiatives like AfricAIED, and open-source textbook generation using LLM. LLMs such as BERT and GPT-4 have demonstrated the potential to address some of the general challenges that Boateng outlines. However, these tools have predominantly been deployed and evaluated in Western settings, with limited or no consideration of the unique needs and challenges faced by students in Africa.

Boateng describes several linked projects which aim to address the issues by developing and deploying AIED tools tailored to the African context. These are (1) SuaCode, an AI-powered smartphone app enabling Africans to learn coding, (2) AutoGrad for automated grading of coding assignments, (3) a code plagiarism detection tool, (4) Kwame—a bilingual AI teaching assistant for coding courses, (5) Kwame for Science providing instant answers to students' questions, and (6) Brilla AI—an AI contestant for a national science and math competition. Despite the numerous challenges he outlines, Boateng is optimistic about AI's potential to transform education in Africa and enable equitable, high-quality learning opportunities for millions across the continent.

Bohne and colleagues consider ways in which AI may offer solutions to a modern challenge of lifelong learning, namely that workplace technologies are changing at such a rapid pace that the labour market is failing to keep up, and it has become very hard for organisation to hire a sufficient number of skilled technicians. In line with Guralnick's call for AI to be used to create practical and immersive project-based, 'learning-by-doing' experiences, they suggest that AR and VR technologies, in combination with AI, can provide additional training and address these types of skills gap.

They introduce a number of AR and VR terms and technologies before going on to outline the results of their study in which computer vision, in combination with a Hololens headset and a 3D model, was used to create a training experience in which machine operators were asked to fix a 3D printer. This project was developed using the Unity game engine and the Microsoft mixed reality suite. The author compared performance using desktop VR and head-mounted VR (using the Hololens headset) and found little difference in the time to completion and number of errors between the two approaches.

They acknowledge that the existing technology may still be relatively difficult to use for learners, although further advances may address this in time. Additionally, there is now the potential to use LLM to further enrich such simulations, leading to a number of avenues for future research.

Roy considers the innovative use of generative AI in English as a Foreign Language (EFL) teaching, specifically on applications of prompt engineering. This chapter involves project-based language learning within a business communication course. The primary objective of this work was to explore the ways in which AI might cultivate an entrepreneurial mindset in students, encouraging them to develop critical business thinking and problem-solving skills, shifting the focus from merely teaching the mechanics of language to guiding students in constructing logical arguments and producing relevant documents. Roy presents several examples that indicate how graduate students in technical disciplines were able to use Google search and ChatGPT for complex assignments, such as employment profile creation, customer profile creation, job applications with video resumes, customer journey mapping, employer profile evaluation, and industry 4.0 presentations.

Roy explains how common prompt types, such as question prompts, completion prompts, story prompts, dialogue prompts, and creative prompts, were applied in the context of business communication. He also discusses more advanced prompting techniques, such as zero-shot prompting, few-shot prompting, and chain-of-thought prompting, which allow for the gradual development of complexity in assignments. This provides a structured approach to implementing prompt engineering in EFL classrooms. He also discusses strategies for tailoring prompt engineering for EFL students, considering factors such as language proficiency levels, vocabulary and grammar support, cultural backgrounds, collaboration, and individualised support. He calls for future research to focus on areas such as adaptive AI-driven prompt generation, multimodal prompts, cross-cultural communication, prompt efficacy assessment, technology integration, and ethical considerations, among others, to further enhance teaching and learning in this domain.

Also focusing on AI and language learning, Ohashi introduces a range of studies that evaluate the strengths and weaknesses of Machine Translation (MT) and GAITs in language education. Ohashi shares insights from her own exploratory work with Alm, which was one of the first wide-scale studies to gauge foreign language educators' response to the technology. The study reveals teachers' strong awareness of ChatGPT, their budding interest in integrating it into their teaching practices, and their concerns about potential negative impacts on students' academic integrity and critical thinking abilities. Ohashi emphasises the importance of teacher training and support in adapting to the challenges and opportunities posed by ChatGPT and provides examples of practical tasks that teachers and learners can try with the technology, such as creating reading passages, generating vocabulary lists, exploring pragmatics, and co-creating fictional stories.

Her chapter also addresses issues of plagiarism and academic integrity and the need for educators to rethink approaches to teaching and evaluating students in light of the widespread availability of generative AI. She argues that while generative AI offers valuable opportunities within language education, adapting will take time and pose a range of challenges that will necessitate comprehensive training for both educators and students. Ohashi calls for educators to unite in supporting one another and their students through this process and for researchers to empirically explore pathways towards pedagogies that deter misuse of AI and help users make the most of its affordances. Her chapter concludes by offering practical recommendations and activities designed to integrate AI tools effectively into language teaching. These aim to enhance language development while maintaining critical language skills.

1.5 Conclusions

The integration of AI into education represents a significant shift in learning and teaching. This chapter has shown that realising AI's potential in education requires understanding both technological capabilities and pedagogical principles. This book's unique contribution is its integration of these perspectives, featuring authors with expertise in both education and engineering or AI development. The following chapters present a range of viewpoints, from theoretical frameworks reconceptualising learning spaces with AI to practical case studies of AI applications in various educational contexts. These contributions highlight the importance of human-centred design, ethical considerations, and using AI to enhance, not replace, human interaction in education.

This interdisciplinary approach has significant implications for education's future. As AI advances, maintaining this integrated perspective will be crucial to ensure technological innovations align with sound educational principles and that pedagogical practices evolve to effectively use AI. This book aims to stimulate further research and collaboration rather than provide definitive answers. It encourages educators, technologists, policymakers, and researchers to engage with key questions about education's future in an AI-enhanced world. This approach can help develop educational practices that maximise human potential, leveraging AI while preserving the essential value of human wisdom and interaction in learning. We hope this volume will be a foundation for ongoing exploration at the intersection of AI and education, with learners at the centre. The integration of AI into education is just beginning, and interdisciplinary efforts like those in this book can shape a learning future that is both technologically advanced and fundamentally human-centred.

References

Academy, K. *Khanmigo*. Available from: https://www.khanmigo.ai/.

Ali, R., et al. (2023). *Performance of ChatGPT and GPT-4 on neurosurgery written board examinations*. medRxiv, pp. 2023.03.25.23287743.

Bronfenbrenner, U. (1979). *The ecology of human development: Experiments by nature and design*. Harvard University Press.

Casbourne, I. (2023). *Should we trust ChatGPT?*. DEFI: DEFI.

Dehaene, S. (2010). *Reading in the brain: The new science of how we read*. Penguin.

Deleuze, G. (1987). *A thousand plateaus: Capitalism and schizophrenia*. U of Minnesota Press.

Jurenka, I., et al. (2024). *Towards Responsible development of generative AI for education: An evaluation-driven approach*. Google.

Kane, M. T. (1992). The assessment of professional competence. *Evaluation & the Health Professions, 15*(2), 163–182.

Latour, B. (2007). *Reassembling the social: An introduction to actor-network-theory*. Oup Oxford.
Mind, M. (2024). *Meet Merlyn Origin, the all-in-one AI assistant for educators*. [cited 2024 May 30]; Available from: https://www.merlyn.org/blog/meet-merlyn-origin-the-all-in-one-ai-assistant-for-educat.
Powell, S., Tindal, I., & Millwood, R. (2008). Personalized learning and the Ultraversity experience. *Interactive Learning Environments, 16*(1), 63–81.
Reimann, P. (2010). Design-based research. *Methodological choice and design: Scholarship, policy and practice in social and educational research* (pp. 37–50). Springer.
Simondon, G. (1958). *DU Mode d'existence des objets techniques*. Aubier.
Stiegler, B. (1998). *Technics and time, 1: The fault of Epimetheus* (Vol. 1). Stanford University Press.
Varanasi, L. (2023). *GPT-4 can ace the bar, but it only has a decent chance of passing the CFA exams. Here's a list of difficult exams the ChatGPT and GPT-4 have passed*. Business Insider.
Wegerif, R., & Major, L. (2023). *The theory of educational technology: Towards a dialogic foundation for design*. Taylor & Francis.
Ye, L., Zhang, H., & Wang, Y. (2023). *Human-in-the-Loop AI practices: Towards transparent and EthOpenAI*. Introducing ChatGPT Edu.

Chapter 2
AI-Enhanced Ecological Learning Spaces

Peter Ilic and Mika Sato-Ilic

Abstract This paper investigates the transformative role of Artificial Intelligence (AI) in creating and enhancing ecological learning spaces. It explores how AI technologies are reshaping educational environments to become more adaptive, interconnected, and personalised. The study begins by tracing the evolution of Generative AI and its application in education, followed by an examination of the concept of ecological learning spaces. The chapter then examines some specific ways AI is impacting these spaces, focusing on three key aspects: dynamic adaptability, interconnectedness, and learner-centred design. It discusses how AI facilitates personalised learning experiences, enables collaborative networks, and supports holistic development and educational accessibility. The paper concludes by outlining the potential of AI in transforming educational experiences and suggesting future directions for research in this field.

Keywords Artificial intelligence in education · Ecological learning spaces · Education · AI · Personalised learning · Adaptive education systems · Holistic educational development

2.1 Introduction

The integration of Artificial Intelligence (AI) into education is revolutionising the way we conceptualise and design learning environments. This chapter explores the emergence of AI-enhanced ecological learning spaces, which represent a significant shift in educational technology. By examining the intersection of AI capabilities

P. Ilic (✉)
University of Aizu, Aizu-Wakamatsu, Japan
e-mail: pilic@u-aizu.ac.jp

M. Sato-Ilic
University of Tsukuba, Tsukuba, Japan
e-mail: mika@risk.tsukuba.ac.jp

P. Ilic et al. (eds.), *Artificial Intelligence in Education: The Intersection of Technology and Pedagogy*, Intelligent Systems Reference Library 261,
https://doi.org/10.1007/978-3-031-71232-6_2

with the principles of ecological learning theory, we investigate how these intelligent systems are reshaping educational practices, fostering personalisation, and creating interconnected, adaptive learning ecosystems that cater to the diverse needs of modern learners.

Alfred North Whitehead's quote, "Civilisation advances by extending the number of important operations which we can perform without thinking about them," (Whitehead et al., 1911) encapsulates a fundamental principle of human progress through automation and task simplification. This idea suggests that as societies develop, they continuously seek ways to automate or simplify complex tasks, allowing individuals to perform them with minimal cognitive effort. By reducing the mental energy required for routine operations, humans can redirect their focus towards more creative, innovative, and challenging endeavours. This process of automation and simplification has been a driving force behind technological advancements, economic growth, and improved quality of life throughout history. Whitehead's insight is particularly relevant to the ongoing debate surrounding AI, as these systems increasingly take on complex tasks that were once the exclusive domain of human cognition. This progression raises intriguing questions about the future role of human intelligence, creativity, and decision-making in an increasingly automated world.

Building on Whitehead's insight, AI has emerged as a powerful tool for extending our cognitive capabilities and automating complex tasks. AI systems possess several key characteristics that make them particularly valuable in this context. These include their ability to automate repetitive or complex processes, provide personalised experiences, adapt to changing circumstances, offer data-driven decision support, scale to handle large volumes of information, improve efficiency across various domains, foster innovation and creativity by augmenting human capabilities, and increase accessibility to specialised knowledge and services. As AI continues to evolve, it exemplifies Whitehead's principle by enabling us to perform an ever-expanding range of important operations with minimal conscious effort, potentially accelerating the advancement of civilisation in unprecedented ways.

2.2 Artificial Intelligence

AI and human reasoning share the fundamental goal of processing information to draw conclusions and make decisions, yet they differ significantly in their approach and capabilities. Human reasoning is a complex cognitive process encompassing context understanding, common sense, creativity, emotional intelligence, and moral considerations (Gigerenzer & Gaissmaier, 2011). In contrast, AI reasoning with language, exemplified by generative models, relies on patterns and associations learned from vast training datasets (Goodfellow et al., 2016).

Generative AI models demonstrate impressive capabilities in language understanding, knowledge representation, contextual reasoning, inference, and language generation. However, they lack the depth and adaptability characteristic of human reasoning. Key differences include humans' superior contextual understanding, common sense reasoning, creativity, emotional intelligence, ability to reason with limited data, and capacity for moral and ethical reasoning (Marcus & Davis, 2019). Additionally, humans possess lifelong learning and adaptability that current AI models struggle to replicate (Csikszentmihalyi, 1997). Despite these distinctions, some researchers argue that the gap between AI and human reasoning may be narrowing. They point to emergent behaviours in sophisticated AI models, improved contextual adaptation capabilities, and the idea of language as a proxy for reasoning. Philosophical debates about the nature of intelligence and the evolving definition of reasoning further complicate this comparison.

The interplay between human and AI reasoning remains a complex and actively researched topic (Skjuve et al., 2021). As AI technology advances, our understanding of both forms of intelligence continues to evolve, challenging traditional distinctions and prompting further discussions about the nature of cognition in humans and machines (Shneiderman, 2020a). This ongoing exploration has significant implications for the development of AI systems and our understanding of human cognition.

2.2.1 History of Generative AI

The history of Generative AI, which encompasses the development of algorithms capable of generating new content, is a fascinating journey. This begins with the early theoretical underpinnings in the mid-twentieth century and extends to the highly advanced models of today. Early theoretical foundations that mark the inception of Generative AI are rooted in the broader field of Artificial Intelligence, which began to emerge in the 1950s (McCarthy et al., 2006; Turing & Copeland, 2004). Early AI research, influenced by pioneers like Alan Turing, was more focused on symbolic AI (Newell & Simon, 1956), which involved rule-based systems (Buchanan & Shortliffe, 1984). However, during this period, the groundwork for neural networks (Rosenblatt, 1958), a key component of modern Generative AI, was laid. Neural Networks and Early Learning Algorithms arrived in the 1980s which saw a surge of interest in neural networks (McCulloch & Pitts, 1943; Schmidhuber, 2015), especially with the development of the backpropagation algorithm (Rumelhart et al., 1986) which allowed networks to adjust their parameters more effectively (Schmidhuber, 2015). This period also witnessed the emergence of basic generative models, although they were limited in capability compared to today's standards. The emergence of deep learning in the 2000s marked a significant turning point (Goodfellow, 2014; Krizhevsky et al., 2012). Key advancements included the development of deep neural networks (Schmidhuber, 2015; Srivastava et al., 2014), which were more effective at processing large amounts of data. This era also saw the introduction of foundational models like Restricted Boltzmann Machines (RBMs) (Hinton, 2012) and Deep Belief Networks (DBNs) (Hinton & Salakhutdinov, 2006; Hinton et al., 2006), which could generate new data after training. Generative Adversarial Networks (GANs) (Goodfellow, 2014) and Variational Autoencoders (VAEs) (Kingma & Welling, 2014;

Rezende et al., 2014) appeared in the 2010s, introduced by Ian Goodfellow and colleagues in 2014 (Goodfellow, 2014), and significantly advanced the field by using two neural networks—a generator and a discriminator—in opposition to generate new, realistic data. VAEs also became prominent, offering a probabilistic approach to generating data. Transformative Language Models in the late 2010s and early 2020s have been dominated by Large Language Models (Boyko et al., 2023) like GPT (Generative Pre-trained Transformer) series from OpenAI (Radford et al., 2018; Yenduri et al., 2023). These models, trained on vast datasets, are capable of generating highly coherent and contextually relevant text, revolutionising natural language processing tasks. Now, the expansion into many diverse domains of Generative AI, including image and music generation, drug discovery, and more (Epstein et al., 2023). This period is characterised by more sophisticated models that are increasingly efficient and capable of handling multimodal data (i.e., data that combines text, images, and other forms of information) (Park et al., 2023). Finally, as the future direction of Generative AI continues to evolve, there's a growing focus on addressing ethical concerns (Zohny et al., 2023) such as bias (McDuff, 2019), misuse (Jo, 2023), and the impact on employment (AbuMusab, 2023). Future developments are expected to involve more efficient and powerful models, broader applications, and continued efforts to address ethical and societal implications (Lim et al., 2023).

Machine Learning (ML) forms the foundation of AI, providing algorithms that enable computers to learn from data and make decisions. Built upon this, Generative AI specialises in creating new data that mimics its training set, using techniques like GANs and VAEs to produce content across various domains. At the pinnacle are Large Language Models (LLMs), a specific application of Generative AI in natural language processing (NLP). LLMs, such as GPT models, use advanced deep learning techniques to process and generate human-like text based on vast language datasets. This hierarchy represents increasing specialisation: ML provides the fundamental tools, Generative AI focuses these tools on data creation, and LLMs apply this generative capability specifically to language tasks. Each layer builds on the previous, resulting in sophisticated models capable of understanding and generating contextually aware, human-like text.

The transformer model, introduced in the paper "Attention is All You Need" by Vaswani et al. (Vaswani, 2017), represents a paradigm shift in NLP by focusing solely on mechanisms called attention, specifically self-attention, to process sequential data. This design allows for parallel processing of sequences and better handling of long-range dependencies compared to its predecessors, such as recurrent neural networks (RNNs) and long short-term memory networks (LSTMs). In the context of transformers in ML, multidimensional space plays a crucial role in representing and processing information. Transformers utilise multidimensional spaces to encode textual information, capture relationships between elements in sequences, and facilitate the learning of complex patterns in data.

2.2.2 Mechanism of the Transformer Process

The transformer architecture has revolutionised the field of NLP by introducing a novel approach to sequence-to-sequence modelling. This architecture relies on a series of intricate steps that enable it to process and understand language effectively. In this section, we will explore the key components of the transformer model, including input embedding, positional encoding, self-attention, multi-head attention, feed-forward networks, residual connections, layer normalisation, and the output layer. By understanding these steps and their interplay within the transformer, we can gain valuable insights into how this powerful architecture achieves state-of-the-art performance across a wide range of NLP tasks.

Input Embedding is the first step in a transformer model, where input tokens (e.g., words, subwords, or characters) are converted into vectors of continuous numbers using embedding layers. This process transforms discrete input tokens into a high-dimensional vector space called the embedding space, which facilitates mathematical operations and allows the model to capture semantic relationships between words. The dimensionality of the embedding space is a hyperparameter of the model. Each dimension can represent different features or aspects of the input data, enabling the model to capture complex relationships between tokens.

To understand this concept better, imagine the embedding space as a vast landscape where each word occupies a unique position based on its relationships and contexts within the language. Words with similar meanings or that frequently appear together will be located closer to each other in this landscape. For example, "cat" and "dog" might be neighbours, along with related concepts like "pet" and "fur." On the other hand, words like "laboratory," "hypothesis," and "data" might occupy a different region of the landscape, representing their semantic similarity. This spatial representation allows AI models to grasp the subtle nuances and relationships between words, enhancing their understanding of language.

In the second step, positional encodings are introduced to address a limitation of transformers. Unlike RNNs or long short-term memory (LSTM) models, transformers process all tokens in the input sequence in parallel, lacking an inherent mechanism to capture the sequential nature of the data. To overcome this limitation, positional encodings are added to the input embeddings to provide information about the position of each token within the sequence. Positional encodings are unique vectors that are added element-wise to the embedding vectors of each token. These encodings are designed to be distinct for each position in the sequence, allowing the model to incorporate the order of tokens into its computations. By adding positional encodings to the input embeddings, the transformer can preserve the sequential nature of the input data within the high-dimensional embedding space.

To illustrate this concept, consider a group of dancers performing a choreographed piece. Each dancer's individual movements are important, but the sequence in which they perform their steps is crucial for conveying the overall story and meaning of the dance. In this analogy, positional encodings in transformers are like assigning each dancer a unique identifier that indicates their position in the sequence. This ensures that even though the transformer processes the entire performance in parallel, the order of the steps is preserved and contributes to the overall meaning of the dance.

Next, at the heart of the transformer architecture lies the self-attention mechanism, the third step, which operates within the multidimensional embedding space. Self-attention allows each token to interact with every other token in the input sequence by computing a set of attention scores. These scores indicate the importance of other tokens when processing a given token. The computation of attention scores involves three vectors for each token: query (Q), key (K), and value (V). These vectors are obtained by projecting the input embeddings into different multidimensional spaces using learned weight matrices. The attention scores are then computed using a scaled dot-product of the Q vector of the target token and the K vectors of all other tokens, followed by a softmax operation to obtain normalised weights that sum to one. The output of the self-attention layer is a weighted sum of these scores and the corresponding V vectors, effectively capturing the relevant information from other tokens.

To provide an analogy for this concept, consider a theatre play with multiple actors on stage, where each actor represents a word in a sentence. A spotlight (the self-attention mechanism) shines on one actor at a time, but it adjusts its brightness based on the relevance of other actors to the overall story. The brightness of the spotlight on each actor represents the attention scores, indicating the importance of each actor to the current focus of the play. The spotlight can also illuminate multiple actors simultaneously to different degrees, representing the weighted sum of values in the self-attention mechanism. This allows the play to convey a richer, more nuanced meaning by highlighting the relationships and interactions between the actors.

Fourth, to capture different aspects of the input data, the transformer employs multi-head attention, which allows the model to jointly attend to information from different representation subspaces at different positions. In this mechanism, the self-attention process is run in parallel multiple times, each with its own set of learned linear projections of the Q, K, and V vectors. These linear projections enable each attention head to focus on different aspects of the input data, effectively creating multiple "self-attention spotlights" that operate simultaneously. The outputs of these parallel attention heads are then concatenated and linearly transformed to obtain the final output of the multi-head attention layer. This approach enables the transformer to capture a richer, more comprehensive understanding of the input sequence by combining the different perspectives learned by each attention head.

To illustrate this more clearly, consider the same theatre play observed by a panel of several critics. Each critic focuses on different aspects of the performance, such as acting, dialogue, or stage design. They might interpret the same scene from contrasting viewpoints, providing diverse insights into the play. Similarly, multi-head attention works like multiple "self-attention spotlights" operating simultaneously, each with a slightly different learned focus. One attention head might concentrate on grammatical relationships, while another might focus on broader themes within the text. These different perspectives are learned by the model during training, allowing it to capture various aspects of the input data. Afterwards, these diverse views are combined to create a richer, more comprehensive understanding of the play.

In the fifth step, after each attention layer, the transformer applies a position-wise feed-forward network (FFN) to the output of the attention mechanism. The FFN consists of two linear transformations with a ReLU activation function in between. The purpose of the FFN is to introduce additional non-linearity and expand the model's capacity to capture complex patterns in the input data. Importantly, the FFN is applied independently to each position in the sequence, allowing the model to capture position-specific patterns and transformations. As information passes through the transformer's layers, it undergoes iterative transformations via the attention mechanisms and feed-forward networks. Each layer projects the input data into new multidimensional spaces, enabling the model to progressively refine and abstract the representations of the input. This iterative process allows the transformer to capture increasingly complex patterns and relationships within the data, building upon the representations learned by the previous layers.

To visualise this process, consider the process of creating a complex painting. The artist starts with initial sketches that form a basic outline of the composition. As layers of paint are added, details, forms, and ultimately themes emerge. The final masterpiece depends on this gradual buildup of complexity, with each layer contributing to the overall representation. Similarly, each hidden layer in a transformer acts like a layer of paint, transforming and refining the input data (the initial sketch) into increasingly abstract and sophisticated representations. Just as each stroke and colour choice builds upon the previous ones, the hidden layers in the transformer build more nuanced understandings of the input text, with each layer expanding the model's capacity to capture complex patterns and relationships.

In the sixth step of the transformer model, both the self-attention and the FFN layers are wrapped with residual connections, followed by layer normalisation. Residual connections help mitigate the vanishing gradient problem (Hochreiter, 1998) and make it easier for the model to learn residual functions with reference to the input. This is achieved by adding the input of each sub-layer back to its output before being normalised. Residual connections ensure that even if a layer makes a less-than-ideal transformation, the original information is preserved, aiding in the stability of the model. Layer normalisation, on the other hand, works to stabilise the training process by normalising the activations across each layer. It ensures that the values within each layer remain well-behaved, preventing them from getting too large or too small. This normalisation step helps to promote smooth training and faster convergence of the model.

To better understand this idea, imagine a team of acrobats performing a daring routine where each acrobat builds upon the moves of the one before, creating increasingly complex formations. Residual connections act like safety nets strategically placed below the acrobats. If someone stumbles, the net allows them to quickly rejoin the routine, preventing the whole structure from collapsing. Similarly, if a layer in the transformer makes a less-than-ideal transformation, the residual connection ensures that the original information is preserved, maintaining the stability of the model. Layer normalisation, in this analogy, works like a levelling tool a worker uses to check the platform between acts. It ensures that the platform remains balanced for the next daring feat, preventing any imbalances that could throw off the performance. In the transformer, layer normalisation keeps the activations across each layer well-behaved, promoting smooth training and faster convergence.

In the seventh step of the transformer model, the decoder side follows a similar structure to the encoder but includes an additional attention layer that focuses on the encoder's output. This additional attention layer allows the decoder to attend to relevant parts of the input sequence at each generation step, enabling it to capture the context and dependencies between the input and output sequences.

The decoder's self-attention layer operates similarly to the encoder's self-attention, allowing each token in the output sequence to attend to the previously generated tokens. However, the additional attention layer enables the decoder to dynamically focus on pertinent segments of the encoder's output as it generates each token in the output sequence. This mechanism is crucial for ensuring that the generated output accurately reflects the key ideas and context of the input sequence.

To clarify this notion, think of the decoder as a translator working on a crucial document. The translator has their own notes and understanding of the target language, which is similar to the decoder's self-attention. Additionally, the translator has access to the original document in the source language, which represents the encoder's output. The decoder's additional attention layer acts like a highlighting tool, allowing the translator to focus on the most relevant parts of the original document as they work on each section of their translation. This dynamic focusing ensures that the final translation accurately captures the context and dependencies of the original text.

In the final step of the transformer model, the output space represents the high-dimensional vector that the model produces at each generation step. In tasks such as language translation or text generation, this output vector needs to be converted into a probability distribution over the vocabulary, indicating the likelihood of each possible output token. The conversion from the high-dimensional output vector to a probability distribution is typically performed using a softmax layer. The softmax layer takes the output vector and maps it to a set of probabilities, where each element corresponds to a specific token in the vocabulary. The resulting probability distribution allows the model to make predictions about the most likely output token at each generation step.

To visualise this process, imagine a sculptor working with a block of clay, where the final sculpture represents the generated text or output. The sculptor's tools are like the model's output mechanism, shaping the clay into different forms. Initially, the block of clay (the high-dimensional output vector) holds endless possibilities. As the sculptor works (the model processes the input), specific features emerge, gradually transforming the clay into a recognisable shape (the predicted output token). The softmax layer acts as the final tool in the sculptor's arsenal, refining the shape of the clay into a specific form. It takes the output vector and moulds it into a probability distribution, assigning higher probabilities to the most likely output tokens based on the learned patterns and relationships from the training data. The final sculpture represents the culmination of the transformation process, where the initially formless clay has been shaped into a precise outcome based on the sculptor's (model's) understanding of the input and the learned patterns.

The use of multidimensional spaces is crucial to the success of transformer models in NLP tasks. These high-dimensional spaces allow transformers to capture and process complex linguistic and contextual information effectively. Throughout the various steps of the transformer architecture, such as input embedding, attention mechanisms, and positional encoding, multidimensional spaces play a vital role. They enable the model to represent words, understand their relationships, and focus on the most relevant information based on the current context. Imagine a multi-layered map where words are positioned based on their semantic and syntactic similarities. The transformer model navigates this map using attention mechanisms, focusing on specific regions to understand the meaning and generate appropriate outputs. By leveraging the power of multidimensional spaces, transformers can learn rich representations of text, capture nuances, and produce coherent and contextually relevant results. This flexibility and capacity to process language in high-dimensional spaces are key to the transformer's state-of-the-art performance across a wide range of NLP applications.

2.3 The Concept of Ecological Learning Spaces: Dynamic, Interconnected, and Learner-Centred Educational Environments

The concept of an ecological perspective on learning spaces is grounded in Bronfenbrenner's ecological systems theory (Bronfenbrenner, 1979), which views development within a complex system of relationships affected by multiple levels of the surrounding environment. Bronfenbrenner is revered as one of the leading world authorities in the field of development psychology. His most important brainchild was the ecological systems theory, where he defines the five concentric systems that are the micro-, the meso-, the exo-, the macro- and the chronosystem (Bronfenbrenner et al., 1994). From a sociocultural perspective, an educational environment consists of individuals and groups engaged in activities that shape and convey their identities. An ecological approach echoes this but considers every aspect of the environment—including physical spaces, tools, technologies, and other people—as resources. This environment is adaptable, adjusting to the evolving interactions between individuals and these resources. These dynamics can be understood as an ecological system (Loi & Dillon, 2006).

Lave and Wenger's Situated Learning Theory (Lave & Wenger, 1991) in 1991 also contributed foundational ideas, emphasising learning as a social process and the importance of context in learning environments. The early 2000s saw initial applications of ecological concepts in education. This period focused on understanding how different environmental factors, including social, cultural, and physical elements, impact learning (Kirschner & Hendrick, 2020). Researchers began to explore the role of technology in these spaces (Siemens, 2005), recognising the growing influence of digital tools in learning environments (Lave & Wenger, 1991). With the advancement of technology, the mid-2000s witnessed a shift towards digital and blended learning environments (Dziuban et al., 2018; Ilic, 2014). This phase marked the recognition of online platforms, virtual classrooms, and digital resources as integral components of learning spaces. It is a way for instructors to strengthen the network of learning formed by collaboration by allowing students to continue the discussion as they move through physical locations (Sharples et al., 2005). For example, mobile phones have created "simultaneity of place" (Traxler, 2009) a kind of bridging of physical space, such as home, school, and work, through the creation of a mobile social space by filling the space in between (Bull, 2005). Mobile technologies transport communities and discussions into physical public and private spaces forcing people to adjust their behaviour to manage a more fluid environment (Traxler, 2009). Private is no longer just what happens when physically alone (Cooper et al., 2002).

The concept of "networked learning"(Harasim, 1995; Jones et al., 2000) emerged, highlighting how digital connections can create new types of learning communities and environments (Cousin & Deepwell, 2005). The last decade has seen the integration of advanced technologies like AI (Crompton & Burke, 2023) into learning spaces. These technologies have been used to personalise learning experiences, provide adaptive learning pathways, and analyse learning data for insights. The focus has been on creating intelligent learning environments (Kinshuk, et al., 2016; Spector, 2016) that adapt and respond to the needs of learners, akin to an ecological system. The most recent literature emphasises the importance of flexibility, adaptability, and learner agency in ecological learning spaces (Esposito et al., 2015).

Current research is exploring the ethical implications of AI in education (Akgun & Greenhow, 2021), the role of data privacy (Komljenovic, 2022; Reidenberg & Schaub, 2018), and the need for equitable access (Gulati, 2008; Haythornthwaite, 2007; Tien & Fu, 2008) to technology-enhanced learning environments. Future directions point towards a more holistic integration of AI into the fabric of ecological learning spaces (Erickson, 2013), ensuring that these technologies serve to enhance rather than dictate the learning process. Throughout these phases, the literature reflects a growing understanding of the complexity and interconnectedness of learning environments (Ilic, 2021). The ecological perspective emphasises the need to consider not just the physical space, but also the social, cultural, and technological dimensions that shape learning experiences.

2.4 AI and the Ecological Aspect of Learning Spaces

This section explores the dynamic role of AI in enhancing ecological learning spaces, emphasising adaptability, interconnectivity, and personalisation. It delves into how AI transforms these spaces into learner-centred environments, facilitating collaborative networks, holistic development, and inclusive education. AI's integration is helpful in personalising learning experiences, engaging students interactively, and providing comprehensive feedback, thereby transforming the educational landscape to be more adaptive and responsive to diverse learning needs.

2.4.1 Adaptability in Ecological Learning Spaces

Ecological learning spaces ideally should be adaptable and responsive to the needs and actions of learners. This adaptability is crucial in a rapidly changing world, where the skills and knowledge required can shift quickly. AI can contribute to this adaptability by enhancing the flexibility and responsiveness of educational environments, thereby enabling the learning environment to evolve in response to student needs. Adaptive learning systems use learning Analytics that systematically collect and analyse data on learners' interactions and progression in education (Siemens, 2013) to tailor learning strategies. The AI uses this data to identify patterns, such as which learning strategies are most effective for individual students (Bienkowski et al., 2012; Ifenthaler et al., 2019). The development and utilisation of these adaptive learning systems are gaining prominence, and they could use AI algorithms to further personalise the learning content and pace (Alavi & Dillenbourg, 2012; Brusilovsky, 2001). For instance, should a student encounter difficulty in grasping a specific concept, these systems have the capability to either furnish additional resources or modify the level of complexity to facilitate better understanding.

Ecological learning spaces encompass holistic development, including social, emotional, and cognitive skills beyond academic knowledge. AI significantly contributes to this approach by tracking and supporting broader aspects of learning. It provides insights into student engagement, emotional well-being, and social dynamics. AI enhances personalisation through adaptive content and difficulty modification, aligning with individual learning paces and preferences. It augments engagement and motivation with interactive tools, making learning more effective and enjoyable.

Inclusion is a fundamental principle in the design and operation of ecological learning spaces. These spaces strive to ensure that all learners, regardless of their background or circumstances, have access to quality education. AI is instrumental in realising this goal, as it can help identify gaps in access and achievement among diverse student populations. Through its analytical capabilities, AI aids in developing strategies to bridge these gaps, thereby promoting a more accessible learning environment (Holstein & Doroudi, 2021; Roshanaei et al., 2023).

Research on the implementation and accessibility of AI systems focuses on their effectiveness in diverse educational contexts and accessibility for all learners, including those with disabilities. A critical focus is on how these innovative systems can be effectively integrated across various educational landscapes and made accessible to a diverse spectrum of learners, including those with disabilities, ensuring fairness in educational opportunities (Burgstahler, 2002; Goldenthal et al., 2021; Lazar et al., 2017).

The rapid advancement of AI in education has the potential to significantly enhance learning experiences through techniques such as adaptive learning systems, Predictive Modelling, student modelling, NLP, and AI-assisted feedback and assessment. These AI-powered systems can personalise learning content and pace, predict learning outcomes, model individual student knowledge and preferences, facilitate intuitive interactions, and provide immediate and targeted feedback. This combination of capabilities can lead to the creation of truly dynamic ecological learning spaces that adapt and respond to the needs and actions of each learner. However, as we integrate these powerful technologies into educational environments, it is crucial to address the ethical and privacy considerations surrounding student data usage and potential biases in AI-driven methodologies. The successful implementation and accessibility of these AI systems across diverse educational contexts, including their ability to accommodate learners with disabilities, will be key to ensuring equal access to educational opportunities for all students. While the path forward presents both exciting possibilities and challenges, the responsible application of AI has the potential to significantly advance and improve the effectiveness of education.

2.4.2 Principle of Interconnectedness

In an ecological learning space, the principle of interconnectedness may be supported through the integration of AI, linking various elements like technology, the physical environment, social interactions, and learning materials. This interconnected approach, enhanced by AI, ensures that learning is viewed not in isolation but as part of a broader system. The emergence of collaborative networks supported by AI enables the effective formation of extensive, interconnected networks that bridge the gap between learners and educators worldwide, facilitating collaboration, idea exchange, and mutual learning (Rosé & Ferschke, 2016; Wenger, 1998). Such networks transcend geographical barriers bringing together diverse perspectives and educational practices, enriching the learning experience.

Furthermore, AI's pivotal role in Data Integration Across Platforms is becoming increasingly crucial (Labarthe et al., 2018; Papamitsiou & Economides, 2014; Siemens, 2005; Siemens & Baker., 2012), and results in a seamless amalgamation of information from diverse educational platforms, enhancing learning experiences. By harnessing AI, educators and institutions can integrate data from a host of educational platforms, creating a seamless flow of information (Ahmad et al., 2023). This integration is pivotal in developing a more comprehensive understanding of individual learning patterns.

The technology also fosters interdisciplinary connectivity, providing access to a wide-ranging web of knowledge that spans different disciplines. This not only deepens subject-specific knowledge but also promotes a holistic and integrated educational approach. The ability of AI in education is its ability to foster this interdisciplinary connectivity (Blikstein & Worsley, 2016; Jacobs, 1989) by acting as a conduit through which learners can access a vast, interconnected web of knowledge that spans various disciplines. This is not just about deepening knowledge in individual subjects; it's about redefining the very approach to education. It encourages a more holistic and integrated understanding, breaking down traditional silos between disciplines. The result is an educational experience that is not only more comprehensive but also more cohesive, fostering a broader perspective and a deeper appreciation of the interconnectedness of knowledge.

These spaces emphasise collaborative and social learning, where AI aids in creating and enhancing interactive educational experiences. Within these ecological learning spaces, collaborative and social learning is greatly emphasised, and AI can assist in this domain by suggesting suitable collaborative projects, which not only enhance academic understanding but also encourage teamwork and communication skills (McLaren et al., 2010; Tan et al., 2022). Additionally, AI can adeptly form study groups by analysing and matching students based on complementary learning styles and skills (Zhai et al., 2021; Zhang et al., 2022). Beyond these practical applications, AI is also capable of simulating social learning environments, providing a platform for students to engage in interactive, socially-rich educational experiences, even in remote or digital settings (Dai & Ke, 2022; Shen et al., 2021).

The integration of AI in ecological learning spaces significantly enhances the principle of interconnectedness by strengthening the links between various elements of the learning environment. AI-supported collaborative networks improve global collaboration and idea exchange among learners and educators, leading to more diverse perspectives and enriched learning experiences. Furthermore, AI bolsters interdisciplinary connectivity by providing learners with better access to a vast, interconnected web of knowledge, promoting a more holistic and integrated understanding of subjects. AI also enhances collaborative and social learning by suggesting more effective collaborative projects, forming study groups based on complementary skills, and simulating engaging social learning environments in both physical and digital settings. By augmenting these aspects of education, AI creates a more interconnected, personalised, and effective learning experience within ecological learning spaces, fostering a deeper appreciation for the interrelated nature of knowledge and skills.

2.4.3 Learner-Centred Ecological Learning Spaces

In learner-centred ecological learning spaces, AI plays a pivotal role in personalising education. It tailors learning experiences to each student's unique needs, preferences, and learning styles by adaptively modifying content and difficulty. This ensures alignment with individual learning paces and enhances student engagement and motivation through interactive tools. AI also provides continuous, personalised feedback and assessment, enabling students to understand their progress clearly and focus on areas needing improvement. This approach not only corrects mistakes but also guides overall learning and development, fostering a more self-directed and reflective educational journey. In the contemporary educational paradigm, personalisation is critical, signifying a shift towards tailored learning experiences (Bartolomé et al., 2018; Skinner, 2016). This often involves using AI to adaptively modify both content and difficulty level (Davies et al., 2021).

In the realm of feedback and assessment, AI provides immediate, personalised responses and aids in assessment creation and grading. Here, AI's contribution is twofold. Firstly, it provides immediate and personalised feedback to students, thereby enhancing their understanding and rectification of mistakes (Giannakos et al., 2019; Shute, 2008; VanLehn, 2006). Secondly, AI aids in the creation and evaluation of assessments (Smolansky et al., 2023), streamlining the grading process and ensuring a more objective evaluation (Owan et al., 2023). However, it is imperative to address issues pertaining to the privacy of student data, ethical usage of such data, and the prevention of any biases that may arise from AI-driven educational methodologies (Holmes et al., 2022; Shneiderman, 2020b; Slade & Prinsloo, 2013).

Similarly, Predictive Modelling forecasts learning outcomes to identify students needing extra support and here AI is adept at forecasting future learning outcomes by analysing students' past performance (Baker et al., 2009; Romero & Ventura, 2010; Xing & Du, 2019). This predictive capability is instrumental in the early identification of students who may require additional support, thereby enabling timely intervention. Moreover, the concept of student modelling is at the forefront of AI education systems that are engineered to construct and continually refine a comprehensive model of each student's knowledge base, skill set, and learning preferences (Santos & Boticario, 2015; VanLehn, 1988; Woolf, 2010). This model evolves in real-time, adapting to the student's interactions with the educational material. While student modelling creates evolving profiles of each learner's knowledge and preferences, NLP enriches student interactions with the systems. NLP enables the system to comprehend and respond to student queries in a manner that is both natural and intuitive, thereby enhancing the overall learning experience and making it more interactive (Graesser et al., 2014, 2018; Perez-Marin & Pascual-Nieto, 2011).

Alongside personalisation, engagement and motivation are equally paramount in the educational process. The use of interactive and immersive AI tools has become increasingly prevalent in maintaining student interest and motivation (Alam & Mohanty, 2022; Chiu, 2023; Huang et al., 2023). By making learning an interactive and enjoyable experience, AI helps sustain student interest and motivation (Fredricks et al., 2004), which are crucial for effective learning and long-term educational success. Furthermore, feedback and assessment are integral components of this evolved educational approach (Sadler, 1989). AI provides continuous and personalised feedback, enabling students to gain a clear understanding of their progress and areas requiring improvement (Darvishi et al., 2022; Richardson & Clesham, 2021; Vittorini et al., 2021). This feedback is not just about correcting mistakes; it's about providing a roadmap for learning and development. AI systems can analyse student responses and learning patterns, offering insights that are tailored to each student's needs. This helps students in identifying their strengths and weaknesses, fostering a more self-directed and reflective approach to learning.

Furthermore, AI enriches feedback and assessment with continuous, personalised insights that guide learning and development. It aids in identifying and addressing achievement gaps among student populations, promoting a more comprehensive learning environment. By monitoring social, emotional, and cognitive aspects, AI enables educators to tailor their approach to each student's needs. These AI-driven enhancements transform learning spaces into more personalised, engaging, inclusive, and holistic environments, fostering optimal learning and development for all students. Ultimately, AI helps create learning environments that nurture the whole student, not just their academic abilities (McStay, 2020; Salas-Pilco, 2020).

2.5 Conclusions

The integration of AI into ecological learning spaces offers transformative possibilities for educational technology and pedagogy. AI could reshape the educational landscape by creating dynamic, interconnected, and learner-centred environments that adapt to individual needs and promote collaborative learning. This aligns with the broader trend of automating complex cognitive tasks, potentially allowing people to focus on higher-order thinking and innovation. Generative AI's evolution suggests the possibility of more personalised and effective learning experiences. By leveraging AI capabilities in adaptive learning systems, predictive modelling, and NLP, educational institutions might create more responsive, engaging, and inclusive learning environments. AI could enhance interconnectedness, transforming learning spaces into rich ecosystems where knowledge is shared, created, and applied across disciplines. This might facilitate global collaboration and promote a holistic approach to education, encompassing social, emotional, and cognitive skill development.

AI-enabled learner-centred approaches could tailor education to individual needs, preferences, and learning styles. This personalisation, combined with continuous feedback and assessment, might empower students to take greater control of their learning journey and potentially achieve better outcomes. However, addressing ethical considerations and challenges is crucial. Issues such as data privacy, algorithmic bias, and equal access to AI-enhanced education must be carefully considered to ensure potential benefits are available to all learners. The potential of AI in education remains vast and largely unexplored. Future research could focus on improving AI algorithms to better respond to diverse learning needs, developing advanced interdisciplinary learning tools, and exploring seamless AI integration while maintaining the essential human element of teaching and learning.

When considering the future, it is crucial to consider the emerging role of AI agents in this evolving educational landscape. These sophisticated entities represent the next frontier in AI, offering capabilities that align closely with the principles of adaptive, personalised learning environments. AI agents are designed to operate autonomously within complex systems, perceiving their environment, processing information, and taking goal-oriented actions. In the context of ecological learning spaces, they could serve as personalised tutors, adaptive assessment tools, and virtual teaching assistants, potentially reshaping the interactions among students, teachers, and other stakeholders in the educational ecosystem. What sets these agents apart is their ability to learn and improve over time, incorporating advanced ML algorithms and NLP. This adaptability makes them particularly well-suited to support the dynamic, learner-centred approach of ecological learning spaces. As we look towards the future of education, the integration of AI agents promises both exciting opportunities and significant challenges in our quest to create more responsive, personalised, and effective learning environments.

In conclusion, AI-enhanced ecological learning spaces offer a promising avenue for potentially more effective, engaging, and accessible education. By continuing to innovate and responsibly explore these technologies, we may be able to create learning environments that not only meet the demands of our rapidly changing world but also nurture the full potential of every learner. This possible progression towards AI-supported learning environments reflects a broader trend in human advancement, where the automation of complex cognitive tasks could free human potential for higher-order thinking, creativity, and innovation in education and beyond.

References

AbuMusab, S. (2023). *Generative AI and human labor: Who is replaceable?* AI & SOCIETY, pp. 1–3.

Ahmad, K., et al. (2023). Data-driven artificial intelligence in education: A comprehensive review. *IEEE Transactions on Learning Technologies, 17*, 12–31.

Akgun, S. & Greenhow, C. (2021). *Artificial intelligence in education: Addressing ethical challenges in K-12 settings.* AI and Ethics, pp. 1–10.

Alam, A. & Mohanty, A. (2022). *Facial analytics or virtual avatars: competencies and design considerations for student-teacher interaction in AI-powered online education for effective classroom engagement.* In *International Conference on Communication, Networks and Computing.* Springer.

Alavi, H. S., & Dillenbourg, P. (2012). An ambient awareness tool for supporting supervised collaborative problem solving. *IEEE Transactions on Learning Technologies, 5*(3), 264–274.

Baker, R. S. J. D., & Yacef, K. (2009). The state of educational data mining in 2009: A review and future visions. *Journal of Educational Data Mining, 1*(1), 3–17.

Bartolomé, A., Castañeda, L., & Adell, J. (2018). Personalisation in educational technology: The absence of underlying pedagogies. *International Journal of Educational Technology in Higher Education, 15*, 1–17.

Bienkowski, M., Feng, M., & Means, B. (2012). *Enhancing teaching and learning through educational data mining and learning analytics: An issue brief.* Office of Educational Technology, US Department of Education.

Blikstein, P., & Worsley, M. (2016). Multimodal learning analytics and education data mining: Using computational technologies to measure complex learning tasks. *Journal of Learning Analytics, 3*(2), 220–238.

Boyko, J., et al. (2023). *An interdisciplinary outlook on large language models for scientific research.* arXiv preprint arXiv:2311.04929.

Bronfenbrenner, U., Husen, T., & Postlethwaite, T. (1994). *Ecological models of human development*, In *International encyclopedia of education.* pp. 37–43.

Bronfenbrenner, U. (1979). *The ecology of human development: Experiments by nature and design.* Harvard University Press.

Brusilovsky, P. (2001). Adaptive Hypermedia. *User Modeling and User-Adapted Interaction, 11*(1), 87–110.

Buchanan, B.G., & Shortliffe, E.H. (1984). *Rule based expert systems: The mycin experiments of the stanford heuristic programming project (the Addison-Wesley series in artificial intelligence).* Addison-Wesley Longman Publishing Co., Inc.

Bull, M. (2005). No Dead Air! The iPod and the culture of mobile listening. *Leisure Studies, 24*(4), 343–355.

Burgstahler, S., (2002). *Universal design of distance learning.* Information Technology and Disabilities, **8**(1).

Chiu, T.K., et al., (2023). *Teacher support and student motivation to learn with Artificial Intelligence (AI) based chatbot.* Interactive Learning Environments, pp. 1–17.

Cooper, G. (2002). The mutable mobile: Social theory in the wireless world. In B. Brown, N. Green, & R. Harper (Eds.), *Wireless world: Social and interactional aspects of the mobile world* (pp. 19–31). Springer.

Cousin, G., & Deepwell, F. (2005). Designs for network learning: A communities of practice perspective. *Studies in Higher Education, 30*(1), 57–66.

Crompton, H., & Burke, D. (2023). Artificial intelligence in higher education: The state of the field. *International Journal of Educational Technology in Higher Education, 20*(1), 22.

Csikszentmihalyi, M. (1997). *Flow and the psychology of discovery and invention.* HarperPerennial.

Dai, C.-P., & Ke, F. (2022). Educational applications of artificial intelligence in simulation-based learning: A systematic mapping review. *Computers and Education: Artificial Intelligence, 3*, 100087.

Darvishi, A., et al. (2022). Incorporating AI and learning analytics to build trustworthy peer assessment systems. *British Journal of Educational Technology, 53*(4), 844–875.

Davies, H. C., Eynon, R., & Salveson, C. (2021). The mobilisation of AI in education: A Bourdieusean field analysis. *Sociology, 55*(3), 539–560.

Dziuban, C., et al. (2018). Blended learning: The new normal and emerging technologies. *International Journal of Educational Technology in Higher Education, 15*, 1–16.

Epstein, Z., et al. (2023). Art and the science of generative AI. *Science, 380*(6650), 1110–1111.

Erickson, G., et al. (2013). *Using the ecological approach to create simulations of learning environments.* In *Artificial Intelligence in Education.* Berlin, Heidelberg: Springer Berlin Heidelberg.

Esposito, A., Sangrà, A., & Maina, M. F. (2015). Emerging learning ecologies as a new challenge and essence for e-learning. *International Handbook of E-Learning* (Vol. 1, pp. 331–342). Routledge.

Fredricks, J. A., Blumenfeld, P. C., & Paris, A. H. (2004). School engagement: Potential of the concept, state of the evidence. *Review of Educational Research, 74*(1), 59–109.

Giannakos, M. N., et al. (2019). Multimodal data as a means to understand the learning experience. *International Journal of Information Management, 48*, 108–119.

Gigerenzer, G., & Gaissmaier, W. (2011). Heuristic decision making. *Annual Review of Psychology, 62*(1), 451–482.

Goldenthal, E., et al. (2021). Not all AI are equal: Exploring the accessibility of AI-mediated communication technology. *Computers in Human Behavior, 125*, 106975.

Goodfellow, I., et al. (2014). *Generative adversarial nets.* Advances in Neural Information Processing Systems, **27**.

Goodfellow, I., Bengio, Y., & Courville, A. (2016). *Deep learning.* MIT Press.

Graesser, A. C., Li, H., & Forsyth, C. (2014). Learning by communicating in natural language with conversational agents. *Current Directions in Psychological Science, 23*(5), 374–380.

Graesser, A. C., McNamara, D. S., & VanLehn, K. (2018). Scaffolding deep comprehension strategies through Point & Query, AutoTutor, and iSTART. In R. Azevedo (Ed.), *Computers as Metacognitive Tools for Enhancing Learning* (pp. 225–234). Routledge.

Gulati, S., (2008). *Technology-enhanced learning in developing nations: A review.* The international review of research in open and distributed learning, **9**(1).

Harasim, L. M. (1995). *Learning networks: A field guide to teaching and learning online.* MIT Press.

Haythornthwaite, C., (2007). *Digital divide and e-learning.* The Sage handbook of e-learning research, pp. 97–118.

Hinton, G.E., (2012). A practical guide to training restricted Boltzmann machines. In Neural Networks: Tricks of the Trade: Second Edition. Springer, pp. 599–619.

Hinton, G. E., Osindero, S., & Teh, Y.-W. (2006). A fast learning algorithm for deep belief nets. *Neural Computation, 18*(7), 1527–1554.

Hinton, G. E., & Salakhutdinov, R. R. (2006). Reducing the dimensionality of data with neural networks. *Science, 313*(5786), 504–507.

Hochreiter, S. (1998). The vanishing gradient problem during learning recurrent neural nets and problem solutions. *International Journal of Uncertainty, Fuzziness and Knowledge-Based Systems, 6*(02), 107–116.

Holmes, W., et al. (2022). Ethics of AI in education: Towards a community-wide framework. *International Journal of Artificial Intelligence in Education, 32*(3), 504–526.

Holstein, K. & Doroudi, S. (2021). *Equity and Artificial Intelligence in Education: Will" AIEd" Amplify or Alleviate Inequities in Education?* arXiv preprint arXiv:2104.12920.

Huang, A. Y., Lu, O. H., & Yang, S. J. (2023). Effects of artificial Intelligence-Enabled personalized recommendations on learners' learning engagement, motivation, and outcomes in a flipped classroom. *Computers & Education, 194*, 104684.

Ifenthaler, D., Mah, D.-K., & Yau, J.Y.-K. (2019). *Utilising learning analytics for study success: Reflections on current empirical findings.* Utilizing learning analytics to support study success, pp. 27–36.

Ilic, P., (2014). *The impact of mobile phones on collaborative learning activities*, In *Education.* University of Exeter.

Ilic, P. (2021). The challenge of information and communications technology in education. *SHS Web of Conferences, 102*, 01009.

Jacobs, H. H. (1989). *Interdisciplinary curriculum: Design and implementation.* ERIC.

Jo, A. (2023). The promise and peril of generative AI. *Nature, 614*(1), 214–216.

Jones, C., Asensio, M., & Goodyear, P. (2000). Networked learning in higher education: practitioners' perspectives. *Research in Learning Technology, 8*(2), 18–28.

Kingma, D. P., & Welling, M. (2014). Auto-Encoding Variational Bayes. *Stat, 1050*, 1.

Kinshuk, et al. (2016). Evolution is not enough: Revolutionizing current learning environments to smart learning environments. *International Journal of Artificial Intelligence in Education, 26*, 561–581.

Kirschner, P., & Hendrick, C. (2020). *How learning happens: Seminal works in educational psychology and what they mean in practice.* Routledge.

Komljenovic, J. (2022). The future of value in digitalised higher education: Why data privacy should not be our biggest concern. *Higher Education, 83*(1), 119–135.

Krizhevsky, A., Sutskever, I., & Hinton, G.E. (2012). *Imagenet classification with deep convolutional neural networks.* Advances in Neural Information Processing Systems, **25**.

Labarthe, H., Luengo, V., & Bouchet., F. (2018). *Analyzing the relationships between learning analytics, educational data mining and AI for education.* In *14th International Conference on Intelligent Tutoring Systems (ITS): Workshop Learning Analytics.*

Lave, J., & Wenger, E. (1991). *Situated learning: Legitimate peripheral participation.* Cambridge University Press.

Lazar, J., Feng, J. H., & Hochheiser, H. (2017). *Research methods in human-computer interaction.* Morgan Kaufmann.

Lim, W. M., et al. (2023). Generative AI and the future of education: Ragnarök or reformation? A paradoxical perspective from management educators. *The International Journal of Management Education, 21*(2), 100790.

Loi, D., & Dillon, P. (2006). Adaptive educational environments as creative spaces. *Cambridge Journal of Education, 36*(3), 363–381.

Marcus, G., & Davis, E. (2019). *Rebooting AI: Building artificial intelligence we can trust.* Vintage.

McCarthy, J., et al. (2006). A proposal for the dartmouth summer research project on artificial intelligence, august 31, 1955. *AI Magazine, 27*(4), 12–12.

McCulloch, W. S., & Pitts, W. (1943). A logical calculus of the ideas immanent in nervous activity. *The Bulletin of Mathematical Biophysics, 5*, 115–133.

McDuff, D., et al., (2019). *Characterizing bias in classifiers using generative models.* Advances in Neural Information Processing Systems. **32.**

McLaren, B. M., Scheuer, O., & Mikšátko, J. (2010). Supporting collaborative learning and e-discussions using artificial intelligence techniques. *International Journal of Artificial Intelligence in Education, 20*(1), 1–46.

McStay, A. (2020). Emotional AI and EdTech: Serving the public good? *Learning, Media and Technology, 45*(3), 270–283.

Newell, A., & Simon, H. (1956). The logic theory machine–A complex information processing system. *IRE Transactions on Information Theory, 2*(3), 61–79.

Owan, V. J., et al. (2023). Exploring the potential of artificial intelligence tools in educational measurement and assessment. *EURASIA Journal of Mathematics, Science and Technology Education, 19*(8), em2307.

Papamitsiou, Z., & Economides, A. A. (2014). Learning analytics and educational data mining in practice: A systematic literature review of empirical evidence. *Journal of Educational Technology & Society, 17*(4), 49–64.

Park, J.S., et al. (2023). *Generative agents: Interactive simulacra of human behavior*. In *Proceedings of the 36th Annual ACM Symposium on User Interface Software and Technology.*

Perez-Marin, D. & Pascual-Nieto, I. (2011). *Conversational agents and natural language interaction: Techniques and effective practices: Techniques and effective practices.* IGI Global.

Radford, A., et al. (2018). *Improving Language Understanding by Generative Pre-Training.*

Reidenberg, J. R., & Schaub, F. (2018). Achieving big data privacy in education. *Theory and Research in Education, 16*(3), 263–279.

Rezende, D.J., Mohamed, S. & Wierstra. D. (2014). *Stochastic backpropagation and approximate inference in deep generative models.* In *International Conference on Machine Learning.* PMLR.

Richardson, M., & Clesham, R. (2021). Rise of the machines? The evolving role of Artificial Intelligence (AI) technologies in high stakes assessment. *London Review of Education, 19*(1), 1–13.

Romero, C., & Ventura, S. (2010). Educational data mining: A review of the state of the art. *IEEE Transactions on Systems, Man, and Cybernetics, Part C (applications and Reviews), 40*(6), 601–618.

Rosé, C. P., & Ferschke, O. (2016). Technology support for discussion based learning: From computer supported collaborative learning to the future of massive open online courses. *International Journal of Artificial Intelligence in Education, 26*(2), 660–678.

Rosenblatt, F. (1958). The perceptron: A probabilistic model for information storage and organization in the brain. *Psychological Review, 65*(6), 386.

Roshanaei, M., Olivares, H., & Lopez, R. R. (2023). Harnessing AI to foster equity in education: Opportunities, challenges, and emerging strategies. *Journal of Intelligent Learning Systems and Applications, 15*(04), 123–143.

Rumelhart, D. E., Hinton, G. E., & Williams, R. J. (1986). Learning representations by back-propagating errors. *Nature, 323*(6088), 533–536.

Sadler, D. R. (1989). Formative assessment and the design of instructional systems. *Instructional Science, 18*, 119–144.

Salas-Pilco, S. Z. (2020). The impact of AI and robotics on physical, social-emotional and intellectual learning outcomes: An integrated analytical framework. *British Journal of Educational Technology, 51*(5), 1808–1825.

Santos, O. C., & Boticario, J. G. (2015). Practical guidelines for designing and evaluating educationally oriented recommendations. *Computers & Education, 81*, 354–374.

Schmidhuber, J. (2015). Deep learning in neural networks: An overview. *Neural Networks, 61*, 85–117.

Sharples, M., Taylor, J., Vavoula, G. (2005). *Towards a theory of mobile learning*, in *mLearn 2005 – 4th World Conference on mLearning.* Cape Town, South Africa.

Shen, Y., et al. (2021). An AI-based virtual simulation experimental teaching system in space engineering education. *Computer Applications in Engineering Education, 29*(2), 329–338.

Shneiderman, B. (2020a). Human-centered artificial intelligence: Reliable, safe & trustworthy. *International Journal of Human-Computer Interaction, 36*(6), 495–504.
Shneiderman, B. (2020b). Bridging the gap between ethics and practice: Guidelines for reliable, safe, and trustworthy human-centered AI systems. *ACM Transactions on Interactive Intelligent Systems, 10*(4), 26.
Shute, V. J. (2008). Focus on formative feedback. *Review of Educational Research, 78*(1), 153–189.
Siemens, G. (2005). *Connectivism: A learning theory for the digital age.* International Journal of Instructional Technology & Distance Learning, **2**.
Siemens, G. & Baker, R.S.D. (2012). *Learning analytics and educational data mining: towards communication and collaboration.* In *Proceedings of the 2nd International Conference on Learning Analytics and Knowledge.*
Siemens, G. (2013). Learning analytics: The emergence of a discipline. *American Behavioral Scientist, 57*(10), 1380–1400.
Skinner, B.F., (2016). *The technology of teaching.* B. F. Skinner Foundation.
Skjuve, M., et al. (2021). My Chatbot companion—a Study of Human-Chatbot relationships. *International Journal of Human-Computer Studies, 149*, 102601.
Slade, S., & Prinsloo, P. (2013). Learning analytics: Ethical issues and dilemmas. *American Behavioral Scientist, 57*(10), 1510–1529.
Smolansky, A., et al. (2023). *Educator and student perspectives on the impact of generative AI on assessments in higher education.* In *Proceedings of the Tenth ACM Conference on Learning@ Scale.*
Spector, J.M. (2016). *Smart learning environments: Concepts and issues.* In *Society for Information Technology & teacher Education International Conference.* Association for the Advancement of Computing in Education (AACE).
Srivastava, N., et al. (2014). Dropout: A simple way to prevent neural networks from overfitting. *The Journal of Machine Learning Research, 15*(1), 1929–1958.
Tan, S. C., Lee, A. V. Y., & Lee, M. (2022). A systematic review of artificial intelligence techniques for collaborative learning over the past two decades. *Computers and Education: Artificial Intelligence, 3*, 100097.
Tien, F. F., & Fu, T.-T. (2008). The correlates of the digital divide and their impact on college student learning. *Computers & Education, 50*(1), 421–436.
Traxler, J. (2009). Learning in a mobile age. *International Journal of Mobile and Blended Learning (IJMBL), 1*(1), 1–12.
Turing, A. (2004). Computing Machinery and Intelligence (1950). In J. Copeland (Ed.), *The Essential Turing: Seminal Writings in Computing, Logic, Philosophy, Artificial Intelligence, and Artificial Life: Plus The Secrets of Enigma* (pp. 433–464). Oxford University Press.
VanLehn, K. (1988). Student Modeling. In M. C. Polson & J. J. Richardson (Eds.), *Foundations of Intelligent Tutoring Systems* (p. 24). Psychology Press.
VanLehn, K. (2006). The behavior of tutoring systems. *International Journal of Artificial Intelligence in Education, 16*(3), 227–265.
Vaswani, A., et al. (2017). *Attention is all you need.* Advances in neural information processing systems, **30**.
Vittorini, P., Menini, S., & Tonelli, S. (2021). An AI-based system for formative and summative assessment in data science courses. *International Journal of Artificial Intelligence in Education, 31*(2), 159–185.
Wenger, E. (1998). *Communities of practice: Learning, meaning, and identity.* Cambridge University Press.
Whitehead, A. N. (1911). *An Introduction to Mathematics* (p. 251). Henery Holt and Company.
Woolf, B. P. (2010). *Building intelligent interactive tutors: Student-centered strategies for revolutionizing e-learning.* Morgan Kaufmann.
Xing, W., & Du, D. (2019). Dropout prediction in MOOCs: Using deep learning for personalized intervention. *Journal of Educational Computing Research, 57*(3), 547–570.

Yenduri, G., et al. (2023). *Generative Pre-trained Transformer: A Comprehensive Review on Enabling Technologies, Potential Applications, Emerging Challenges, and Future Directions.* arXiv preprint arXiv:2305.10435.

Zhai, X., et al. (2021). A review of Artificial Intelligence (AI) in education from 2010 to 2020. *Complexity, 2021*, 8812542.

Zhang, Q., Lee, M.L., & Carter, S. (2022). *You complete me: Human-ai teams and complementary expertise.* In *Proceedings of the 2022 CHI Conference on Human Factors in Computing Systems.*

Zohny, H., McMillan, J., & King, M. (2023). Ethics of generative AI. *Journal of Medical Ethics, 49*(2), 79–80.

Chapter 3
Reimagining Learning Experiences with the Use of AI

David Guralnick

Abstract Advances in artificial intelligence (AI) provide the opportunity for the creation of new approaches to learning. Progressive educational methods such as personalized, project-based learning have been effectively used for many years, but are labor-intensive for teachers and expensive to provide. With the aid of artificial intelligence, we can create personalized, engaging, active, meaningful learning experiences that can easily scale up and be made available to large audiences. Learning experiences that take advantage of AI can put learners in realistic environments where they can explore, make decisions, and receiving coaching guidance for example, a student who wished to learn about ancient Greece could potentially explore a simulation of ancient Greece, with an AI coach guiding them. AI-based approaches to learning have the potential to create educational systems that foster creativity and critical thinking skills, and can change the structure of both traditional education and workplace learning.

Keywords Learning by doing · Simulation · Learning experiences · Artificial intelligence · AI · Virtual reality · VR · Augmented reality · AR · Metaverse · Coaching guidance · Feedback · Critical thinking

3.1 Introduction

When I was young, I was fortunate enough to go to a fantastic school[1] during my formative years, from preschool through elementary school (until I was around age 12). The school was very non-traditional—rather than having lots of quizzes and tests, we had projects and learned by doing, and we were encouraged to be experimental, to follow our interests (even at very young ages!), and that "school" was a place of collaborative enjoyment, not a place where you competed for the highest score and that was the only focus. The school actually didn't have student grades at all

D. Guralnick (✉)
Kaleidoscope Learning, Columbia University, New York, USA
e-mail: dguralnick@kaleidolearning.com

P. Ilic et al. (eds.), *Artificial Intelligence in Education: The Intersection of Technology and Pedagogy*, Intelligent Systems Reference Library 261,
https://doi.org/10.1007/978-3-031-71232-6_3

(no number or letter grades summarizing your performance, even among the older students), and instead had detailed evaluations of each student that teachers would write and then go over with the student's parents. The evaluations were exceptionally detailed and often covered the student's growth as a person, not just their academic performance.

One of my most memorable learning experiences ever was at that school, when I was 10 years old and in the fifth grade. We had a section in a class in which we learned about U.S. state governments, and rather than simply reading about state government history and laws, we were given a project to do: each student was asked to choose a state, and the goal was to create a scrapbook showing key elements of that state's government—everything from the state bird and flower to interesting state laws. To accomplish this, we had a lot of freedom—we'd meet with our teacher regularly and individually, but her role was much more that of an advisor than that of a traditional instructor. We were encouraged to experiment and try new approaches—not just do library research (this was in the pre-Web/Google days), but to write letters to government agencies, maybe make phone calls, whatever we wanted to try as a way to achieve our goals.

This was a fascinating project and we all enjoyed it; it was individualized in so many ways, even from getting to choose your state to focus on instead of simply being assigned a state (in my case, I chose Pennsylvania, the state I was born in and lived for all of one year, so I didn't exactly have a lot of memories of it but was interested in learning more about my birth state). Also, since the overall structure and vibe of the class, and the school, were collaborative, we, as students, naturally and happily shared information with each other, often informally in conversation—for example, one of my friends had good luck reaching someone at the Chamber of Commerce of the state capital city of his state, so he mentioned that to me and I tried the same thing with my state (my friend had better luck than I did with the Chamber, though).

Overall, this was an experience that was memorable and fun, and we also learned a lot more about how state governments work from all of our varied research and discussions (and there were lots of discussions, including outside of class—we were into this project!) that we would have learned simply by memorizing facts from a textbook.

A few of the key advantages of this approach included:

- *Learning by Doing:* We learned not just by reading, but by doing something—now, we didn't get to practice being a state governor and really learning by doing in an ideal way, but the project involved active learning rather than only passive reading (the use of learning by doing dates back centuries; see Dewey (1938), among many others, and various studies have demonstrated its effectiveness—see, for example, Knight et al. (2010); also, see Schank (1997) for descriptions of a variety of early workplace learning-by-doing simulations, including many that we designed by this author).
- *Constructivism:* We really created, or constructed, something: the scrapbook that was the end goal of the project (constructivism's origin is generally attributed to Piaget (1967) and others).

- *One-on-One Coaching/Mentoring:* Our teacher played the role of a coach or mentor, rather than that of a traditional teacher, which was incredibly valuable.
- *Collaboration with other students:* We worked with our peers—the projects were individual, but the community worked together to share information and run ideas by each other.
- *Critical Thinking:* Working on this project involved a fair amount of decision-making and initiative on the part of the student; we weren't told what to do, so we needed to look at different opportunities, explore them, and think critically about our project, its progress and goals, and how we might best achieve our goals.

This was an incredibly successful project in my view, and one that followed a number of principles of progressive learning and theories from cognitive science such as situated learning (Brown et al. 1989; Collins et al., 1989).

But as successful as it was, this project had one drawback, and it was a major one: it was very, very expensive and, related, limited in access. The school's tuition was high, classes were limited in size, and this class only had one teacher, who was presumably paid well by teacher standards. She was fantastic—but even if there had been more budget, finding other teachers who were this good at all aspects of their job and were available and interested and lived close enough to the school to work there every day—well, that would have been a challenge.

Perhaps the greatest barrier to experiences like the one I had being more common and widespread is that of scalability—due to cost and resource limitations, this type of experience doesn't scale up well. We'd need a lot of money and a lot of available great teachers in a specific location to reach more students with this experience.

And that's where technology, and especially artificial intelligence (AI), can really make a difference—with thoughtful use of technology, we can design experiences that make use of the concepts we've discussed above and can be experienced by a large-scale audience. That's where we have to potential for a true revolution in education and workplace learning and the opportunity to take an approach to learning that's personalized, respectful, and often practical, and make it available to everyone. In 1983, the writer and futurist Isaac Asimov (see Asimov, 2018) predicted that by 2019:

> *Education, which must be revolutionized in the new world, will be revolutionized by the very agency that requires the revolution—the computer....*
> *There will be an opportunity finally for every youngster, and indeed, every person, to learn what he or she wants to learn, in his or her own time, at his or her own speed, in his or her own way....*
> *Education will become fun because it will bubble up from within and not be forced in from without.*

As I write this in 2024, we're not there yet, either in education or in workplace learning, but we're getting closer, especially with appropriate uses of AI—if we do it right.

3.2 Motivation and Creating a Culture of Learning

Educational courses and workplace learning experiences are frequently designed around *extrinsic* motivational factors, usually quiz and test scores—learners are motivated to do well in order to achieve a high score and the benefits that come with that. One motivational process that has gained popularity in recent years, particularly in corporate learning, is a form of *gamification* (see, e.g., Kapp, 2012) in which a leaderboard is posted and learners compete with their peers for the highest score; sometimes a prize is given to the person with the high score. Research (e.g., Malone (1981) has shown that *intrinsic* motivation—when the person is motivated from within due to finding the task enjoyable or meaningful—is more powerful than extrinsic motivation.

On my fifth-grade state government project, I was primarily motivated by intrinsic factors: I was interested in the state I was born in, I enjoyed the project process and found some parts of it especially cool—getting to write letters to, and sometimes hear back from, government offices was a fun thing for a 10-year-old and felt very "adult"—and putting together a scrapbook was a fun, personal, creative task. I also had pride in my work and wanted to do a good job, but I wasn't receiving any of the common external motivators—I wasn't receiving a letter or number grade and wasn't at all thinking about whether the project would be covered on my teacher's next evaluation that went to my parents. But the intrinsic motivating factors were powerful, plus the school culture added to the motivation as well—everyone else seemed enthusiastic about the project and about school in general. And overall my fifth-grade project seemed meaningful and relevant to me and I felt emotionally connected to it. Plus the structure of the project and the culture of the school helped us consistently feel comfortable and confident.

In contrast, much of traditional education, both in school and in workplace learning, puts the learner in a position where they feel they're being told what to do, whether it's by an instructor or by a machine. So much current online learning follows a traditional classroom format and involves watching a video clip or reading text, then answering quiz questions. These methods are typically not nearly as effective as well-designed learn-by-doing approaches, and they're not very motivating, relying almost entirely on extrinsic motivational factors.

AI and new technologies can allow us to create intrinsically-motivating, effective learning experiences on a large scale. In the next section, I'll discuss some online learning experiences we can create.

3.3 Human-Centered, Technology-Based Learning Experiences

How can we best use AI and new technologies to create learning experiences that embody the progressive learning theories of my fifth-grade project and connect with their audience, and are built in a way that can scale up?

3.3.1 A Sample Future Learning Experience

Here's one potential AI-infused experience: imagine a student who's interested in learning about ancient Greece. In a future world that's not so far away, the student might project a 3-D world on the wall of their room, immersing them in ancient Greece, as shown in Fig. 3.1. (This would be a form of virtual reality, and could be accomplished in today's world already in a slightly different form with VR goggles or glasses).

The student in this world would have the ability to explore at their own pace and based on their own interests, and would also have an AI coach to communicate with, via text or audio. The coach would always be available to answer any questions that the student had, and could also provide advice and suggestions based on the coach's knowledge of the student's interests, which would grow over time (see Figs. 3.2 and 3.3).

The student could also potentially interact with other students who were "in" the same ancient Greek world, either synchronously or asynchronously, forming and

Fig. 3.1 A student projecting an immersive image on their wall

Fig. 3.2 A student asks a question

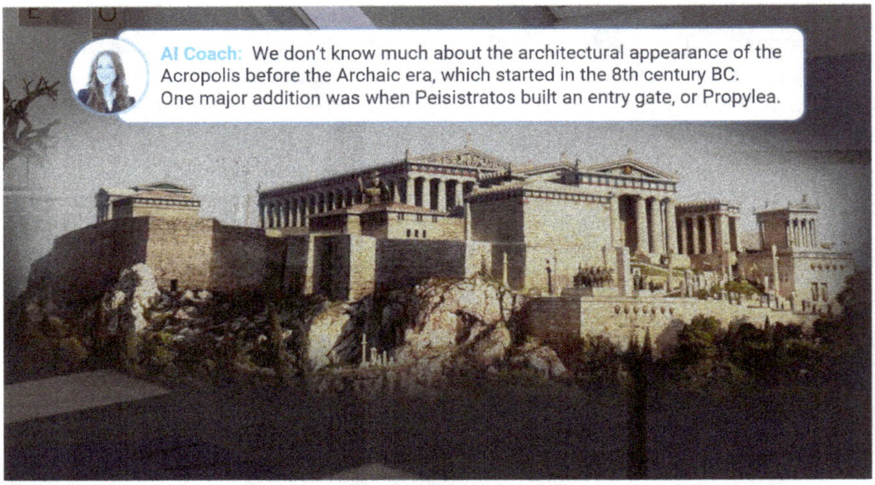

Fig. 3.3 The AI coach answers the student's question

furthering bonds with other students (I can imagine this as an experience that a cohort or class of students take together, or something that people do individually and then they can decide how much they want to interact with others).

This experience includes a number of key elements that we would expect to contribute to its success:

- *Learning by doing, in a (very!) realistic context:* This is a true learning experience, involving learning by doing in various forms—the student can explore ancient

Greece on their own, have a conversation with a simulated version of Socrates, or discuss what they see with their coach or other students.

- *Coaching guidance:* The AI coach can answer questions from the student and also play a more active role in guiding the student—making suggestions based on things that are of general interest or things that are related to what the student has shown interest in, for example.
- *Feedback (from a coach and from peers)*: The AI coach can interject to provide advice and feedback on actions the student takes in the simulated world. And the students' peers can potentially provide their thoughts as well—a variety of different activities and experiences could be designed within the simulated virtual world.
- *Learner control, but the learner is assisted and advised*: Another key feature of this experience is that the learner has a lot of control—they're not just in lockstep with other students, but can explore things of interest. At the same time, the AI coach is watching them and watching over them, jumping in with suggestions and guidance and there to help when needed. Different students may prefer, or need, different levels of guidance. Also, we can design activities in such a virtual world that do involve students doing something together when needed—e.g., perhaps everyone takes a virtual tour of the Acropolis together and then has freedom to explore after that.
- *An immersive, personalized experience:* The immersive nature and high level of personalization of this experience are at the heart of what can make this approach successful.

The learning environment and sample experience above show one way in which we can reimagine learning experiences using new technologies—rather than just taking elements of traditional education and training and mimicking them in a new-technology world or with an AI teacher, we can instead redesign, reinvent, and reimagine learning experiences, based on research, practical experience, and creativity—all in ways that can scale up and reach a large-scale audience.

3.3.2 The Potential Future of Workplace Learning: An Example

Workplace learning experiences typically need to be skills-focused and provide transfer to the actual job. In an ideal world, new employees would be able to gain experience via practice in a realistic way, thus preparing them for their role in ways that give them at least some qualities of an experienced employee before they even start.

Online training simulations (and before that, computer-based simulations that ran locally, using hard drives and other technologies such as videodisk players and CD-ROMs), have been used in the professional world for some time; one of the earliest documented examples was the Boston Chicken Cashier Trainer from 1991 (Guralnick, 1996), a learn-by-doing simulation that I created in which learners practiced using a specialized cash register to process customer orders quickly and efficiently. In this simulation, customers appeared on video with their order and the learner needed to ring up each order as quickly and accurately as they could. A text-based "tutor" provided guidance by answering specific, pre-programmed questions, and feedback when the learner took an incorrect action, and the tutor used AI techniques to provide a realistic interaction at a time when such interactions were rarely seen. Meters on the screen showed the learner's speed and accuracy scores, serving as a motivating challenge because, though the meters are an extrinsic motivational factor, they did show the learner's performance on authentic criteria, the same criteria that would define whether they did the real job well.

This simulation was successful in its time and later, similar experiences focused on soft skills, such as customer service skills (Guralnick, 2007). In soft-skill contexts, the limitations of the technology were more noticeable, particularly in terms of the learner's interaction with another character, such as a customer. The learner would have to choose from a set of options as far as what to say to a character, or sometimes piece together a sentence from components—in either case, the options were limited and risked making it too obvious what the reasonable approaches were. While many such simulations have been successful (e.g., Guralnick, 2007), a fully-realistic, immersive environment with free-form, verbal communication would be ideal. And that's the environment we can envision with the use of new AI approaches and advances in virtual reality.

Here's an example of how such an immersive simulation experience might work: Imagine a new retail customer service employee, either at their home or in a room in the back offices of a store, projecting an immersive environment on the wall in the same way that the student in the previous example projected ancient Greece. In this immersive simulation, the employee/learner could see a retail store environment with a customer near them, browsing (see Fig. 3.4).

The learner might decide to offer to help the customer and to initiate a conversation with her. As the learner does so, they can see the customer's response in realistic video, with text captions available as an option (see Fig. 3.5), and also have an AI coach watching their interaction and deciding when and whether to interrupt with feedback. At the end of the interaction, the AI coach might provide a summary of the learner's performance in key areas, even including scores for each area (see Fig. 3.6).

This experience includes a number of key elements, many of which are similar to those in the Greece experience above.

- *Learning by doing, in a realistic context:* The learner's experience with a customer in this environment looks and feels incredibly similar to the real experiences the learner will have once they're on the job—with the educational advantage that the customers and situations in this environment can be carefully designed and programmed so that learners will encounter the situations they need to learn to deal with, and in a sequence that suits them. This type of learning should result in the clear transfer of skills to the real job.

Fig. 3.4 A trainee enters an immersive world

Fig. 3.5 A sample customer interaction

- *Personalized learning:* This experience could be designed so the learner has some control over which customers to assist, but also so that the system, or the AI coach, determines which customers and situations the learner works with based on the learner's performance so far—e.g., to take a simple example, a learner who is having particular trouble dealing with angry customers might get more angry customers to practice on, or might intentionally get a non-angry customer as a "break" and then another angry customer, all at the system's discretion.

Fig. 3.6 A performance summary

- *Coaching guidance:* The AI coach can answer questions from the learner and also interject to guide them as needed.
- *Feedback (from a coach, the environment, and potentially from peers)*: The AI coach can also interject to provide feedback on actions the learner takes in the simulated world, plus the learner will get feedback of some sort from the customer's reaction. And the students' peers can potentially be involved as well—this example was designed as an individual experience, but peer feedback could easily be incorporated.
- *Learner control, but the learner is assisted and advised*: While the learner's world is fairly restricted relative to the Greece example, the learner still has a lot of control over what they do and say to a customer. At the same time, the AI coach is watching them and watching over them, jumping in with suggestions and guidance and there to help when needed.
- *An immersive, personalized experience:* As with the Greece example, the immersive nature and high level of personalization of this experience are at the heart of what can make this approach successful.

One broader key advantage of this approach is that it feels meaningful and relevant to the learners in the context of their jobs. Traditional online learning often feels boring to the learner, but even more critically, it's often hard for learners to appreciate the direct application of the training to the job that they'll perform. The immersive environment described here leverages the strong points of previous learn-by-doing simulations and goes far beyond that, to create a learning experience that feels very much like doing the real job but in a safe, simulated environment where the learner's mistakes don't have real-life consequences, and where they can have guidance and feedback from a coach.

3.3.3 Improving Workplace Performance Without the Need for Training

In workplace situations, the employee's goal is to perform their job well—effectively and efficiently—and "learning" in the traditional sense fits in terms of being in service of the overall performance goal. AI and new technologies can play a very critical and useful role in improving performance via the technology that's "just-in-time"—the employee uses the system when they need it while performing the job. Such a "performance support" approach (Gery, 2002; Rossett & Schafer, 2007) works well and often minimizes the need for training, and works best in cases where the timing of the interactions works—for example, a retail customer service employee needs to be able to work with customers well before they actually do so—at least with today's technology, there's no appropriate way for an in-person customer service employee to consult a system to help them interact with customers while they're in the middle of the conversation, other than for specific, appropriate things such as looking up information about a product or price.

Traditionally, performance support experiences are often straightforward—an information reference that an employee can search, or, as a bit more complex example, a system that augments a piece of software with suggestions about how to complete a particular task. These are often very useful systems; with new technologies, including virtual reality (VR), augmented reality (AR), and sensors, "just-in-time" performance support experiences can reach an entirely new level of effectiveness for people in a wide variety of jobs. Let's take a look at an example of a potential such system.

Imagine a veterinarian, working in their office, examining a dog. In this new world, an AI coach might speak to the veterinarian, either via voice or text, and say "Based on information from the sensors, you might want to check her right back paw to see if she has a cut there." (see Fig. 3.7). And then as the exam continues, the veterinarian might communicate with a colleague who could also, through the magic of technology, see the dog who's being examined. For example, the veterinarian might notice a spot on the dog's ear and ask an oncologist colleague for their opinion (see Fig. 3.8). Of course they could do this today by sending or texting a photo to the colleague, but this vision allows a more seamless way for primary veterinarians and specialist veterinarians to work together.

This experience has several key elements:

- *Just-in-time performance support:* The entire experience is designed not to train the veterinarian (or the dog), but to support the veterinarian to make their work better, easier, and more efficient, while still leveraging—and not diminishing—their expertise.
- *Coaching guidance, in real time: The* AI coach is there to help—for questions the veterinarian may have and also to interject, as we saw in the example.
- *The integration of sensors/other technology into the experience:* This example demonstrates a way in which additional technology—in this case a sensor that can indicate a potential cut that would be easy to miss otherwise—can be integrated

Fig. 3.7 Assistance from an AI coach and information from sensors

Fig. 3.8 Collaborating with a colleague, a veterinary specialist

into a performance support experience. The sensor information could be made available to the veterinarian directly or, as in this example, via the AI coach.

- *Real-time collaboration with colleagues:* the technology here was designed to support and facilitate easy, efficient interaction with a colleague, integrated into a natural workflow.

AI plays a role in the coaching component, and it's easy to imagine the sensors and other technologies using AI to learn about the specific animal being examined and create a completely new model of an animal's medical chart. For example, there's already technology that examines a dog's DNA and returns a report noting the dog's potential risk for certain diseases, which can be used to make medical decisions, such as adding medication or changing a dog's diet, to lessen the chances of the dog ever having those diseases. Information, in this situation, really is power, and AI can take this type of analysis to a new level, in the veterinary field or in nearly any field.

3.3.4 A Holistic Approach to Experience Design

When new technologies begin to gain widespread use, a natural tendency is for people to employ them in ways that resemble the uses of previous technologies. This has certainly occurred in the world of educational technology, where a large number of online learning experiences have been designed to mimic the classroom environment in many ways, either by focusing on lectures via video, live virtual sessions, or the online "read and test" model that remains common in corporate learning. These models essentially take existing classroom approaches and use technology as a way to reach a larger audience but with a similar learning experience.

Following this approach misses the opportunity to take advantage of the capabilities of each new technology to reimagine learning experience design—and even more, many of the existing classroom models gained popularity out of necessity, as a way to educate large numbers of students and allow efficient grading and evaluation of students in a pre-internet age—these were never the best approaches, just the most practical—it was impossibly expensive, at the very least, to provide individualized instruction and exciting experiences on a large scale.

But with new technologies, the world of possibilities changes—we can, or will soon be able to, create magical immersive worlds using virtual reality (VR) and holograms; use AI to learn about a particular student and provide individualized learning, including not just working on their weaknesses but supporting their strengths and providing ways for them to explore and pursue their interests. In order for learning experience designers to best take advantage of what new technologies can provide, there's a need to focus on the experience—starting with the learning and performance goals and the audience, being creative in terms of what the experience can be like—rather than viewing information as the key currency in education and workplace learning, as it so often is.

Ideally, in my view, the design of learning experiences should be not just about information or "content," but should instead follow a creative approach that takes advantage of different uses of technology—and not just cutting-edge technologies, either—for example, creating video clips or asking students to create videos could be a part of such an experience and keeps the overall learning and performance goals and the learners' interests, emotions, and time commitments in mind as well. In my ideal world, learning experience design will often involve designers coming up with new experiences that employ technology in new and different ways, for individual and/or collaborative use, following the path shown by the examples above in this section but continuing to go far beyond those.

3.4 AI's Uses in Education: The Past

Artificial intelligence has a long history in terms of its involvement and potential involvement in learning experiences. A few key areas in AI history that relate to learning experience design and laid the groundwork for many of the ideas in the examples above. Those areas are summarized below.

3.4.1 Eliza

Joseph Weizenbaum's ELIZA (Weizenbaum, 1966) was a program that was able to hold a realistic-feeling natural language conversation with a user. The ELIZA program's conversations involved ELIZA playing the role of a psychotherapist, and relying on keywords and a set of rules to formulate its responses. A sample conversation excerpt from Weizenbaum's seminal article is below (the computer system that Weizenbaum used did not allow question marks to be used in the generated text):

User: Men are all alike.
Eliza: In What Way
User: They're always bugging us about something or other.
ELIZA: Can You Think of a Specific Example
User: Well, my boyfriend made me come here.
ELIZA: Your Boyfriend Made You Come Here
User: He says i'm depressed much of the time.
ELIZA: I AM Sorry To Hear You Are Depressed

The conversations often felt realistic, and ELIZA is sometimes referred to today as the first chatbot. The ELIZA system didn't truly understand what the user was saying, but it was nonetheless able to simulate the feel of a real conversation in the role and context that Weizenbaum chose (which were no accident! The role of a therapist, in which it's often appropriate to repeat parts of the other person's statement or ask a general question, was perfect for ELIZA. ELIZA's conversations often felt realistic, and Weizenbaum's hope (Weizenbaum, 1966) at the time of ELIZA's first release was that ELIZA could evolve over time to better learn from its conversations and eventually "prove an interesting and even useful conversational partner" (Weizenbaum, 1966). While ELIZA may not have evolved as Weizenbaum dreamed that it might, and his approach differed significantly from that of ChatGPT and other Large Language Models we see today, the program did prove to be extraordinarily influential over time.

3.4.2 Intelligent Tutoring Systems

In the late 1960's and 1970's (and continuing after that), *intelligent tutoring systems* that could play the role of a teacher became a popular area of research and experimentation, perhaps most notably at Carnegie Mellon University, initially due to the work of Jaime Carbonell (1970). These systems were built to teach students specific content areas and to have helpful, rich dialogue with students, often helping them learn to solve problems within a particular content domain. A later well-known example of an intelligent tutoring system is the LISP tutor from Carnegie-Mellon's John Anderson and his colleagues (Anderson, Conrad, & Corbett, 1989), which taught students how to program in the LISP programming language. Creating an intelligent tutoring system, or ITS, typically required substantial effort in terms of the "knowledge representation" work needed to define the rules, data structures, and data that would drive the ITS for a particular content domain. ITSs have certainly had some successes over the years but have never really become widespread in their use.

3.4.3 Expert Systems

During a similar time period, in roughly the mid-1960's and 1970's, work on *expert systems*, artificial-intelligence systems that could make decisions, such as in medical areas. Edward Feigenbaum was considered the major pioneer in this area; the DENDRAL system that he created along with other colleagues as a way to do chemical analysis (Lindsay et al., 1993). Expert systems had the goal of advising people rather than teaching them, and were, in a sense, kindred spirits to the later electronic performance support systems (Gery, 2002) that helped people perform a task rather than necessarily teaching them how to perform the task themselves, though sometimes expert systems' "advice" was really the answer to the problem rather than just advice.

3.4.4 Learn-BY-Doing Simulations

In the late 1980's and early 1990's, there was significant work on learn-by-doing simulations that were created using AI techniques (e.g., see Kass & Guralnick, 1991; Schank, 1997), in which learners practiced a role or skill in a safe, simulated environment, with teaching guidance and feedback. In the corporate learning world, the program that was considered the first corporate learning-by-doing simulation was from my own work as a Ph.D. student of Roger Schank at his Institute for the Learning Sciences at Northwestern University (see Guralnick, 1996). This program, completed in 1993, helped employees of Boston Chicken, a fast-casual food chain, learn how to ring up customers' orders quickly and accurately using the specialized, custom cash register that Boston Chicken restaurants used at the time.

In this simulation, created for new cashier employees as part of their initial training, learners play the role of a cashier and have to process the order of a customer who appears in video with their food. The goal is to process each order as quickly and accurately as possible by using the on-screen cash register and pressing the correct keys. This is perhaps more difficult than it sounds; the keyboard was designed for efficient use by trained employees, but some training was definitely necessary—as one "tricky" case, a customer might order a half chicken with spinach and potato salad, and that would be correctly rung up using the "1/2 WITH SIDES" key on the keyboard rather than ringing up the half chicken, the spinach, and the potato salad individually, even though each of those did have their own dedicated key on the keyboard. The notion of "correct" was not just due to speed, but to accuracy as well—the "1/2 WITH SIDES" key correctly applies a discounted price relative to the prices of all the items individually.

At any point in the simulation, the learner can make use of three buttons that allow them to conduct a simulated mini-dialogue with a tutor; these buttons are labeled NOW WHAT?, HOW DO I DO THAT?, and WHY?, and a sample interaction might look like the following:

> *[A customer appears on video with a half chicken with spinach and potato salad; the customer tells you what they have and you, as the learner, can also see the order in the video].*
> *Learner: Now What?*
> *Tutor: You need to ring up the customer's order.*
> *Learner: How Do I Do that?*
> *Tutor: Click the "1/2 with sides" key, which is flashing now [key flashes on keyboard]*
> *Learner: Why?*
> *Tutor: You should click the "1/2 with sides" key because that's the most efficient and accurate way to ring up a half chicken with spinach and potato salad—those are the sides.*
> *Learner: [clicks "1/2 with sides" key, cash register display updates accordingly].*

The learner can also ask a few follow-up questions of the customer using on-screen buttons, and as the learner progresses through the scenario, the speed and accuracy meters update constantly. If the learner takes an incorrect action, the tutor will interrupt and explain why it was incorrect.

The learning experience included a variety of different scenarios, and learners start with simpler cases and gradually take on more difficult situations. The tutoring component used AI techniques and was created using a set of rules plus data for each situation, similar in some ways to the ELIZA approach.

Overall, this simulation was very well-received; the skills learned clearly transferred to the real job, and the on-screen meters added a nice "challenge" element in an authentic way—this was a realistic game, not an artifice, in that speed and accuracy, the measures showing on the screen, are the same measures that indicate performance on the real job.

Looking back at key uses of AI in education over the past decades, we can see how the foundation was created for a lot of the experiences we can now create—as technology has improved—and will soon be able to create.

3.5 AI's Uses in Education: New and Evolving Experiences

As AI improves and technologies such as AR, VR, and holograms become more widespread, new learning experiences that resemble the "future" learning experiences shown earlier in this chapter are becoming possible and practical. Examples include:

- Medical simulation using holograms and other technologies: Fernando Salvetti and Barbara Bertagni's e-REAL Labs has created a series of interactive simulations for the medical field, in which learners can learn by doing in a mixed-reality environment using sensors, holograms, and more. One example is a "phygital" classroom at the International Red Cross "Luigi Gusmeroli" Learning Center in Bologna, Italy, in which students learn medical procedures in a hybrid simulation (Salvetti et al, 2022).
- Uses of humanoid robots as teaching assistants in a university environment: Ilona Buchem's group at the Berlin University of Applied Sciences built learning experiences around robots that "could communicate via speech, non-verbal sounds, visual pattern recognition, gestures, and touch" (Buchem, 2023).
- A company called Interflexion has created AI-based role-play simulations that have been used in sales, ethics training, and other area (Hack, 2023).

Other examples include uses of VR and AR to create simulated and mixed-reality worlds that can help people learn in ways that have not been possible before, and particularly not possible to do on a large scale. Advances in mainstream technology from Unity, Meta (particularly via the Quest 3, a mixed-reality headset with advanced gesture control) provide more and more opportunities for the creation and deployment of new immersive learning experiences.

One key to the future success and acceptance of technology-based learning is to ensure that it's truly built around learning principles—it's not sufficient to just create a simulation environment, but to create a learning experience that incorporates guidance and feedback, and is structured in a way that puts learners in situations that are challenging and help them truly learn. The examples in this section are just a handful of experiences that demonstrate the range of possibilities.

3.6 The Potential Impact of AI-Based Learning Experiences

AI and other new technologies have the potential to revolutionize learning; the examples above provide a feel for how AI-based experiences can work and allow us to effect real change in the worlds of education and workplace learning, by moving away from the traditional models of instruction and instead taking advantage of new technologies to create different learning experiences and different pedagogical methods that can reach a large-scale audience. Thoughtfully-designed learning experiences that use technology in creative ways can completely change the world of education by marrying progressive learning theories with new technologies to create learning that's truly personalized, meaningful, engaging, and fun.

It seems plausible to imagine that if standard education shifts from a lecture-based, teacher-centered format to one that focuses on individuality, on finding your strengths, on doing things and experimenting—along the lines of my own fifth-grade experience—and can do so affordably on a large scale, thanks to technology, this could also change the learning culture of many schools. Many students often consider education today to be a chore rather than an enjoyable experience, which is understandable considering the current structure of many courses. In my own experience, my family moved not that long after my fifth-grade learning experience, so I needed to switch schools and ended up in a more traditional setting. As one memorable seventh-grade experience in my new school—but not in a good way—we were asked to memorize and write down, as a quiz, the rules of capitalization and punctuation. I believe there were 12 such rules. Now, I was a disciplined student with a good memory, so I did this task and that was that. But even as a seventh-grader, or perhaps especially as a seventh-grader, the task struck me as silly—it was clearly just a memorization test and it seemed easy to argue both that a student could be good at capitalizing and punctuating without being able to list the rules, and also that a student could recite the rules but not be able to put them into practice. Given tasks such as this one, it was easy for students to consider school something to get through rather than something to embrace. But with new, progressive learning experiences that use technology so they can scale up, we can create new generations of students—and workplace learners—who feel engaged in learning and that learning is natural and fun. It's also challenging and hard work a lot of the time—but working hard at something you're excited about feels very different than working hard at something that feels boring or irrelevant.

One key risk as education and workplace learning involve even more technology in the future is that education could become less human rather than more human. The approaches I suggest above all use technology in ways that will help people feel emotionally connected to the experience, even if it's an experience that doesn't involve interaction with other people, and the experiences above all put the learner in a position where they can make decisions, explore their interests, and feel respected, often much more so than in a traditional classroom. But with more technology comes the risk of schools and organizations focusing on standardization and scoring rather than on personalization and exploration, as it becomes easier to automate detailed performance evaluations. In my view, while there's certainly a role for evaluation, including in helping personalize the learning experience, we also need to be sure to foster experimentation, confidence, and emotional connections, as we saw in my fifth-grade example. It's all too easy for the technology to take over and for future learning to feel more rigid and dehumanizing, when the feel that we want is just the opposite.

3.7 What's Next

As time goes on, current technologies will tend to improve and become more affordable, and new technologies will emerge. There will continue to be more opportunities to imagine and create new types of learning experiences; one key to the widespread adoption of new experiences is for there to be a practical way for people to create high-quality learning experiences, so it may be ideal if we see new and different tools that help in the creation process. AI could certainly end up playing a critical role in this process. Change is always difficult, and having tools that help make it easier for people to create new learning experiences greatly increases the chances that we can see real change and a real movement to new models of education and workplace learning.

The potential of AI to help revolutionize learning experiences—with the right philosophy and thoughtful, creative approaches to experience design that incorporates AI—is tremendous. By leveraging progressive learning theories and a willingness to rethink what educational experiences can be like, we have the opportunity to create a new world of learning, through school, work, and life.

Acknowledgements This work would not have been possible without contributions from Christy Levy, Veira Petersen, Lara Ramsey, Anelise Spyer, and Marcos Guisande. Photo credits to Juan de Lara, www.archaeological-reconstructions.com, for the Acropolis image (modified by Marcos Guisande) and kjpargeter/Freepik for the bedroom image in the "future learning experience" Greece example.

References

Anderson, J.R.., Conrad, F.G., & Corbett, A.T. (1989) Skill Acquisition and the LISP Tutor. Co

Asimov, I. (2018). 35 years ago, Isaac Asimov was asked by the Star to predict the world of 2019. Here is what he wrote. (Originally published in 1983). Toronto Star, December 27, 2028. https://www.thestar.com/news/world/35-years-ago-isaac-asimov-was-asked-by-the-star-to-predict-the-world-of/article_4c25735d-c2e4-5bde-8774-21cc2e5a5561.html

Buchem, I. (2023). Designing learning experiences with humanoid robots as teaching assistants. From "Creativity in Learning Design," The Learning Ideas Conference's Winter Event.

Brown, J. S., Collins, A., & Duguid, P. (1989). Situated cognition and the culture of learning. *Educational Researcher, 18*, 32–42.

Carbonell, J. R. (1970). AI in CAI: An artificial-intelligence approach to computer-assisted instruction. *IEEE Transactions on Man-Machine Systems, 11*, 190–202.

Collins, A., Brown, J. S., & Newman, S. E. (1989). Cognitive apprenticeship: Teaching the craft of reading, writing, and mathematics. In L. B. Resnick (Ed.), *Knowing, learning, and instruction: Essays in honor of Robert Glaser* (pp. 453–494). Lawrence Erlbaum Associates.

Dewey, J. (1938). Experience & education. New York, NY: Kappa Delta Pi.

Gery, G. (2002). Performance support—driving change. In Rossett, A. The ASTD E-Learning Handbook. New York: McGraw-Hill.

Guralnick, D. (1996). An authoring tool for procedural-task training. Ph.D. dissertations, Northwestern University's Institute for the Learning Sciences.

Guralnick, D. (2007). Online learning by doing for soft skills: A methodology and an authoring tool. In the Proceedings of European Conference on E-learning (ECEL) 2007, Copenhagen, Denmark.

Hack, J. (2023). Developing conversational fluency at scale with AI-driven interactive role-play. From The learning ideas Conference 2023, New York.

Kass, A. & Guralnick, D. (1991). Environments for incidental learning: Taking road trips instead of memorizing State Capitals. In the Proceedings of the International Conference on the Learning Sciences, Evanston, IL.

Kapp, K. (2012). *The gamification of learning and instruction: Game-based methods and strategies for training and education.* John Wiley & Sons.

Knight, J. F., Carley, S., Tregunna, B., Jarvis, S., Smithies, R., de Freitas, S., Dunwell, I., & Mackaway-Jones, K. (2010). Serious gaming technology in major incident triage training: A pragmatic controlled trial. *Resuscitation, 81*(9), 1175–1179.

Lindsay, R. K., Buchanan, B. G., Feigenbaum, E. A., & Lederberg, J. (1993). DENDRAL: A case study of the first expert system for scientific hypothesis formation. *Artificial Intelligence, 61*, 209–261.

Malone, T. W. (1981). Toward a theory of intrinsically motivating instruction. *Cognitive Science, 4*, 333–369.

Piaget, J., & Inhelder, B. (1967). *The Child's conception of space.* W. W. Norton & Co.

Rossett, A., & Schafer, L. (2007). *Job aids and performance support: Moving from knowledge in the classroom to knowledge everywhere.* John Wiley & Sons.

Salvetti, F., Garnder, R., Minehart, R., & Bertagni, B. (2022). Effective extended reality: A mixed-reality simulation. In Guralnick, D., Auer, M., and Poce, A., Innovative Approaches to Technology-Enhanced Learning for the Workplace and Higher Education: Proceedings of 'The Learning Ideas Conference' 2022. Cham, Switzerland: Springer Nature.

Schank, R. (1997). *Virtual learning: A Revolutionary approach to building a highly skilled workforce.* McGraw-Hill.

Weizenbaum, J. (1966). ELIZA: A computer program for the study of natural language communication between man and machine. Communications of the ACM, 9(1).

Chapter 4
Generative AI Integration in Education: Challenges and Approaches

Steven Watson and Shengpeng Shi

Abstract Generative AI, exemplified by models, such as ChatGPT, Gemini, and Midjourney, carry transformative potential for educational personalisation and inclusion. Yet, its integration into educational systems raises concerns about its ethical use, plagiarism, and bias. In this chapter, we explain what generative AI is, how it functions and discuss how it can be integrated into educational programmes and practices. We advocate for approaches that are human-centred, where educators and learners have direct access to and are encouraged to use generative AI models to develop generative AI literacy–learning by doing . Although it is a challenging mode of integration, we set out an approach for researching the ethical and effective integration of generative AI in education. This involves conducting participatory transdisciplinary research using a design-based research methodology (DBR). This is to facilitate iterative but sustainable transformation in educational practices, advocating for systemic change over superficial adoption.

Keywords Generative AI · Generative AI integration · Design-based research · Generative AI literacy

4.1 Introduction

Since the release of ChatGPT by the US tech company OpenAI in 2022, there has been considerable speculation and interest in generative AI's potential for education as well as in other aspects of society. Many have gone as far as suggesting that it has the potential to revolutionise education (e.g., Bahroun et al., 2023; Gao et al., 2024; Prather et al., 2023), but with both opportunities and challenges (e.g., Fengchun &

S. Watson · S. Shi
Faculty of Education, University of Cambridge, Cambridge CB2 8PQ, UK
e-mail: sw10014@cam.ac.uk

S. Watson · S. Shi (✉)
College of Education for the Future, Beijing Normal University, Zhuhai, China
e-mail: ss2619@cam.ac.uk

P. Ilic et al. (eds.), *Artificial Intelligence in Education: The Intersection of Technology and Pedagogy*, Intelligent Systems Reference Library 261,
https://doi.org/10.1007/978-3-031-71232-6_4

Holmes, 2023; Michel-Villarreal et al., 2023; Yan et al., 2023). While generative AI is impressive in its capacity to generate new content based on extensive training data, how it can be integrated into education is less clear (e.g., Kaplan-Rakowski et al., 2023; Michel-Villarreal et al., 2023). Indeed, since the release of ChatGPT, and the subsequent release of DALL-E, Gemini (previously Google Bard), Claude and Midjourney, for example, the use of generative AI in educational contexts has been quite limited[1]. It has attracted widespread media attention, both positive and negative, in terms of its impact on education (Fütterer et al., 2023). This has created concern that the educational use of generative AI has the potential to undermine the integrity of the educational system and its assessment processes (e.g., Woolcock, 2023).

Concerns are around Generative AI's potential for contract plagiarism, wherein students simply prompt generative AI models to produce assessed work (Grassini, 2023). This concern is exacerbated by the fact that content produced by generative AI is not reliably detectable by plagiarism detection systems (Chaka, 2023). This is because the output produced by generative AI in response to a user prompt is based on a sophisticated recursive probabilistic model and therefore each response is unique. It has been further suggested that learners may become over reliant on the technology and that it might undermine their critical thinking. There have also been widespread concerns raised about ethical and regulatory issues. The extensive training sets used in developing Large Language Models (LLMs) inherently contains structural biases. Consequently, output from models like ChatGPT may reflect these biases, leading to content that predominantly aligns with perspectives such as white, anglophone, elite, and patriarchal viewpoints. It may contribute to societal issues like perpetuating gender stereotypes and even promoting Western cultural imperialism. In addition, there are unresolved regulatory issues in relation to the copyright and attribution to the original content creator in training datasets. Concerns also persist about data privacy and the protection of personal data in both training data and user prompts (Kasneci et al., 2023).

The response of educational leaders in some cases has been to restrict the use of and sometimes even ban its use (e.g., Johnson, 2023). Yet, its educational potential remains, in terms of its potential for individualised and personalised learning and for educational inclusion and accessibility (e.g., Nixon et al., 2024). Thus, advocates for the use of generative AI in education have been frustrated at the resistance and the control of the revolutionary potential of generative AI.

The central challenge in exploiting its capabilities for personalisation and inclusion comes down to integration. How can generative AI be introduced into educational contexts so that it can be used ethically, safely, effectively and in regulatory compliant ways? This chapter put forwards an approach to doing this through contextualized participatory transdisciplinary research and development approaches to realise the potentials of the technology in sustainable ways, in other words that lead to systemic change in educational practices.

In what follows, we begin with an explanation of what generative AI is and its capabilities beyond content generation. We then consider the potential use of generative AI in educational contexts. We go on to reflect on the challenge of educational change and how the autonomy and integrity of the education system are highly dependent on its continuity. The challenges in integrating technological innovations into existing educational programmes and practices are thus outlined. We conclude by proposing an approach to the integration of generative AI using a design-based research (DBR) methodology which we argue can contribute to its sustainable and systemic integration. Moreover, by using an analytic generalisation strategy from specific iterative contextualised research endeavours, knowledge and theory can be developed to inform other generative AI integration contexts and that can contribute to educators' and learners' generative AI literacy—that is an awareness of the rationale behind the technology and how to use it ethically, safely, creatively and effectively for teaching and learning, while being regulatory compliant in terms of copyright and data privacy.

4.2 What is Generative AI?

The broad definition of generative AI is that it is an artificial intelligence that produces human-like content in response to user prompts, in text, image, audio and video, for example. However, this definition of generative AI does not fully account for the technology's capability. To comprehend this, it is necessary to consider the architecture and functioning of generative AI models. There are two main types of generative AI, all using deep learning technology based on neural networks. Text-based models, i.e., LLMs such as ChatGPT, Gemini and Claude, use Transformer architecture that employs self-attention giving them the capability to produce contextually specific and nuanced responses in text (Vaswani et al., 2017). For many image, video, and audio applications, for example, the underlying technology uses a Diffusion model (Ramesh et al., 2022). There are similarities in the overarching function of Transformer architecture and Diffusion models, but they are substantially different in the way in which they create content. However, for our purposes here, we are going to focus on text based LLMs that use Transformer architecture. This is because text remains the dominant medium in educational settings, even though other multi-modal media have become and will continue to become increasingly important educationally. Furthermore, the overarching principles and capabilities that we explain in terms of LLMs and Transformers can be used to understand image, audio and video generative AI based on Diffusion models (see Watson, In review).

An LLM is initially trained on an extensive data set of publicly available text on the Internet. This allows it to quantify its millions or even billions of parameters. These are weightings within the neural network architecture that allow it to identify relational features in the training data. This begins with tokenisation, which differentiates the training data into elements, often this is at the word level, but can also be phrases, syllables, characters, and symbols. The different approaches to tokenisation, word, syllable or symbol, permit the model not only to deal with multiple languages but also different symbolic systems such as computer code or mathematics.

The relational dependencies can be explained in terms of *word embeddings*. This is an approach that predates LLMs and is a way in which the model can establish and quantify relationships between words in a multi-dimensional vector space. These multi-dimensional relationships are also used with other tokenised symbolic systems. This, although very difficult to imagine in terms of the scale of its multidimensionality and vastness of the numbers of relationships, is relatively simple to comprehend when one considers an example.

In word embedding the word 'jump' will be spatially close to similar words like 'leap', 'hop', 'bounce', 'skip' etc. (see Fig. 4.1). Words that are similar but less close to 'jump' in the vector space of word embeddings might include terms that share a broad theme of movement or transition but do not directly describe the action of jumping. These words may reflect related concepts or physical activities where jumping might be a component or associated action, such as, 'run', 'climb', 'slide', 'glide' or 'fly'.

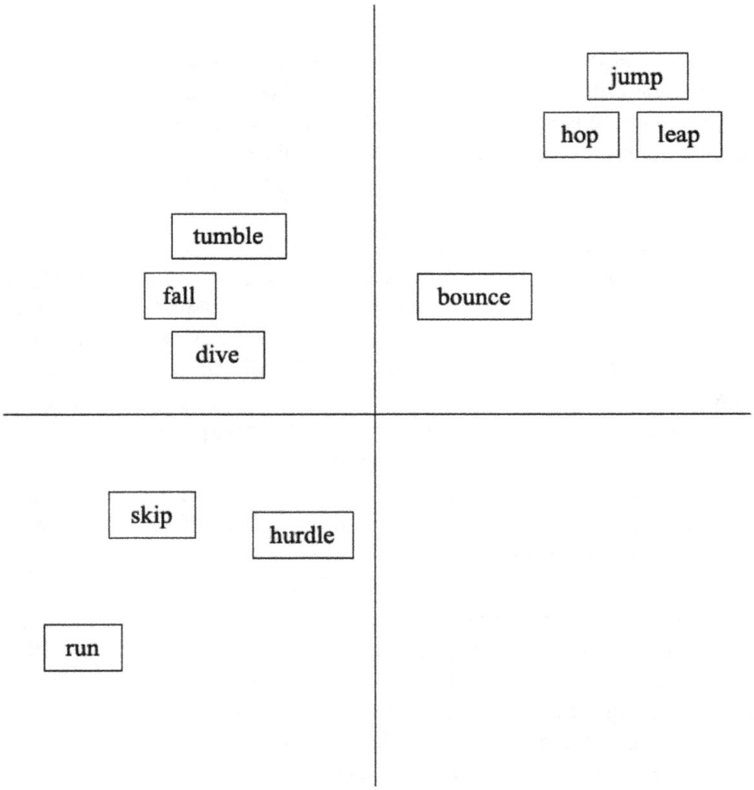

Fig. 4.1 Simplified illustration of word embeddings in 2-dimensions using proximity based on meaning

This is where it becomes difficult to visualise the vast multidimensional vector space of all possible relationships between words with proximity and distance reflecting closeness or variance in meaning or syntactic function. This multidimensionality allows for words having different meanings in various contexts[2]. The possibility of imagining the complexity of this when extended to multiple languages and different symbolic systems is further stretched. But what is ultimately mind bending is to consider that LLMs do this at the tokenisation level, where the multidimensional vector space is a representation of parts of words, syllables, or symbols, or parts of characters such as in Chinese. The approach to tokenisation is dependent on the model.

What should be understood from this is that LLMs have a vast record of possible contextualised meaning and their relationship to semantic and syntactic structures. This is why a response to a prompt is s meaningful or plausible but is not necessarily factually correct. It does not work like a search engine which uses keywords to find information and knowledge but works entirely on meaning and syntactic relationships. This is the basis on which users observe hallucinations, i.e., responses such as fictitious references that look real but do not exist. This is also why understanding generative AI is important, as its capability is not in finding facts but in exploring and processing contextualised meaning. But we shall return to this as there is another important dimension of most contemporary LLMs and that is Transformer architecture and self-attention.

Word embeddings provide a powerful means of generating a model of meaning and syntactic relationships. But this offers only a partial explanation of the way in which LLMs generate plausible and contextually meaningful responses to user input and their Natural Language Processing (NLP) capability. NLP means that users input prompts in natural language, the model translates this into other forms while maintaining the meaning of the original input, seemingly understanding the input. The response might be paragraphs of text in response to a stream of consciousness set of notes or ideas, a poem from a piece of academic writing, an explanation of the steps in a mathematical proof, producing computer code from everyday language or even turning an angry email response into something that is more nuanced and less emotional.

The key to this capability lies in the way LLMs can process sequential data such as text. A piece of text has meaning as a whole and the elements (e.g., words or phrases) within it all contribute to the whole meaning as well as being related to each other. The breakthrough in NLP capability in LLMs was the development of Transformer architecture which uses *self-attention*. This means that a user prompt is not processed sequentially, self-attention allows the model to identify which elements of the text are important in the overall meaning of the text and to weight aspects of the text in terms of their contribution to the overall meaning of the text[3] This utilises the model's parameters, or in simple terms, the multidimensional vector space of word embeddings. This allows the development of a coherent response to a prompt that is consistent with the meaning and semantic structure of the user input.

Consider the sentence: 'The bank of the river was flooded'. In earlier models without self-attention, each word would be processed in sequence, potentially leading to ambiguity about the meaning of 'bank'—whether it refers to a financial institution or the side of a river. The Transformer architecture, through self-attention, assesses the entire sentence, recognizing that 'river' and 'flooded' are key to understanding and processing 'bank' in this context as the side of a river.

This can also be illustrated with another example of how an LLM writes computer code. When prompted to write a Python function for the Fibonacci sequence up to a certain number 'n', the LLM discerns the task as coding, focusing on 'Fibonacci sequence' and 'up to n' as key elements. Through self-attention and word embeddings, it 'understands' the Fibonacci sequence as numbers formed by the sum of the two preceding ones, with 'n' marking the generation limit. Recognizing the need for an input parameter and sequence generation method within the limit, the LLM constructs code to fulfil the request based entirely on the semantic and syntactic relationships between elements of the programming language, although it has no database of syntax or logic capability. This explains why the code is not always syntactically correct or will function correctly.

To really understand the educational potential of generative AI, it is important to comprehend generative AI not in terms of its generative capability but in terms of its *semantic transduction* capability (see Watson, In review). This reflects its capacity to deal with contextualised meaning and transform and translate (i.e., transduce) text from one form to another while keeping the meaning same. Or, to change the meaning and form of a text as instructed by the user. This can be interlingual or intralingual translation, summarising text, rephrasing, providing alternative perspectives, or explaining mathematical reasoning without the model 'knowing' mathematics.

The key then to the capability of LLM generative AI and its potential for education is in its semantic transduction capability over its ability to simply generate content.

Three Categories of Generative AI Integration

Much has been said about the potential for generative AI to revolutionise education through personalisation as well as, of course, the widespread concerns about its use educationally. Here, we will consider this in terms of how it is being understood and presented in the literature and how personalisation, individualisation and inclusion might be facilitated through generative AI's semantic transduction capability.

Personalised learning has been an educational consideration since antiquity, from Socrates to Rousseau to Montessori to Vygotsky, Gardner and Bloom, the idea that pedagogy and curriculum should be adapted to individual needs, perspectives and learning styles, representing a prominent theme throughout history. The tension between personalised and standardised education reflects broader debates about the goals of education, equity, and the practicalities of implementing individualised learning at scale. While personalised learning emphasises the development of individual potential and aligns with constructivist theories of education, standardized education focuses on equality, accountability, and the logistical realities of educational delivery in large and diverse societies (Tyack & Cuban, 2000).

The role of technology in personalised learning can be traced back to correspondence education from the late nineteenth century, through broadcast technologies in the mid twentieth century (Anderson, 2008), to computer-assisted learning and online learning in the latter part of the twentieth century. This evolution reflects a shift from teacher-centred methods to learner-centred approaches, facilitated by technology's capacity to tailor educational experiences to individual learners' needs, abilities, and interests. The latest phase includes AI-based adaptive learning technologies (Xie et al., 2019). These systems analyse learners' interactions, performance, and preferences in real-time, adapting content, pacing, and learning strategies to the individual. The use of data analytics and learning analytics has become integral to personalised learning, enabling educators and education systems to make informed decisions based on data from learner interactions.

We identify three broad categories of generative AI integration for personalisation and inclusion (see Table 1.1). The first type, *Autonomous Generative AI Tools* (GAITs) include AI autonomous generative AI systems like ChatGPT, Gemini or Claude, which are designed to create content or data from inputs. In these systems, users directly interact with the model; users input prompts and receive AI-generated outputs, functioning independently without integration into other technological systems. The chatbot presentation allows the user to develop output iteratively through the course of a 'chat'. The second category of generative AI integration involves the *integration of generative AI into applications* (IGAA), exemplified by Microsoft Copilot and more recent developments with Google Gemini. IGAAs augment the functionality and usability of existing applications, thereby facilitating efficiency and user experience. The generative AI components enhance existing software applications, such as word processing, email and spreadsheets, by embedding augmentative features within existing functionalities and make use of NLP to improve productivity. The third category of integration are *Generative AI-based Automatic Decision-Making Systems* (GADMS), these employ AI for data analytics and logic processes to facilitate autonomous decision-making. GADMS analyse data and generate insights while relying on logical systems for decision-making, using generative AI to augment input and output functionality. They extend on existing adaptive learning systems to enhance adaptive learning capabilities. GADMS and are also utilized across various sectors like finance, healthcare, and logistics to automate complex decision-making processes. An example of an experimental educational GADMS is Khan Academy's Khanmigo.

This categorization delineates the spectrum of autonomy in generative AI applications, from fully user-dependent tools to completely autonomous decision-making systems. GAITs, IGAAs and GADMSs all facilitate individualisation but in different ways and that align with different educational principles. Relative to each other they have advantages and disadvantages (summarised in Table 1.1).

The GAIT approach has most potential in terms of the possibilities for flexible inclusion and access. Learners have near direct access to the generative AI model which means a high level of learner orientation. The semantic transduction capability of the technology can be used by the learner to translate their ideas into the forms required for assessment. Furthermore, it can translate ideas in the curriculum into forms that have meaning for the learner and translate curriculum content materials into forms that the learner can make meaning of in terms of their own perspectives, experience and understanding. This might include learner prompted explanations

Table 1.1 Three categories of generative AI integration

Generative AI (GenAI) integration category	Advantages	Disadvantages
Autonomous GenAI tools (GAITS) e.g., ChatGPT, Gemini, Claude	• Potential for personalised inclusion and access • Direct access to model • Opportunities to develop GenAI literacy • Flexible and versatile	• Challenging to integrate • Need to develop high level of user knowledge and skill in GenAI
Integrated GenAI Applications (IGAA) e.g., Microsoft Copilot	• Increased productivity and workflow • Some degree of personalisation • Some inclusion and access benefits	• Moderate integration challenges • Limited access to model • Limited development of user GenAI literacy
GenAI-based Automatic Decision-Making Systems (GADMS) e.g., intelligent or smart tutoring systems	• Easy to integrate • Improved efficiency in existing educational programmes and practices • Some inclusion and access benefits • Cost effective at scale	• Black box decision making (limited human-centricity) • Very limited development of generative AI literacy • Inflexible • Limited human/user orientation • Personalisation in orthodox procedural knowledge and skill development

of ideas in text, mathematics, or computer code. Using image-to-text capability can further extend this capability to providing alternative explanations of visual material. There are similar potentials for advanced speech-to-text systems. For learners with specific needs, generative AI using the GAIT approach, can allow improved mediation of those needs to facilitate adaptation of curricula, pedagogy, and the learning environment by educators to support learner access. The major challenge of the GAIT approach is that it does require both learners and educators to have a high level of skill and knowledge in using generative AI i.e., a high level of generative AI literacy. This, from our experience, requires considerable user learning and experimentation to comprehend the semantic transduction capability of generative AI. Users appear to start from a point of understanding and using generative AI as a tool for content generation, or as a type of search tool, with time and within a supported learning environment, users begin to understand and utilise the semantic transduction capability of generative AI. Educators and learners engaging with generative AI through GAIT integration has the potential to facilitate contextual and learner-oriented development of generative AI literacy.

The second category of integration, IGAA, involves generative AI augmentation of existing applications. This provides a structured means by which users are supported in the use of word-processing, email and spreadsheet packages through NLP. They can give instructions to support the development of specific content and functionality in the integrated applications. For example, they may support writing in word processing in a 'smart' way that goes beyond existing add-ons such as grammar and spell checking. This may include suggesting alternative text or supporting the development of functionality and automation in spreadsheets. At the time of writing this, IGAA approaches are being developed and increasingly rolled out by technology organisations. This looks set to be an important future feature of the generative AI educational integration landscape and has many benefits in terms of inclusion and accessibility though with less flexibility and user orientation than the GAIT approach. While familiarity with generative AI through IGAA can contribute to generative AI literacy, there are limitations since the learner or educator is not directly accessing and instructing a model as much as they are with the GAIT approach. The main advantage of the IGAA approach to generative AI integration is in increased productivity and accessibility with existing software applications. Since, IGAAs potentially provide a revenue model for technology companies through additions to existing application subscriptions they are likely to become a significant a generative AI integration. Microsoft, for example, has made significant investment in Copilot and advertised it extensively. Furthermore, it is likely that there will be significant demand for generative AI augmentation of existing ubiquitous education software applications.

The final category of integration builds on existing forms of technology-based adaptive learning systems. These involve learning analytics and logic systems for automatic decision-making approaches. Smart or intelligent tutoring systems, draw on a body of data about learning trajectories to provide an automatic response to learner input and response based on learners' previous responses and interactions with the system. Automatic decision-making systems (ADMS) are effective in developing procedural and factual knowledge and procedural competency in ways that are adapted to the individual learner, through data mining and big data analytics of learning trajectories and curriculum content. Effectively, ADMS adaptive learning systems identify learner misconceptions and misunderstandings based on data analytics and can suggest questions or prompts to develop the required knowledge and skill, thus addressing learner misconceptions and misunderstandings in an individualised and personalised way. ADMS intelligent tutoring systems are increasing integrating machine learning and deep learning into data analytics which improves the capability of previous database-based systems. Generative AI is increasingly being integrated into intelligent tutoring systems, in what we refer to here as GADMS. The NLP capability of generative AI provides a more conversational interface for the learner and can address knowledge and skills development across a range of subjects and on a range of educational issues that are confronted by the learner.

While GADMS introduce more flexibility and greater personalisation than previous forms of ADMS intelligent tutoring systems, they are limited in the extent to which learners can develop deep conceptual and relational understanding of content and ideas. This is a result of the largely logic-based automatic decision making that remains central to automated systems. Deep and relational conceptual understanding are dependent on greater learner orientation and in facilitating more exploratory, inquiry and problem-based and collaborative approaches to learning, where the methods and approaches required are multifaceted and non-standard. They also limit the social dimension of learning that is a key part of the interaction between learners and between learners and educators.

It is likely that GADMS will become an increasing feature of education, to supplement and perhaps even replace in some cases, interactive educative experiences in classrooms or seminars. They are efficient in developing procedural knowledge and skills and most importantly they can be deployed at scale and generate considerable revenue streams for technology companies. They are relatively easy to integrate as they reflect the more traditional aspects of knowledge and skill proficiency that has been central to the educational system for centuries. Advances in learning and data analytics facilitated by AI alongside the NLP capability of generative AI means that they have significant potential in personalised learning. However, since the decision making within these systems uses logic systems and not based on human judgement, they have limited human centricity.

4.3 Toward Human-Centred Generative AI Integration in Education

While it is probable that the integration of generative AI into education is likely to be dominated by IGAA and GADMS, as result of cost and efficiency, we argue that the GAIT approach has most value educationally. This is a because of its flexibility for inclusion and accessibility and learner orientation. However, as we have emphasized, GADMS and IGAA are more likely to be on the path of least resistance in terms of educational integration and that their development will be underpinned by the potential global market for them and revenues they are likely to generate. Yet, the GAIT approach must be developed and supported educationally. This is not to say that GAITs do not provide revenue for technology companies, but the cost of integration, i.e., the resources required for educators and learners to become proficient in their use and integrated into educational programmes and practices, means that IGAA and GADMS will have economic advantages.

However, generative AI seems likely to be here to say and will be an increasing feature of societal communication and the human-technology interface. It is therefore incumbent on education systems and for policy to support the development of generative AI literacy. The world that younger learners will live in and work in will be underpinned by AI and generative AI based systems and comprehending them is going to be essential.

Generative AI literacy encompasses a comprehensive set of knowledge and skills essential for understanding, interpreting, and engaging with generative AI technologies. It involves a technical awareness of generative AI, deep learning, LLMS, Transformer and Diffusion models. Additionally, ethical awareness is a crucial component. This extends to issues related to data privacy, the potential biases in AI outputs due to biased training data, and the risks of misuse in creating deepfakes or spreading misinformation.

Generative AI literacy also extends to the creative application of generative AI tools in various fields such as writing, STEM, art and design, interlingual and intralingual communication centred on a comprehension of the semantic transduction capability of generative AI. Furthermore, critical evaluation skills are vital, enabling individuals to comprehend the implications of hybrid human-technology generated content and assessing the accuracy and relevance of AI responses, and identifying potential biases. Lastly, an awareness of the societal impact of generative AI, i its potential to automate jobs, influence media and entertainment, and shape public opinion, must be integral to generative AI literacy. This understanding is crucial for responsibly leveraging these technologies and navigating the rapidly evolving digital and technological landscape. Moreover, this implies developing a philosophical perspective on technology, its societal role and its relationships to individuals in society.

Central to developing deep and relational conceptual understanding of generative AI, i.e., generative AI literacy, is the GAIT approach to generative AI integration in education. While conceding, in this, there is a high level of technological mediation even within GAIT approaches, educators and learners do have opportunity to engage directly with generative AI models and therefore it offers a more human centred approach to AI integration in education than IGAA and GADMS integration.

However, the major challenge of the GAIT approach is its integration into existing educational practice. This is largely because generative AI is not well understood by both educators and learners and therefore approaches to research and development are required that are contextual and participatory so that knowledge, skills, and new kinds of educational practice can be developed and disseminated. In the remainder of this chapter, we will consider how to address the challenge of integrating generative AI into education premised on the GAIT approach and describe a research and development approach based on Design-Based Research (DBR) (Watson, In review).

4.4 Participatory Co-Design in Human-Centred Generative AI Integration in Education

We have already outlined some of the challenges in integrating human-centred generative AI in education in the context of the GAIT approach. The use of autonomous generative AI tools in educational settings, such as generative AI chatbots, present several challenges in terms of the effective, regulatory compliant and ethical use of such flexible and versatile technologies. Not least of which is in developing educators', educational decision makers' and learners' generative AI literacy and in understanding the fundamental capability of the technology in terms of its capacity for semantic transduction. However, the challenges for integration are not limited to learning about the technology and how to use it. There is a profound integration challenge related to the education system itself and the nature of educational discourse and its autonomy on society (Watson, In review).

To understand this, we consider education as an autonomous discourse in society, that is paradoxically interconnected with wider societal discourse, such as science and knowledge, politics and policy making, economics and finance, media and information, and law and regulation. Education is a discursive system in that its discourse provides a basis for meaning making for educational programmes and practices as well as individual learner and education actions and behaviours. At the same time education as a discursive system is subject to contingency, the contingent experience of educators and learners and the contingency of other discursive systems such as policy making, law, media, science and economics (Watson & Romic, 2024).

Education maintains is autonomy and integrity, while remaining interconnected with external contingency, through its self-referential meaning making through educational programmes and practices. In this, it operates with a generalised discursive medium of the learner and differentiates its programmatic operations through the progress of the learner. Thus, the integrity and autonomy of education is highly dependent on the continuity of its programmes and practices i.e., for example, curriculum, pedagogy, practices and assessment processes. This means, and that has been borne out by educational reform and change literature, educational programmes and practices are difficult to change as a result of policy or research interventions or indeed technological innovations. This is not to say that educational programmes do not change and that it is not possible to reform programmes and practice, but it is important to recognise that change must be systemic, and time and resource are required to precipitate reforms in order that meaning making in relation to contingent experience can adapt to change. This also implies that reform or change initiatives cannot easily be anticipated or predicted in terms of how new forms of programmes and practice and emerge (Watson, In review; Watson & Romic, 2024).

This complexity was initially acknowledged in part by pragmatist philosophy but notably operationalised by Kurt Lewin in the period immediately after the Second World War. Lewin's Action Research approach introduced rigour to contextualised iterative participatory research and development with cycles of formative evaluation and knowledge making that led to emergent programmatic change and understanding of the nature of the systemic change (Lewin, 1946).

This principle was developed by Ann Brown in the 1980s with her seminal work on design experiments in educational settings (Brown, 1992). This was further developed into design based research (DBR) methodologies (Barab & Squire, 2004; Christensen, 2018). This involves the introduction of new ideas or new programmatic designs that might include technological innovation, through cycles of contextualised and participatory formative evaluation. This methodology provides a means by which innovations can be introduced into complex contextual education milieu and through which new programmes that are systemically coherent can emerge and that are more likely lead to sustained change and that can be adapted and scaled to similar contexts.

Moreover, although challenging because of the complexity of the research and development contexts as well as the extensive data produced (Dede, 2004), knowledge through analytic generalisation can be precipitated. At the cutting edge of this transdisciplinary approaches to theory development are used to develop theory drawing on a network of disciplinary concepts that are analysed abductively with data and in relation to the semantic structuration of their disciplinary origins e.g., psychology, social psychology, psychology, anthropology, philosophy and technology (Marschall et al., In review). This can also be understood as presenting two fundamental research questions in relation to the research and development context and case: *what is this a case of?* And *what is behind the case?* Thereby creating a theoretical relationship between the particular and general (Watson, In review).

This is the approach that we advocate in GAIT integration of generative AI in education. This allows, during the iterative research and development process, a high level of participation, co-design and formative evaluation, involving learners, educators and decision makers in a structured and rigorous approach that allows the emergence of new educational programmes and practices. Emergence is key in this, as this acknowledges that changes cannot be predicted based on a linear summation of contributing parts, that is the technology itself, generative AI; the perspectives of participants; existing practices; and the decisions of education leaders and policy makers as well as the effects of wider societal dynamics. But in reminiscence of the work of Lewin and Brown, allows new unpredicted forms to evolve and be understood.

4.5 Concluding Remarks

In this chapter, we have advocated for a human-centred approach to integrating generative AI into education. This approach entails utilizing generative AI tools that enable direct interaction between educators and learners and the AI models themselves, thereby fostering the development of Generative AI literacy. By embracing this technology, educators and learners can effectively adapt to the educational and societal changes brought about by its advancement. Additionally, we have outlined an approach to contextual and participatory research and development using DBR. Given the sustained presence of generative AI, it becomes imperative for education to collaboratively develop programmes and practices with educational stakeholders. These initiatives should support the ongoing development of generative AI literacy through iterative research and the utilization of generative AI models.

Endnotes

1. There is little actual data on the use of generative AI in education. However, based on the authors' experiences in schools and higher education, there is little use of generative use of the technology in educational contexts for teaching and learning. This is partly on account of fears about the technology, and anxieties

about using it safely and ethically. Surveys from the US and the UK suggest that students in higher education are making use of it more that educations, while in schools the teachers are using it more than students, but mainly for productivity rather than teaching and learning (Hennessey, 2023; Laird et al., 2023; Tyton Partners, 2023; Walton Family Foundation, 2023).
2. For a visualization of the word embeddings, see, http://projector.tensorflow.org.
3. For visualization and further explanation see, https://ig.ft.com/generative-ai/.

References

Anderson, T. (Ed.). (2008). *The theory and practice of online learning* (2nd ed.). AU Press.

Bahroun, Z., Anane, C., Ahmed, V., & Zacca, A. (2023). Transforming education: A comprehensive review of generative artificial intelligence in educational settings through bibliometric and content analysis. *Sustainability, 15*(17), 12983. https://doi.org/10.3390/su151712983

Barab, S., & Squire, K. (2004). Design-based research: Putting a stake in the ground. *Journal of the Learning Sciences, 13*(1), 1–14. https://doi.org/10.1207/s15327809jls1301_1

Brown, A. L. (1992). Design experiments: Theoretical and methodological challenges in creating complex interventions in classroom settings. *Journal of the Learning Sciences, 2*(2), 141–178. https://doi.org/10.1207/s15327809jls0202_2

Chaka, C. (2023). Detecting AI content in responses generated by ChatGPT, YouChat, and Chatsonic: The case of five AI content detection tools. *Journal of Applied Learning and Teaching, 6*(2), 12. https://doi.org/10.37074/jalt.2023.6.2.12

Christensen, K. (2018). The development of design-based research. In R. E. West & R. E. West (Eds.), *Foundations of Learning and Instructional Design Technology*. EdTech Books. https://edtechbooks.org/lidtfoundations/development_of_design-based_research

Dede, Chris. (2004). If design-based research is the answer, What is the Question? A commentary on collins, Joseph, and Bielaczyc; diSessa and Cobb; and Fishman, Marx, Blumenthal, Krajcik, and Soloway in the JLS special issue on design-based research. *Journal of the Learning Sciences, 13*(1), 105–114. https://doi.org/10.1207/s15327809jls1301_5

Fengchun, M., & Holmes, W. (2023). *Guidance for generative AI in education and research*. UNESCO.

Fütterer, T., Fischer, C., Alekseeva, A., Chen, X., Tate, T., Warschauer, M., & Gerjets, P. (2023). ChatGPT in education: Global reactions to AI innovations. *Scientific Reports*. https://doi.org/10.1038/s41598-023-42227-6

Gao, L., López-Pérez, M. E., Melero-Polo, I., & Trifu, A. (2024). 2024 Ask ChatGPT first! Transforming learning experiences in the age of artificial intelligence. *Studies in Higher Education*. https://doi.org/10.1080/03075079.2024.2323571

Grassini, S. (2023). Shaping the future of education: Exploring the potential and consequences of AI and ChatGPT in educational settings. *Education Sciences, 13*(7), 692. https://doi.org/10.3390/educsci13070692

Hennessey, M. (2023). *Exclusive: Almost half of Cambridge students have used ChatGPT to complete university work*. Varsity Online. https://www.varsity.co.uk/news/25463

Johnson, A. (2023). *ChatGPT in schools: Here's where it's banned—and how it could potentially help students*. Forbes. https://www.forbes.com/sites/ariannajohnson/2023/01/18/chatgpt-in-schools-heres-where-its-banned-and-how-it-could-potentially-help-students/

Kaplan-Rakowski, R., Grotewold, K., Hartwick, P., & Papin, K. (2023). Generative AI and teachers' perspectives on its implementation in education. *Journal of Interactive Learning Research, 34*(2), 313–338.

Kasneci, E., Sessler, K., Küchemann, S., Bannert, M., Dementieva, D., Fischer, F., Gasser, U., Groh, G., Günnemann, S., Hüllermeier, E., Krusche, S., Kutyniok, G., Michaeli, T., Nerdel, C., Pfeffer, J., Poquet, O., Sailer, M., Schmidt, A., Seidel, T., & Kasneci, G. (2023). ChatGPT for good? On opportunities and challenges of large language models for education. *Learning and Individual Differences, 103*, 102274. https://doi.org/10.1016/j.lindif.2023.102274

Laird, E., Dwyer, M., & Grant-Chapman, H. (2023). *OFF TASK! EdTech threats to student privacy and equity in the age of AI*. The Center for Democracy & Technology (CDT). https://cdt.org/wp-content/uploads/2023/09/091923-CDT-Off-Task-web.pdf

Lewin, K. (1946). Action research and minority problems. *Journal of Social Issues, 2*(4), 34–46. https://doi.org/10.1111/j.1540-4560.1946.tb02295.x

Marschall, G., Watson, S., Kimber, E., & Major, L. (In review). A transdisciplinary study of a novice mathematics teacher's instructional decision making. *Journal of Mathematics Teacher Education.*

Michel-Villarreal, R., Vilalta-Perdomo, E., Salinas-Navarro, D. E., Thierry-Aguilera, R., & Gerardou, F. S. (2023). Challenges and opportunities of generative AI for higher education as explained by ChatGPT. *Education Sciences, 13*(9), 856. https://doi.org/10.3390/educsci13090856

Nixon, N., Lin, Y., & Snow, L. (2024). Catalyzing equity in STEM teams: Harnessing generative AI for inclusion and diversity. *Policy Insights from the Behavioral and Brain Sciences*. https://doi.org/10.1177/23727322231220356

Prather, J., Denny, P., Leinonen, J., Becker, B. A., Albluwi, I., Craig, M., Keuning, H., Kiesler, N., Kohn, T., Luxton-Reilly, A., MacNeil, S., Petersen, A., Pettit, R., Reeves, B. N., & Savelka, J. (2023). The robots are here: Navigating the generative AI revolution in computing education. In *Proceedings of the 2023 Working Group Reports on Innovation and Technology in Computer Science Education*, pp. 108–159. https://doi.org/10.1145/3623762.3633499

Ramesh, A., Dhariwal, P., Nichol, A., Chu, C., & Chen, M. (2022). *Hierarchical text-conditional image generation with CLIP Latents* (arXiv:2204.06125). arXiv. https://doi.org/10.48550/arXiv.2204.06125

Tyack, D. B., & Cuban, L. (2000). *Tinkering toward Utopia: Century of public school reform*. Harvard University Press.

Tyton Partners. (2023). *Generative AI in higher education: Fall 2023 update of time for class study*. Turnitin.

Vaswani, A., Shazeer, N., Parmar, N., Uszkoreit, J., Jones, L., Gomez, A. N., Kaiser, L., & Polosukhin, I. (2017). *Attention Is All You Need* (arXiv:1706.03762). arXiv. https://doi.org/10.48550/arXiv.1706.03762

Walton Family Foundation. (2023). *Survey finds majority of teachers, parents report positive impact of ChatGPT on teaching and learning*. Walton Family Foundation. https://www.waltonfamilyfoundation.org/about-us/newsroom/survey-finds-majority-of-teachers-parents-report-positive-impact-of-chatgpt-on-teaching-and-learning

Watson, S., & Romic, J. (2024). ChatGPT and the entangled evolution of society, education, and technology: A systems theory perspective. *European Educational Research Journal*. https://doi.org/10.1177/14749041231221266

Watson, S. (In review). *Emergent discourses on generative AI in education*. Cambridge Univerissy Press.

Woolcock, N. (2023). AI 'is clear and present danger to education'. *The Times*. https://www.thetimes.co.uk/article/ai-is-clear-and-present-danger-to-education-3sk09ftlf

Xie, H., Chu, H.-C., Hwang, G.-J., & Wang, C.-C. (2019). Trends and development in technology-enhanced adaptive/personalized learning: A systematic review of journal publications from 2007 to 2017. *Computers & Education, 140*, 103599. https://doi.org/10.1016/j.compedu.2019.103599

Yan, L., Sha, L., Zhao, L., Li, Y., Martinez-Maldonado, R., Chen, G., Li, X., Jin, Y., & Gašević, D. (2023). Practical and ethical challenges of large language models in education: A systematic scoping review. *British Journal of Educational Technology, 55*, bjet.13370. https://doi.org/10.1111/bjet.13370

Chapter 5
Navigating AI in Education—Towards a System Approach for Design of Educational Changes

Li Yuan, Tore Hoel, and Stephen Powell

Abstract As artificial intelligence (AI) continues to advance, recent developments of Generative AI (GenAI) have sparked great interest, posing questions for policy-makers, technology innovators, educators, and EdTech researchers about possible paradigm changes for education. This chapter critically examines the development of Artificial Intelligence in Education, which promised to revolutionise educational practices by providing effective, personalised, learning at scale—first through rudimentary teaching machines and subsequently via advanced adaptive learning systems. We argue that Adaptive Intelligent Tutoring Systems (ITS) reinforce the content delivery model and restrict pedagogic opportunities in teaching and learning when adapted to existing educational models. The chapter examines the relevance and value of existing theories of learning in the development of educational technology and a need for new theories when an AI agent becomes an active partner in teaching and learning process and discusses the complexity of educational innovation from interdisciplinary perspectives. Finally, we offer an analytical assessment of the opportunities and limitations of GenAI in education using the cybernetic principle of variety, and propose a framework to address organisational, pedagogical, and technological challenges for using GenAI to support new formal learning and pedagogical practices.

Keywords Artificial Intelligence in Education (AIED) · Generative AI (GenAI) · Pedagogical agents · Teaching machine · Viable systems · Conversation theory

L. Yuan (✉) · T. Hoel
Beijing Normal University, Beijing, China
e-mail: l.yuan@bnu.edu.cn

S. Powell
academyEX, Auckland, New Zealand

P. Ilic et al. (eds.), *Artificial Intelligence in Education: The Intersection of Technology and Pedagogy*, Intelligent Systems Reference Library 261,
https://doi.org/10.1007/978-3-031-71232-6_5

5.1 Introduction

Over the past 60 years, researchers and developers in educational technology areas have deployed artificial intelligence in education (AIED) to develop systems and tools to promote flexibility, inclusion, personalisation, engagement, and effectiveness (Luckin & Holmes, 2016). According to Holmes et al., (2019), most pedagogical models, and the accompanying tools, adopted in AIED systems represent knowledge-mastering approaches in teaching and learning. For example, a teaching model incorporating Vygotsky's (1978) Zone of Proximal Development aims to present knowledge that is appropriately challenging for the student (the 'more knowledgeable other'), neither too simple nor too difficult. Additionally, by implementing individualised formative feedback, the model ensures that students receive timely feedback that is conducive to their learning progress. The recent development of Generative AI (GenAI) tools with advanced ability in producing text, graphics, audio, and video has thrust "technical disruption" and "new education paradigms" into the mainstream conversation regarding the future of education.

Technological innovation creates uncertainty that poses challenges both to institutions and individuals as they struggle to adapt to the environment. The proliferation of technological options lead to increasing difficulty in choosing and coordinating effective action (Johnson et al., 2022).

> Whether technology is seen as an 'extension' of existing capabilities [..], amplifications [..], reordering of nature [..], or changes to the means of production and social relations [..], an additional option means that the complexity of choosing whether to use a new method or stick with the old method increases. (Johnson et al., 2022)

The evolution of educational technology indicates that a simplistic perspective, which posits that mere technology integration results in educational transformation, is inadequate. Scanlon et al. (2021) reviewed over 100 projects, products and programmes in Technology Enhanced Learning (TEL), they concluded that TEL was a complex ecosystem which includes communities, technologies and practices that are informed by pedagogy. It is the interaction between these different facets that influence design and development, and embed technology innovations successfully into educational interventions. Pedagogy can be described as "a theorised approach to teaching and learning" (Ferguson et al., 2016). The utilisation of technology in educational contexts is strongly influenced by underlying theoretical frameworks and the pedagogical strategies employed by instructors, as evidenced by various models and theories (Millwood, 2013). For instance, Skinner's (1968) behaviourist teaching machine aimed to supplant teachers, whereas Laurillard's (2002) conversational framework for learning design, enhanced interactions between educators and students. More recently, Siemens and Downes' theory of connectivism has been spurred by the advent of Massive Open Online Courses (MOOCs), which provide access to global learning opportunities. One of the critiques on educational technology adoption is that it is too narrow and prescriptive, and limited only to influences from behavioural sciences. It fails to recognise the development and influence of broader theoretical perspectives on teaching and learning. Technologies are often used to reinforce existing instructor-centred, lecture-driven approaches, and rote memorisation. This is, in many cases, contrary to the promises of the technology advocates who market interactive, student-centred learning environments.

Despite previous challenges with technology integration in education, the rapid development of AI, especially the continued advancements of Generative AI has offered new opportunities for education. These developments potentially transform teaching and learning by creating innovative educational platforms and tools, generating human-like texts and realistic learning content, questions and exercises and providing personalised feedback to individual learners. The continued development of AI technology has captured public imagination globally and opened up new avenues could empower learner agency, support personalisation, enable learners to reflect on learning and develop high-order thinking skills. However, as Dawson et al. (2023) pointed out, there is a need to examine the relevance and value of existing theories of learning when an AI agent becomes an active partner rather than a simple technology in the teaching and learning process. They call for new frameworks, models, and ways of thinking to revise existing team-based and collaborative theories of learning.

In this chapter, an initial review is conducted on the evolution of AIED in relation to various pedagogical paradigms, encompassing behaviourism and constructivism, as well as an emerging framework that employs AI to enhance collective intelligence. Subsequently, to give conceptual tools which analyse and design educational technologies for the AI age, the chapter revisits Gordon Pask's (1975) Conversation theory and Diana Laurillard's (2002) conversational frameworks, which have had a considerable impact on the development of educational technology, which we expect to continue to influence the formulation of new educational paradigms involving advanced AI technologies. The Viable Systems Model (VSM) is also examined to articulate the complexities associated with technological innovation in education. Finally, the chapter outlines a system approach to assess and design AI technologies in facilitating the development of complex skills and higher-order thinking within educational contexts.

5.2 Entangled Technology and Pedagogy in AIED

Various AI technologies have been implemented to create intelligent learning environments for behaviour detection, predictive modelling, learning recommendation, etc. Using these technologies, researchers and developers have been designing AIED tools, such as intelligent tutoring systems, teaching robots, learning analytics dashboards, etc. AI has been considered a powerful tool to facilitate educational paradigm changes that are otherwise impossible to achieve in the traditional education modes (Holmes et al., 2019; Hwang et al., 2020). Understanding the relationship between AI and established educational and learning theories is crucial. Equally important is examining the extent to which AI technologies affect learning and instruction (Hwang et al., 2020). This necessitates a critical reflection on the theoretical, pedagogical, and computational dimensions of AIED, particularly how various AI technologies are employed to address educational challenges in teaching and learning.

5.2.1 From "Teaching Machine" to "Learning Machine"–A Behaviourist Paradigm

The idea of a technology that would allow students to "go at their own pace" has a long history, from Sidney Pressey's mechanised positive-reinforcement provider to B. F. Skinner's behaviourist bell-ringing box. Watters (2021), in her book "Teaching Machines: The History of Personalized Learning", documented how textbook publishers and early advocates for computerised learning promoted machines with "bite-sized content" and "individualised instruction". Since the 1920s, psychologists have explored ways to automate teaching. In the 1950s, the psychologist B. F. Skinner used the techniques he had developed for training rats and pigeons for teaching a Harvard University course in natural sciences. Skinner's "programmed learning" was refined and adopted in many classrooms in the 1960s (Skinner, (1968).

Skinner's teaching machine was taken up by researchers and designers in AIED to develop Intelligent Tutoring Systems (ITS) or later Intelligent Adaptive Learning Systems (or Learning Machine, which has recently been popularised in the Chinese educational market[1]). Adaptive learning systems are designed to dynamically adjust to the level or type of course content based on an individual student's abilities or skill attainment, in ways that accelerate a learner's performance with both automated and instructor interventions. However, as Holmes et al. (2019) pointed out, in adaptive learning systems, questions were pre-scripted, although students could proceed at their own pace, they went through the same list of questions as every other student and in the same order.

By incorporating behaviourism-supported approaches, AIED systems provide learners with personalised and adaptive learning experiences, reinforcing positive behaviours and encouraging mastery of the material. ITS has raised concerns that knowledge has been delivered through a narrow, instructivist pedagogical model and focused on skills mastery and information recall through 'drill and practice'. Therefore, it is essential to balance behaviourist methods with other pedagogical approaches to create a well-rounded and effective learning environment.

[1] Education tech firm Squirrel AI on market prospects, see https://www.chinadaily.com.cn/a/202109/16/WS6142fa9ca310e0e3a6822120.html.

5.2.2 *Inquiry-Based, Exploratory AIED Experiments—A Constructivist Paradigm*

The focus of early AIED, developing ITS, was primarily on cognition and the cognitive aspects of learning; however, in recent years, more and more AIED researchers have realised that education goes beyond the cognitive; and that metacognition, affect, and motivation play important roles in learning (Holmes et al, 2019; Rebolledo-Mendez et al., 2022). One of the experiments developed by the AIED community was to use Pedagogical Agents in learning environments via inquiry-based and exploratory learning that encourage active engagement and investigation (Richards & Dignum, 2019). Methods such as virtual labs and simulations, case-based learning, data exploration and visualisation, project-based learning and gamified learning environments were used. Overall, integrating inquiry-based and exploratory learning strategies into AIED systems enhances learners' motivation, critical thinking skills, and problem-solving abilities, leading to more meaningful and engaging learning experiences.

Grounded upon a social constructivism view, learning occurs when a learner interacts with people, information, and technology in socially situated contexts (Vygotsky, 1978). The dialogue-based tutoring systems (DTSs) or the exploratory learning environments (ELEs) have been developed to achieve mutual interactions between the system and the learner (Holmes et al., 2019). On the one hand, the AI system collects and analyses emerging, multimodal data from the learner to assess the learner's learning status. For example, Biswas and his colleagues (Biswas et al., 2016) created "Betty's Brain", a computer system to help promote students' understanding of metacognitive skills and to reinforce knowledge as part of a science curriculum in an Open-Ended Learning Environment (OELE). Betty's Brain diverges from a classic intelligent tutoring system (ITS) and adopts the learning by teaching (LBT) paradigm where computer agent interactions are focused around completing a primary task unrelated to the acquisition of domain content knowledge.

In this way, the AI system and the learner build mutual interactions, resulting in learner-centred learning through mutual interaction and sustained collaboration between the learner and the AI system. Nonetheless, numerous technical limitations impede the seamless and synergetic interactions between humans and computers. These constraints arise from the intricate nature of the interactions, as neither the learner's information and data nor the state of the system remain static or uncomplicated (Holmes et al., 2019). It is critical for the AI systems to offer real-time data analysis and immediate feedback to learners, and for the learner to use that feedback to enhance the ongoing, emergent learning processes. Therefore, it would be beneficial if the AI system maintains continuous learner-generated data collection and analysis, and provides learners with real-time, exploratory opportunities to make decisions about learning.

5.2.3 Human–AI Collaborative Adaptive Learning Systems—an AI Augmented Collective Intelligence Paradigm

In the collective intelligence paradigm, learner agency is the core of AIED and AI is viewed as a tool to augment human intelligence. Education is a complex adaptive system, where a synergetic collaboration between multiple entities in the system is essential to ensure the acquisition of desired knowledge and the development of abilities and skills for the learner. AIED need to be designed and implemented with the recognition that AI technology is part of a broader system, which includes learners, teachers, administrators, policy-makers, parents and other stakeholders. For education, concepts like human–machine collaboration, human-centred AI, and augmented intelligence are gaining prominence. The learners lead their own learning and manage the risks of automated AI decisions, and develop more flexible and engaging approaches to learn with peers and teachers with the support of AI agents.

Following Dewey's 'social intelligence' (Dewey, 2004), Wegerif (2007, 2013) views dialogic education as a way to collectively respond to the following challenge: the ability of a society to think together, learn together and respond appropriately in a global context. The development of students' higher-order thinking is also an important goal for education. Wegerif defines dialogue as interaction that supports learning by opening up a shared dialogic space of reflection, within which new insights and understandings can emerge. Education takes place through dialogues, and dialogues require the opening of a dialogic space. Wegerif and Major (2023) emphasise the crucial role that technology plays in "opening, widening, deepening and sustaining dialogic spaces", which make learning possible. They offer a "foundation of design" to bring the collective dialogue of educational theory and educational technology together, to enrich and orchestrate teaching and learning, by designing for engagement, connection, expansion, and the unknown—all as an ethical practice. The concept of expanded dialogic space and time has significant implications for developing and designing future teaching and learning environments and processes using AI agents in education.

The ongoing design and evaluation of AI systems to support complex problem solving, collaboration, critical and collective thinking, hold considerable potential, particularly given the rapid and ongoing development of GenAI and applications in numerous fields, not least art and design. So far it is still open as to how AI can be used to effectively support developing metacognition and higher-order thinking skills and capabilities in schools, homes and organisations. The ultimate goal (Tegmark, 2018) is that future AIED should be designed and developed to empower learners to take full agency of learning, optimise AI technology to provide real-time feedback for multiple stakeholders about emergent learning, and rethink learning changes brought by AI in complex, interconnected learning systems. In the new education paradigm, AI will be used to develop higher-order thinking with a focus on the collective dimension of thinking: thinking as an aspect of collaborations and dialogues. Technology enters into thinking represented by dialogue in two distinct ways: (1) as the medium of the dialogue and (2) as a voice in the dialogue (Casebourne et al., 2023). Collective intelligence could be achieved in expanded dialogic spaces with the support of AI agents and orchestrated by a teacher or a more knowledgeable other, who can support processes of learning to learn together and ease induction into communities that are learning together, in addition to supporting communities in ongoing learning projects.

5.3 Reviewing Conversational Theories and Design Frameworks for Educational Technology

Education systems embody complexity, serving multiple functions outlined by Biesta (2009) as qualification, socialisation, and subjectification. These functions influence the selection, teaching, and assessment of knowledge. Beyond the inherent social and cultural intricacies, the complexity is further amplified by the disparate yet interrelated roles of three additional components: curriculum, pedagogy, and assessment. These elements operate across all levels of the education system but manifest in distinct ways. Recent developments in GenAI and Large Language Models (LLMs) are widely seen to challenge existing education processes and practices. According to Dawson et al. (2023), AI usage in education could produce two different outcomes: (1) AI could revolutionise and enhance existing teaching methods, assessment and learner engagement; (2) it could create an entirely new education system that fosters creative thinking and problem solving, which are critical in modern and highly complex settings. Questions have been raised on the relevance of existing education theories and practices, for example, whether the adoption of AI in education requires modifications or revisions in how we learn, whether a complete restructuring of the education system is required, and whether there is a need for new theories. It may be fair to say though, as AI continues to impact education, there is a need for new frameworks, models, and ways of thinking as we observe non-human agents being incorporated as active partners in teaching and learning. It is essential for advanced education and learning theories to ensure that established principles, values, and trusted constructs shape the use of AI in education (Dawson et al., 2023; Wegerif & Major, 2023).

The UNESCO's *Guidance for generative AI in education and research* calls for support of teachers' "unique roles in facilitating higher-order thinking, organizing human interaction, and fostering human values" (Miao & Holmes, 2023, p. 26).

> The most fundamental perspective of the long-term implications of GenAI for education and research is still about the complementary relationship between human agency and machines. One of the key questions is whether humans can possibly cede basic levels of thinking and skill-acquisition processes to AI and rather concentrate on higher-order thinking skills based on the outputs provided by AI. (Miao & Holmes, 2023, p. 37).

Although existing theoretical lenses and frameworks offer insights into organisational disruptions and the integration of technology in learning, it is uncertain terrain when it comes to the impact of GenAI on higher-order thinking, social interaction, and human values. It is essential to apply existing or new pedagogical theories and design frameworks, which will give directions for future exploration for new paradigm changes in education. In the following, we will revisit Gordon Pask's (1975) Conversation theory and its applications in education; we will then discuss Diana Laurillard's (2002) conversational model and its influence on the development of eLearning and Virtual Learning Environments and its pedagogical and technical limitations.

5.3.1 Conversation Theory and Its Application for Learning

The core idea of Conversation theory is that learning occurs recursively through "conversations". It is based upon Gordon Pask's interest in cybernetics, where he conceived human–machine interaction as a form of conversation, a dynamic process, in which the participants learn from each other (Pask, 1975). It presents an experimental framework heavily utilising human–computer interactions and computer theoretic models to explain how conversational interactions lead to the emergence of knowledge between participants who are engaged in conversation with each other.

In order to facilitate learning, Pask (1975) argued that subject matter should be represented in the form of structures that show what is to be learned. These structures exist in a variety of different levels depending upon the extent of the relationships displayed. The critical method of learning according to Conversation theory is "teachback" in which one person teaches another what they have learned.

Conversation theory is the general study of how knowledge is constructed through interactions between humans. Learning is a continual conversation: with the external world and its artefacts, with oneself, and also with other learners and teachers. The most successful learning comes when the learner is in control of the activity, is able to test ideas by performing experiments, asking questions, collaborating with other people, seeking out new knowledge, and planning new actions (Sharples, 2005).

After having looked at conversations from a learning perspective we now turn to how the conversational perspective has influenced the design of learning technologies.

5.3.2 A Conversational Framework to Inform Learning Design

Laurillard (2002) has applied Conversation theory to address pedagogical challenges in the design of learning technology and support collaborative learning. She developed a 'conversational framework' in which technology may provide or enrich the environment in which conversations take place either online or face to face incorporating the full range of subject areas and topic types; the theory is discipline, or subject, agnostic. Associated pedagogic strategies have to consider different forms of communication and associated mental activities: discussion, adaptation, interaction and reflection.

The Conversational Framework has been used as a pedagogic design tool for learning platform, developers and learning designers to provide a description of the components of a collaborative process and how the different components of the pedagogic design interrelate to motivate the learner to learn. The technology provides a shared conversational learning space, which can be used not only for single learners but also for learning groups and communities. Technology can also demonstrate ideas or offer advice at the level of descriptions, as with the worldwide web or online help systems, or through specific tools to negotiate agreements, such as concept maps and visualisation tools. Even though technology for learning conversations, such as virtual learning environments, discussion forums, online communities and help systems, have had some success in mediating learning, their value is limited. In part their ability to give practical advice is limited because they cannot support the full range of conversations, and they do not share the learner's context (Sharples, 2015).

Building on the conversational perspectives discussed in both the previous and current sections, we will shift to a systems-oriented viewpoint in the ensuing section. Here cybernetic concepts of variety, feedback and other terms come to use to help us navigate in the increasingly complex landscape of AIED.

5.4 Viable Systems and Feedback Loops

The cybernetic concept of variety can be used as an analytical tool to evaluate the pedagogical relationships and communications in an educational system (Millwood & Powell, 2011). This concept was identified by Ross Ashby as The Law of Requisite Variety and was used by Stafford Beer in his Viable Systems Model (1985) to explain the management and coordination of organisations. A useful definition of variety is given by Britain et al. (2007) as "the number of states of which a system is capable of attaining". If we consider the pedagogical organisation of an educational system such as a school or an institution, we can see that in the relationships between teachers (controllers) and students (controlled) there is a great deal of complexity in communications required for learning to take place. The theory goes on to explain that if the system is to be stable (functioning effectively) the controllers need to have sufficient ways of responding to the system they are controlling and this is achieved by attenuation and amplification. For example, in a classroom context, a teacher may attenuate or reduce variety by asking students to raise their hand before answering a question, and they may amplify themselves by producing worksheets that explain a task to students so that they are not constantly being asked questions.

Two further key concepts of the VSM, useful for this analysis, are viability and recursion. Viability refers to the ability of something to maintain a separate identity within a particular environment (Beer, 1985). A state school could be an example of a viable system that survives in a context including the local community, funding streams, etc. It can be said to be viable because it would be theoretically possible to move that school to a different similar environment where it would continue to thrive. Other examples of viability in sense of the meaning could be Human Resource departments that could be lifted out of one company environment and set down in another and be able to function well. Returning to the state school, we can see that it exists in a horizontal context of many other schools, and vertically within an educational system. The educational system, the school, and the subjects taught within the schools are part of the same system but can be analysed through different levels of recursion (Beer, 1985). Again, viability is key and we can see that at the level of the subject, say geography, it would be possible to extract a geography department and place it in a different school environment and it would remain viable.

This rudimentary depiction of the complexities introduced by pedagogical diversity within an educational system highlights the potential contributions of technology to the learning environment. Britain et al., (2007) explored this in terms of the Virtual Learning Environment (VLE) whose tools offer multiple opportunities for attenuation and amplification of the teacher. GenAI technologies have the potential to offer an infinite number of states or responses to students to meet their learning needs that go way beyond that of a VLE; and then the challenge of variety is eliminated. The key questions then become qualitative ones related to the language models being used, the technology-mediated experience of the learner, and the purpose to which the GenAI is being used. For example, there are issues around information source biases that may be discriminatory to groups in society. If the purpose of the GenAI interaction is for summative assessment products, there is the challenge of ensuring academic integrity in terms of the veracity of the work and the need for submitted work to be that of the student or properly attributed to sources. In considering which technologies to use, it is questions like these that will determine which of the specialised GenAI platforms best meet the needs of the learner and the institution.

The learning theories discussed above (Behaviourism, Constructivism, Conversational Theory and Conversational Framework) all have feedback loops as a central mechanism for learning. For higher order skills and development of capabilities the feedback will be rich; for example, evaluative feedback, motivational, thought provoking, etc. It is these types of feedback that are currently best performed by humans.

Figure 5.1 situates the theories discussed above in a GenAI world. We can see that GenAI technology encompasses the whole educational system, but that its application will be context specific related to the level of recursion for a given system in view (Macro, Meso, Micro) and the actors involved. It is in the process of learning (the controlled) that the value of human interaction comes to the fore, where the value of relationships between different participants cannot be replaced by technology, and where the variety presented by learners can be absorbed by other controlled actors.

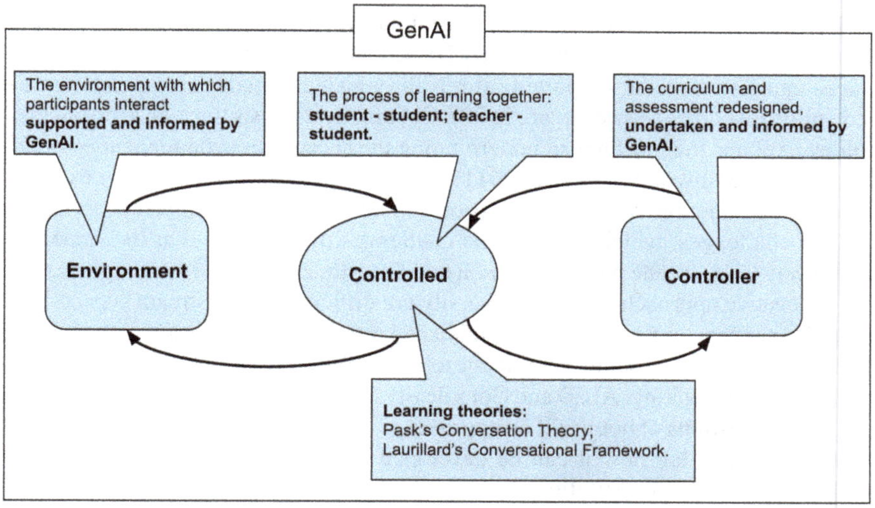

Fig. 5.1 A system perspective on understanding how GenAI will impact different parts of education

5.5 Developing an Analytic Framework for Designing and Adapting AIED

Education is interconnected with other systems and consists of multiple subsystems which interact with each other. Studying AIED, cybernetic system theory and complexity theory are particularly useful to understand and explain how technology and pedagogy are entangled.

A cybernetic systems-thinking approach views education as a network of self-regulating feedback loops between elements at different levels of recursion (sub systems) in a given system in focus. AIED strengthens our understanding of these feedback loops by offering an analysis of learning through digital technology as an integral connectivity of complex educational systems, influencing how all the parts interact. For example, the use of student data to help to explain and predict the impact of interventions across the system, approaches employed by those seeking to measure educational gain in higher education. These increasingly real time 'actionable insights' can be used to inform decisions about teaching methods and strategies offering a personalised and nuanced understanding of what works for individuals in which circumstances aiding student self-regulation and moving us beyond relatively simplistic pedagogic driven design models. Moreover, Complexity Theory reminds us that the effects of technology in education are not predictable or linear, that the use of AIED can lead to unexpected learning outcomes but the system will be able to adapt to non-linear learning paths that better align with the complex nature of individual learning processes.

The Ultraversity project at Anglia Ruskin University developed a wholly online undergraduate degree programme from 2004, with 300 students undertaking workplace inquiries for credit, supported entirely through facilitated learning communities and online resources (Powell et al., 2008). Cybernetic analysis through a lens of technology explains the mechanism underpinning the successful pedagogic approach of Ultraversity (Millwood & Powell, 2011) that are applicable to GenAI. For example, the use of Authentic Assessment approaches focussed on workplace activities or societal challenges, which mitigate the challenges of GenAI and at the same time offer a way to utilise the benefits (Powell and Forsyth, 2024). However, organisationally innovative approaches like Ultraversity are difficult to mainstream successfully, struggling as they must against the established models of higher education to secure the necessary resources and investment to grow (Powell et al., 2015). The application of systems thinking, AIED and Complexity Theory offers much to understanding evaluative questions of not just if an approach is effective at the level of an individual or a programme, but how it can be developed to challenge the existing dominant models at the scale of institutions and nationally.

In the rapidly evolving landscape of AI technologies, emphasising a perspective centred on complexity and emergence serves as an indicator that the quest for comprehensive evaluation and design frameworks remains an active and continuous effort. For example, Wegerif and Mansour (2010), and Wegerif and Major (2023) propose a dialogic foundation for the design of education technology as a way of thinking about education which will inform the development of different design frameworks in different contexts. In the present chapter, foundational elements have been set forth for the development of a system approach for assessing pedagogical implications for the development of AIED and a design framework for educational changes (Cuban, 2018). This has been achieved by: (1) delineating the historical trajectory through which technologies have entered educational settings, impacted teaching and learning modalities, and have simultaneously been situated within and framed by the dominant pedagogical approaches of their periods; and (2) revisiting old and tested theoretical approaches (systems and complexity theories; cybernetics and design-based educational theories) to see how the new challenges of powerful and more general AI tools, such as GenAI, might be understood and met with new policies and practices. In the end, this approach should lead to the design of models that should be evaluated, revised, and further developed through iterations of design-based educational research.

While developing this framework we anticipate a process involving numerous diagrams being drawn and populated with diverse perspectives for evaluation. The swift advances of GenAI presents a twofold application: it can generate disruptive educational scenarios for consideration, while simultaneously serving as an autonomous tool to accelerate the design process itself. Figure 5.2 gives an example of a template of three educational sub-systems and three organisational levels depicting some hypothetical answers to disruptions caused by GenAI. The suggested measures include redesign of educational organisations at all levels—from the course level to the national system level—to allow potential benefits to be realised. There are also suggested amendments to curriculum (what should we teach and learn), to pedagogy

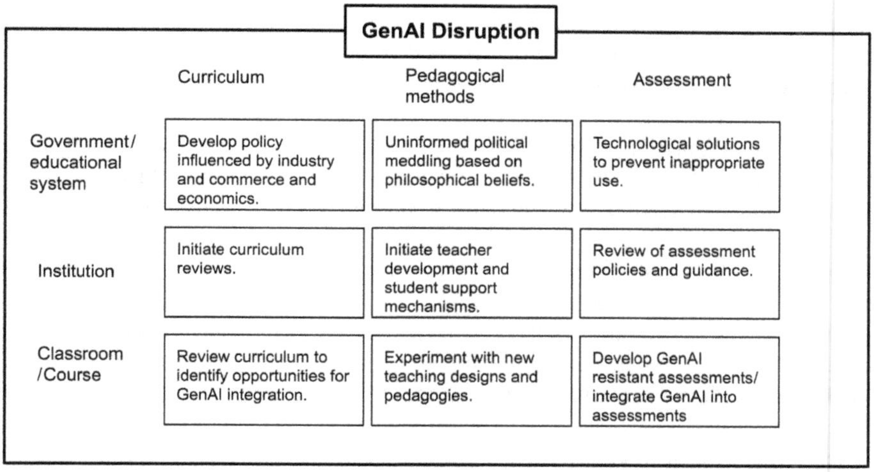

Fig. 5.2 A system approach towards educational changes and a framework for designing AIED

(how should we teach and learn), and to assessment (how should we assess what we have taught and learned?). Those elements and questions are interconnected and they could be understood, interpreted and implemented differently by various stakeholders at different levels in educational systems. Without sufficient consideration of the aims or established practices of teachers and students, it will leave educators susceptible to an inadequate appreciation of the complexity and how it is interwoven in educational activities.

In the current landscape of AI innovation, predicting the performance of new technologies remains an uncertain endeavour. Our most realistic expectation is to cultivate a sufficiently detailed possible scenarios for collective stakeholder discussion, Fig. 5.2.

When GenAI hit education, it was the certification subsystem that reacted. This is the system that does assessment and accreditation, and may from a business point of view be seen as vital to education. Educational cybernetics gives us some tools to analyse how vital certification is, and how it may be affected. First, the principle of recursion implies that every viable system contains and is contained by other viable systems. According to the Viable Systems Model each recursive level has a similar structure, it has a management or regulatory aspect and an operational aspect. What becomes clear, is that larger systems often have tensions between subsystems.

To enable effective reflection, teachers need a clear understanding of both their own beliefs and alternative beliefs about technology and pedagogy (Sandholtz et al., 1997). The shift from focus on declarative knowledge to the development of the whole person requires new literacies, new capabilities, and new methods and approaches to develop these aspects of our faculty and our students. Therefore, sociocultural considerations should be at the forefront when discussing how technology and pedagogy are interwoven. Entangled pedagogy is collective; learner agency is negotiated between teachers, students and other stakeholders; and outcomes are contingent on complex relations that cannot be determined in advance. Uncertainty, imperfection, openness and honesty should be embraced when the aim is to develop pedagogical knowledge that is collective, responsive and ethical.

Given the significant impact GenAI has had, the ethical implications of implementing AI in education is a natural starting point. For example, AI may provide dynamic adaptive feedback to the learner through deeper profiling of the learners from the capture of 'stealth' assessments as the student works to 'robust' learner profiling including additional data on their preferences, interests and biases which may well raise some ethical issues as time goes on (Gardner et al., 2021). We would advise a human-centred approach being open to all questions and viewpoints different stakeholder groups have. This is an emergent discussion as there is little actual data on the use of GenAI in education as usage differs between age groups, types of education, countries, etc. However, we know that use of GenAI by students and to a lesser extent is widespread as reported through professional networks, institutional policy responses and the media.

Using cybernetics as a theoretical lens, we observe that education systems adeptly navigate the complexities introduced by GenAI. In Table 5.1 we have included the concerns Rienties et al. (2024) identified amongst distant education students from different disciplines regarding the design and use of a hypothetical AI Digital Assistant. The concerns are broad in scope and resonate well with the current state of discourse we observe worldwide. In the table, we have given some keywords describing a potential response to these concerns driven by our system approach to addressing this complex field.

Cybernetic analysis works in part by attenuating external variables that could potentially destabilise the system, such as the threats and ethical dilemmas identified by the distance education students represented in the table above. Furthermore, the approach amplifies external elements that contribute to the system's core objectives, such as personalised learning and new and more efficient assessment mechanisms, etc. This dual approach enables the system and subsystems to maintain stability and adaptability amid various phases of technological disruption. In the case of assessment, we have seen this play out in several countries, starting by banning GenAI in universities, and ending up with establishing national guidelines for how to use these new tools safely, ethically and effectively (see Australia as an example[2]).

5.6 Conclusion

The rapid development of AI provides new opportunities for education, but by simply adding technology to existing education systems, there is a risk that applications, such as teaching machines and ITSs, which were designed and developed based on

[2] https://education.nsw.gov.au/teaching-and-learning/education-for-A-changing-world/guidelines-regarding-use-of-generative-ai-chatgpt

Table 5.1 Concerns about GenAI (building on Rienties et al., 2024) and potential system approach responses

Concerns about GenAI	Cybernetics and system framework response
Academic Integrity: use in assignments, including plagiarism risks and maintaining academic standards	Use authentic forms of assessment adaptive to work and societal contexts, with a focus on grading the process of thinking and learning. In Cybernetic terms, enabling a wider variety of students' meaningful engagement with the assessment task
Operational Challenges: effects on reliability and validity of student learning, emphasizing the need for tool accuracy, and equality of access	A systemic understanding of teaching and learning to ensure that all aspects of AI's impact on education and society are addressed in the learning environment
Ethical and Social Implications: changing balance between technology and human-centric approaches to teaching	Embed ethical reasoning and social context considerations into the curriculum, ensuring learners develop the capacity to critically assess AI technologies' roles and impacts, fostering responsible use and innovation
Data Privacy and Use: handling of student data by AI, highlighting the importance of transparency and obtaining consent	Implement stringent data governance frameworks, ensuring transparent data practices, securing consent, and maintaining strict controls over access and use, to safeguard privacy and build trust in AI-enhanced educational environments
Future of Education: adaptations in teaching methods and learning expectations resulting from AI adoption	Adapt curricula to include skills essential for the AI era, emphasize interdisciplinary learning, and promote flexible, lifelong education models to prepare learners for continuous evolution in the workforce and society

behaviourism, might be used to "scale up bad pedagogical practices" (Tuomi, 2018, p. 4). Constructivism theories and conversational frameworks have been exploited in various learning contexts with limited technology mediation. There is a need to research and discuss existing and new educational theories and design principles to fully realise the potential of advanced AI technologies and GenAI applications in the teaching and learning process. More critically, with the global impact of AI in society, there is a need to redesign education systems at all levels, from the class, the school (institution) to the national policy and strategies, from curriculum, assessment and new pedagogy with the support of AI agents. For instance, creative thinking and problem solving are critical in modern and highly complex settings. When AI begins to serve as an active partner in sustained social, creative and intellectual pursuits, the impacts on existing practices remain unknown. It is certainly problematic to introduce technology without sufficient consideration of the aims or an inadequate appreciation of complexity relating to how it is entangled in educational activity. An AI-enabled education system should encourage collaboration across sectors and stakeholders globally and facilitate open, flexible, interactions between learners, teachers and automated agents and systems.

Currently it is not possible to be definitive about the impact of AI and particularly GenAI on global educational systems. Much is claimed by futurologists presenting multiple scenarios, predictions, and possibilities; however, at this time it is unknown how the new technologies will play out. We have been in this position before with predictions about how computers would change education in the 1980–90's, the world wide web in the 2000's and more recently MOOCs in 2013. Although these technological innovations have had an impact on educational systems, the basic models of education have largely remained the same. The way technology impacts education is unpredictable, to date, certainly, it has not been on the scale of the industrial revolutions in industry. In many ways, fundamental aspects of education—including teaching methodologies, learning processes, and assessment techniques—have undergone minimal transformation over the past two centuries.

In this chapter, an analytic framework has been developed to support designing and implementing AI in education by emphasizing the interconnectedness of education systems and the complex interactions between technology and pedagogy—a holistic system wide view. We argue that adapting AI in education needs to accommodate diverse stakeholder perspectives and to ensure that technology integration aligns with teachers' values, beliefs and attitudes, national policies, and pedagogical innovation. The framework recognises the ethical and pedagogical challenges of AIED and ensures comprehensive evaluation of both benefits and drawbacks. We advocate for AI tools, carefully designed and implemented, that improve teaching and learning while ensuring sufficient stability for educational institutions to function effectively during the adoption of GenAI.

References

Beer, S. (1985). *Diagnosing the system for organizations.* Wiley.

Biesta, G. (2009). Good education in an age of measurement: On the need to reconnect with the question of purpose in education. *Educational Assessment, Evaluation and Accountability (formerly Journal of Personnel Evaluation in Education), 21*(1), 3346.

Biswas, G., Segedy, J. R., & Bunchongchit, K. (2016). From design to implementation to practice—a learning by teaching system: Betty's brain. *International Journal of Artificial Intelligence in Education, 26*, 350–364. https://doi.org/10.1007/s40593-015-0057-9

Britain, S., Liber, O., Perry, S., & Rees, W. (2007). *Modelling organisational factors affecting the development of e-learning in a university using a cybernetics approach.* University of Bolton Institutional Repository.

Casebourne, I., Shi S., Hogan M., Holmes, Y., Hoel, T., Wegerif, R., & Yuan, L. (in press). Using AI to support education for collective intelligence

Cuban, L. (2018). *The flight of a butterfly or the path of a bullet?* Harvard Education Press.

Dawson, S., Joksimovic, S., Mills, C., Gašević, D., & Siemens, G. (2023). Advancing theory in the age of artificial intelligence. *British Journal of Educational Technology, 54*, 1051–1056. https://doi.org/10.1111/bjet.13343

Dewey, J. (2004). *Democracy and education: An introduction to the philosophy of education.* The Free Press.

Ferguson, R., Brasher, A., Clow, D., Cooper, A., Hillaire, G., Mittelmeier, J., Rienties, B., Ullmann, T., & Vuorikari, R. (2016). Research evidence on the use of learning analytics—implications for education policy. In Vuorikari, R., Castaño Muñoz, J. (Eds.). Joint Research Centre Science for Policy Report; EUR 28294 EN; https://doi.org/10.2791/955210.

Gardner, J., O'Leary, M., & Yuan, L. (2021). Artificial intelligence in educational assessment: 'Breakthrough? Or buncombe and ballyhoo?' *Journal of Computer Assisted Learning, 37*(5), 1207–1216. https://doi.org/10.1111/jcal.12577

Holmes W., Bialik M., & Fadel C. (2019). Artificial intelligence in education: promises and implications for teaching and learning, Center for Curriculum Redesign, Boston, MA. 28294 EN. https://doi.org/10.2791/955210

Hwang, W., Xie, H., Wah, B. W., & Gasevic, D. (2020). Vision, challenges, roles and research issues of artificial intelligence in education. *Computers and Education: Artificial Intelligence, 1*, 100001. https://doi.org/10.1016/j.caeai.2020.100001

Johnson, M. W., Suvorova, E. A., & Karelina, A. A. (2022). Digitalization and uncertainty in the university: Coherence and collegiality through a metacurriculum. *Postdigital Science and Education.* https://doi.org/10.1007/s42438-022-00324-1

Laurillard, D. (2002). *Rethinking university teaching: A framework for the effective use of educational technology* (2nd edition) London; Routledge Falmer (0-415-25679-)

Luckin, R., & Holmes, W. (2016). *Intelligence Unleashed: An argument for AI in Education.* UCL Knowledge Lab.

Miao, F.C., Holmes, W., Huang, R.H., and Zhang, H., (2023). *Guidance for generative AI in education and research,* policy. In Vuorikari, R and Castaño Muñoz, J (eds.), Joint Research Centre Science for Policy Report; EUR, https://www.unesco.org/en/articles/guidance-generative-ai-education-and-research

Millwood, R. (2013). *Review of learning theories and their relationship with Technology Enhanced Learning,* https://www.academia.edu/5170757/Review_of_Learning_Theories_and_their_relationship_with_Technology_Enhanced_Learning

Millwood, R., & Powell, S. (2011). A cybernetic analysis of a university-wide curriculum innovation. *Campus-Wide Information Systems, 28*(4), 258–274. https://doi.org/10.1108/10650741111162734

Pask, G. (1975). *Conversation, Cognition, and Learning.* Elsevier.

Powell, S., & Forsyth, R. (2024). Generative AI and the implications for Authentic Assessment. In S. Beckingham, J. Lawrence, P. Hartley, & S. Powell (Eds.), *Using Generative AI Effectively in Higher Education: Sustainable and Ethical Practices for Learning, Teaching and Assessment* (pp. 97–105). Routledge.

Powell, S., Olivier, B., & Yuan, L. (2015). Handling disruptive innovations in HE: Lessons from two contrasting case studies. *Research in Learning Technology, 23*, 1–14. https://doi.org/10.3402/rlt.v23.22494

Powell, S., Tindal, I., & Millwood, R. (2008). Personalized learning and the Ultraversity experience. *Interactive Learning Environments, 16*(1), 63–81. https://doi.org/10.1080/10494820701772710

Rebolledo-Mendez, G., Huerta-Pacheco, N. S., & Baker, R. S. (2022). Meta-affective behaviour within an intelligent tutoring system for mathematics. *International Journal of Artificial Intelligence in Education, 32*, 174–195. https://doi.org/10.1007/s40593-021-00247-1

Richards, D., & Dignum, V. (2019). Supporting and challenging learners through pedagogical agents: Addressing ethical issues through designing for values. *British Journal of Educational Technology, 50*(6), 2885–2901. https://doi.org/10.1111/bjet.12863

Rienties, B., Domingue, J., Duttaroy, S., Herodotou, C., Tessarolo, F., & Whitelock, D. (2024). What distance learning students want from an AI Digital Assistant. *Distance Education.* Online: https://oro.open.ac.uk/96703/1/Distance_learning_19_03_2024_oro.pdf

Sandholtz, J. H., Ringstaff, C., & Dwyer, D. C. (1997). *Teaching with Technology: Creating Student-Centered Classrooms* (p. 211). Teachers College.

Scanlon, E. (2021). Educational technology research: contexts, complexity and challenges. *Journal of Interactive Media in Education,* 2021(1): 2, pp. 1–12. https://doi.org/10.5334/jime.580

Sharples, M. (2005). Learning as conversation: transforming education in the mobile age. In *Proceedings of Conference on Seeing, Understanding, Learning in the Mobile Age* (pp. 147–152). Budapest, Hungary.

Sharples, M. (2015). Seamless learning despite context. In L.-H. Wong, M. Milrad, & M. Specht (Eds.), *Seamless learning in the age of mobile connectivity* (pp. 41–55). Springer. https://doi.org/10.1007/978-981-287-113-8_2

Skinner, B. F. (1968). *The technology of teaching*. Appleton-Century-Crofts.

Tegmark, M. (2018). *Life 3.0: Being human in the age of artificial intelligence*. Vintage.

Tuomi, I. (2018). *The impact of artificial intelligence on learning, teaching, and education. policies for the future,* In Cabrera, M., Vuorikari, R & Punie, Y., (Eds.) EUR 29442 EN, Publications Office of the European Union, Luxembourg, 2018, ISBN 978-92-79-97257-7, https://doi.org/10.2760/12297, JRC113226.

Vygotsky, L. S. (1978). *Mind in Society: The Development of Higher Mental Processes*. Harvard University Press.

Watters, A. (2021). *Teaching Machines: The History of Personalized Learning*. The MIT Press.

Wegerif, R. (2007). *Dialogic Education and Technology*. https://doi.org/10.1007/978-0-387-71142-3.

Wegerif, R. (2013). *Dialogic: Education for the internet age*. London: Routledge (ISBN-10: 0415536782 ISBN-13: 978–0415536783).

Wegerif, R., & Major, L. (2023). *The Theory of Educational Technology: Towards a Dialogic Foundation for Design* (1st ed.). Routledge. https://doi.org/10.4324/9781003198499

Wegerif, R., & Mansour, N. (2010). A Dialogic Approach to Technology-Enhanced Education for the Global Knowledge Society. In M. Khine & I. Saleh (Eds.), *New Science of Learning*. Springer. https://doi.org/10.1007/978-1-4419-5716-0_16

Chapter 6
AI in the Assessment Ecosystem: A Human–Centered AI Perspective

Alina A. von Davier and Jill Burstein

Abstract Artificial Intelligence (AI) is increasingly embedded in the processes of digital, high–stakes educational assessments across test development, measurement, and security. As the use of AI in any setting impacts humans, this brings ethical considerations into play. To that end, *human–centered AI* introduces an ethical perspective. Human–centered AI refers to the development and implementation of AI technologies that prioritize human values, needs, goals, and decision power in system design. This chapter explores assessment drivers that reflect human–centered AI priorities for modern, AI–powered assessments. The chapter explores connections among three critical drivers for digital assessments that draw on human–centered AI principles: (a) a validity argument that considers the use of AI, (b) human-in-the-loop AI practices in assessment, and (c) responsible AI standards as a professional responsibility. The chapter discusses how these three assessment drivers interact within an assessment ecosystem, and contribute to safe AI assessment practices.

Keywords AI · AI ethics · Assessment standards · Digital-first assessment · Educational assessment · Ethical principles · Responsible AI

A. A. von Davier · J. Burstein (✉)
Duolingo, Pittsburgh, USA
e-mail: jill@duolingo.com

A. A. von Davier
e-mail: avondavier@duolingo.com

P. Ilic et al. (eds.), *Artificial Intelligence in Education: The Intersection of Technology and Pedagogy*, Intelligent Systems Reference Library 261,
https://doi.org/10.1007/978-3-031-71232-6_6

6.1 AI in the Assessment Ecosystem: A Human–Centered AI Perspective

In recent years, Artificial intelligence (AI[1]) has revolutionized the landscape of learning and testing. The rapid development of AI technologies has led to the creation of advanced testing platforms and assessment tools that can adapt to individual test–takers' abilities, providing a better testing experience. As AI systems become increasingly integrated into educational contexts, it becomes a priority to explore the commitment to *human–centered AI* and the role of stakeholder trust in these systems.

6.1.1 Human-Centered AI

Human–centered AI refers to prioritizing human values, needs, goals and decision power in the design and implementation of AI technologies (Auernhammer, 2020; Shneidermann, 2020). This is consistent with the ethically-aligned design principle of *Embedding Values into Autonomous Systems* (The IEEE Global Initiative on Ethics of Autonomous and Intelligent Systems, 2017), which asserts that when machines engage with humans, they should be trained to social and moral norms, which must be embedded in the system. In the context of educational testing, this translates to humans creating AI systems that align with assessment science principles, such as validity, ethical concerns of educational stakeholders (such as instructors and administrators) and test–takers, and test–takers' experience and access. Stakeholder trust in AI systems is crucial for its widespread adoption;users must feel confident in the reliability, fairness, accuracy, and security of AI–driven assessments.

6.1.2 Challenges to the Human-Centered AI Perspective

Human-centered AI offers a positive lens through which human values are reflected in the development and implementation of AI systems. However, there are challenges that may conflict with the notion of human-centered AI that need to be addressed. Amariles and Baquero (2023) discuss challenges as they pertain to law. However, their discussion resonates across sectors, and we can imagine that these concerns might also apply in assessment. The authors point out two key challenges. First, they claim that there is a focus on accountability at the design stage of AI systems which does not consider the full life cycle. They provide an example of the need to evaluate *bias* across the AI system life-cycle as discussed in Floridi et al. (2022a). They illustrate that bias can occur across the life cycle in design (historical bias), development (representation bias), evaluation (evaluation bias) and operation (deployment

[1] Note that AI is used broadly and also includes AI–adjacent methods such as machine learning, computational and statistical modeling.

bias). Second, they assert that regulatory initiatives invested in human-centered AI may be market-driven which conflicts with the ability to promote human values. As mentioned earlier, Future of Life's (2023) "Pause Giant AI Experiments: An Open Letter" taps into concerns about market priorities over human values. Garibay et al. (2023) discuss six challenges with respect to adherence to human-centered AI. These assert that AI development should attend to human well-being; be responsible; respect privacy; incorporate human-centered design and evaluation frameworks; attend to governance and oversight; and, respect human cognitive processes with regard to human-AI interaction. Greene et al. (2024) also highlight the need for AI development that addresses these challenges. As precedent, they use the Belmont Report (1979) which operationalized human subjects review practices to protect the rights of individuals participating in human subjects research. While responsible AI frameworks (such as the NIST, 2023) leverage human-centered approaches and ethical principles for system development, recent work by Dotan et al. (2024) highlights challenges associated with AI ethics frameworks. They note that while organizations may develop responsible AI frameworks that lean on ethical principles for AI use, there is a question of the extent to which the principles are operationalized (i.e., translated in practice). Dotan et al. introduce a maturity model that aims to evaluate how ethical principles embodied in AI ethics frameworks are operationalized.

Grounding AI use with human-centered AI for assessment practices (Fig. 6.1) helps to mitigate risks and build stakeholder trust. Recent discourse highlights AI distrust, emphasizing the need for ethical AI practices. For example, such discourse was exemplified by the Future of Life's (2023) "Pause Giant AI Experiments: An Open Letter", and policy initiatives by various entities, including the Department for Science, Innovation & Technology (2023), The White House (2023), Madiega (2023), and Holmes et al. (2022). The computer science community has advocated for systematic AI audits to address diverse risks (Mökander & Axente, 2023; Mökander & Floridi, 2021; Raji et al., 2020; Raji & Buolamwini, 2019), spanning from healthcare trust issues (LaRosa & Danks, 2018) to legal implications of content souring (Foltynek et al., 2023; Levene, 2023) to issues of biased classifications by gender and race (Buolamwini & Gebru, 2018; Nicoletti & Bass, 2023).

This chapter explores human–centered AI in educational testing, addressing three key factors that must be taken into account when developing and implementing AI–driven assessment tools to ensure that the challenges of human-centered AI are addressed: *validity requirements through validity arguments*; *human in the loop AI practices* as part of design principles and psychometrics in an assessment ecosystem; and, *responsible AI practices* as a professional responsibility to guide ethical considerations. Figure 6.1 provides a visualization to illustrate the interactions of these drivers for digital tests, as well as how the human-centered AI approach to assessment is at the core of these interactions.

These factors are addressed in the novel theoretical *assessment ecosystem* proposed by Burstein et al. (2022) for a digital–first *language* assessment. The novelty lies in the acknowledgement that technology–based assessments require multiple frameworks and tools to be thoughtfully integrated in order for the assessment to function smoothly. To that end, their ecosystem is composed of an integrated set of

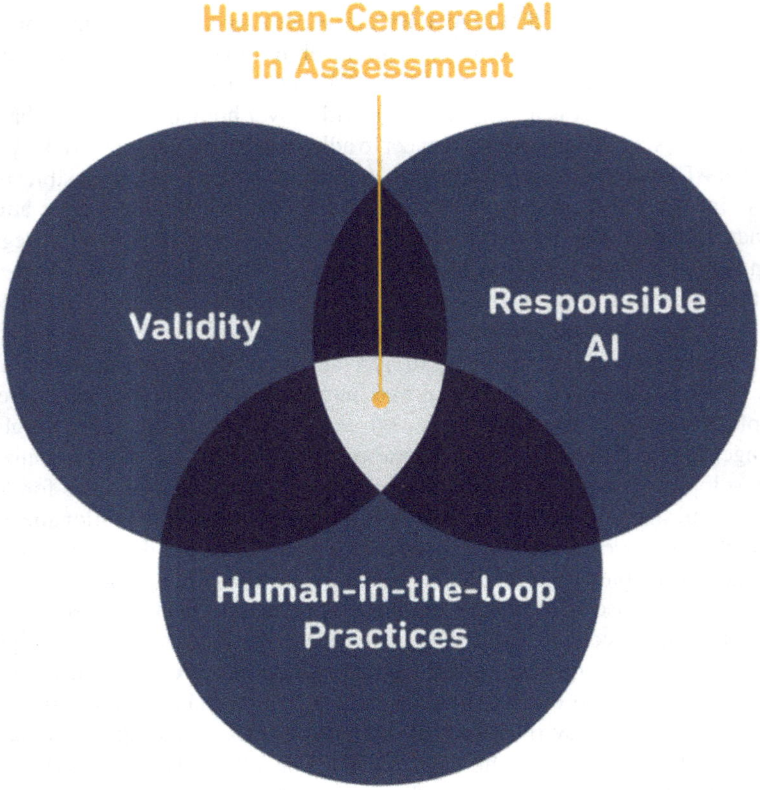

Fig. 6.1 Assessment drivers that reflect human–centered AI for digital assessments

frameworks aligned with the assessment workflow (language assessment design, measurement, and test security), as well as test–taker experience factors (TTX) (such as equitable test access). Langenfeld et al. (2022) generalized the ecosystem to learning and assessment systems beyond language testing. See Fig. 6.2 for an illustration of the assessment ecosystem.

Within Burstein et al.'s (2022) assessment ecosystem, *human–in–the–loop AI* practices support human–centered AI processes and decisions in assessment systems, such as human review of automatically–generated test content. Building on assumptions of traditional *validity* argument approaches (Chapelle et al., 2008; Kane, 1992, 2011, 2013), validity arguments for modern assessment explicitly address AI used on an assessment to produce valid, fair, reliable, and secure test scores. In earlier work, Williamson et al. (2012) address the use of automated essay scoring in high-stakes writing assessment. Chapelle, Cotos, and Lee (2015) used the chain of inferences to frame a validity argument for the feedback effectiveness of a writing instruction application. Burstein et al. (2022) proposed a *digitally–informed chain of inferences* (DCI) that interacts with the ecosystem frameworks and TTX factors. Finally, ethical

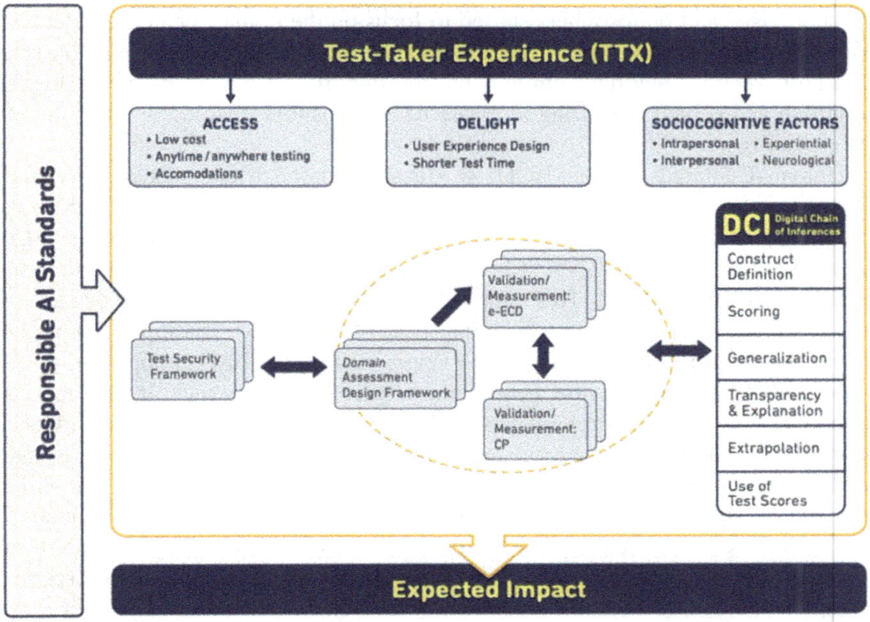

Fig. 6.2 A Domain–agnostic assessment ecosystem governed by responsible AI standards

principles are discussed as captured in *responsible AI (RAI)* standards for a language assessment (Burstein, 2023). RAI standards embed ethical principles to builduser trust, including fairness, explainability, privacy, and transparency and accountability.

This chapter discusses how human–in–the–loop practices, digitally–informed validity arguments, and RAI standards interact within the assessment ecosystem and achieve a human-centered approach to AI developments and applications in assessment. We emphasize the importance of both theoretical frameworks that account for the use of technology, and human–centered and responsible AI practices. Together, frameworks and practices support the development of inclusive and ethically–sound assessment technology that can be safely implemented on assessments, and aim to build human trust in AI systems.

6.2 Human–Centered AI in Assessment

Artificial intelligence (AI) systems and technologies have come a long way since their inception (Crevier, 1993; Grudin, 2009). Machine learning, computer vision, natural language processing, expert systems, and robotics are just a few of the innovations that have emerged in the AI domain, paving the way for increasingly sophisticated interactions between humans and computers. Over time computer scientists, psychologists, and philosophers started to focus on the human–centered aspects of AI, taking into account usability, esthetics, and the overall experience of the user (Markoff, 2005). They have emphasized the importance of understanding human cognition and behavior, striving to create AI systems that seamlessly integrate into our daily lives.

Human–Centered Design (HCD) is a design approach that places individuals at the heart of the development process, taking into account their unique needs, motivations, emotions, behaviors, and cultural context (see Auernhammer, 2020; Schneiderman, 2020; and The IEEE Global Initiative on Ethics of Autonomous and Intelligent Systems, 2017). This approach goes beyond merely involving people in the design; rather, it emphasizes the importance of truly understanding and centering them in every aspect of the design. The language used in different applications can often provide insight into how people are perceived. For instance, in engineering, individuals are frequently referred to as "human factors," signifying their influence on the design and performance of a technological system. Similarly, within management, people are often labeled as "human resources" or "human capital," highlighting their role as assets to be utilized. In contrast, HCD demands that we view humans as people at the center of a system. Each person brings their own unique experiences, needs, desires, and lifestyles, all of which are deeply ingrained within their specific cultural contexts. Through this approach, HCD fosters the creation of solutions that genuinely resonate with users, accommodating their diverse characteristics and ultimately leading to more effective and meaningful outcomes. By centering people and their complexities, Human–Centered Design champions a more empathetic and holistic approach to design and innovation.

In assessment, human–centered AI refers to the design and implementation of AI systems in the context of evaluating human performance, skills, or abilities, cultural background, with a focus on understanding and addressing the needs and experiences of the people involved. HCD in assessments is inclusive and supports diversity, equity and inclusion (DEI) that promote the fair treatment of all people. This approach prioritizes ethical principles that encode human values, and inclusivity when creating or implementing AI–driven assessment tools. In the context of human–centered AI assessment, some key considerations include:

1. *Fairness*: Ensuring that AI assessment tools are unbiased and do not discriminate against certain individuals or groups based on factors like gender, race, or socioeconomic status.
2. *Transparency*: Making the AI algorithms clear and understandable to users, and including human experts in the decision making, so the users can trust the assessment outcomes.
3. *Explainability*: Providing clear explanations of the AI's components, role and recommendations, helping users understand the rationale behind the assessment results.
4. *Privacy and security*: Safeguarding the data used by AI assessment tools, and ensuring that users' personal information is protected.

5. *Customization*: Allowing for personalization of AI assessment tools to better address the unique needs and circumstances of individual users.
6. *Collaboration*: Encouraging collaboration between AI systems and human users, where AI tools support and augment human decision–making, rather than replacing it.

By prioritizing these aspects, human–centered AI in assessment aims to create more ethical, effective, and inclusive evaluation tools that respect human values and improve the overall assessment experience for users.

6.3 Validity

According to the *Standards for Educational and Psychological Testing* (American Educational Research Association, American Psychological Association, & National Council on Measurement in Education, 2014), test validity is measured by the degree to which evidence and theory support the use of a test score for its intended purpose. Related to key factors in human–centered AI, validity includes fairness and explainability as part of a validity argument for an assessment.

6.3.1 Validity Arguments for Traditional Assessment

To support test score interpretation, Kane (1992) proposed a validity argument approach. In this approach, a *chain of inferences* serves as a mechanism to confirm the validity, fairness, and reliability of test score interpretations for the intended purpose (Kane, 1992, 2011, 2013). Earlier discussions of the validity argument approach appeared prior to the emergence of digital assessments and the use of AI on assessment. Chapelle et al. (2008) expanded on and applied the validity argument approach to a large–scale, high–stakes, computer–based English language assessment. In their approach, each inference in the chain has an underlying *warrant* (general principle) and *assumptions* to support the warrant. Satisfying each inference in the chain as part of test development and measurement processes contributes to the validity argument.

This chain of inferences remains valuable for framing validity arguments and is utilized in modern digital assessment frameworks (Burstein et al., 2022; Huggins–Manley et al., 2022; Williamson et al., 2012). However, while Chapelle et al.'s (2008) chain of inferences was designed for a computer–based assessment, it did not formally address digital considerations, such as the use of AI.

6.3.2 Validity Arguments for Digital Assessment

In this section we discuss the role of the *human–in–the–loop AI* practices as part of the human-centered AI and as a building block for the digital chain of inferences in a digital assessment ecosystem. We will start by clarifying the meaning of human–in–the–loop AI, then we describe the theoretical ecosystem for digital assessments, followed by the digital chain of inferences for supporting validity claims.

6.4 Human-in-the-Loop AI Practices

Human–in–the–loop (HITL) AI practices involve integrating human intelligence with AI systems to create a symbiotic relationship that enhances the performance and capabilities of both parties. These practices are associated with human–centered AI fairness, transparency, explainability, privacy, and security factors as part of design, measurement and security practices within the assessment ecosystem (as illustrated in Fig. 6.2; see Burstein et al, 2022).

For purposes of this chapter, we adopt the the definition of HITL as defined in the machine learning. In the machine learning literature, the HITL loop has been defined in terms of control of the outcomes, from humans judgments used to train an AI system, to a system supporting the humans by providing relevant information that facilitates the decision making process (see Mosqueria-Rey, et al., 2023). In addition, HITL AI in educational assessment involves the development and refinement of assessment algorithms. By incorporating human expertise and feedback, AI systems can improve their understanding of complex educational concepts and their ability to contribute to the evaluation of test–takers' performance accurately. This collaboration between human and machine learning algorithms allows for a continuous improvement loop, as human experts can correct the AI system's mistakes, while the AI system can learn from these corrections and update its recommendations. The human weighs in on the final decision, through direct grading or through proctoring and evaluation.

HITL is particularly relevant for automatic item generation, where the content is generated by the AI and evaluated by human experts. HITL AI practices can also be employed to design more personalized and adaptive testing experiences. This level of customization helps improve test–takers' engagement and lead to better testing outcomes. A HITL AI approach also plays a crucial role in addressing potential biases in educational assessment. By having humans actively involved in both the assessment process and the AI applications, it becomes easier to identify and mitigate potential biases that may arise from the use of AI algorithms (Belzak et al., 2023). This can help ensure that all test–takers are assessed fairly and accurately, regardless of their background or personal circumstances.

Another application of HITL AI practices in educational assessment is the development of more effective and efficient feedback systems. With the combined insights of human experts and AI algorithms, we can provide students with timely, specific, and actionable feedback that can help them improve their performance through test preparation. HITL AI practices that are grounded in research and systematically evaluated hold significant potential for enhancing the quality and effectiveness of educational assessment. As AI technology continues to advance, the collaboration between humans and AI will become increasingly important in ensuring the validity of educational assessment practices.

6.5 Theoretical Assessment Ecosystem

Burstein et al.'s (2022) *assessment ecosystem* was designed for the Duolingo English Test (DET)—a digital–first, computer–adaptive, high–stakes English language assessment. The DET's ecosystem is composed of a set of integrated frameworks for test design, measurement and security. HITL AI practices are applied throughout the test's ecosystem. Examples include human review of automatically–generated item content for test design and human labeling to build automated scoring for measurement automated plagiarism detection. The ecosystem interacts with the digitally–informed chain of inferences (DCI) to build a validity argument that considers AI use on the test (Mislevy, 2018; Weir, 2005). As mentioned above, Burstein et al.'s (2022) ecosystem has implications for any technology–based learning and assessment system and can be applied to *any* domain (see Langenfeld et al., 2022). Figure 6.2 illustrates a domain–agnostic version of their ecosystem, governed by the Responsible AI Standards.

The ecosystem is comprised of four complex and integrated frameworks: (a) the *Domain* Assessment Design Framework; (b) the Expanded Evidence–Centered Design (e–ECD) Framework (Arieli–Attali, et al., 2019); (c) the Computational Psychometrics (CP) Framework (von Davier, 2017; von Davier et al., 2019, 2021): and (d) the Test Security Framework. Digital affordances across the ecosystem frameworks support test–taker experience factors associated with equitable test access (such as, lower cost and anytime/anywhere access), and address sociocognitive factors that may play a role in test outcomes.

The Domain Design Framework. This framework focuses on construct definition. It considers constructs relevant to the *domain* of the assessment; identifies potential secondary constructs that may impact test–taker performance, such as sociocognitive factors (e.g., intrapersonal, interpersonal and neurological) (Mislevy, 2018), and experiential (e.g., test item familiarity; Weir, 2005) factors; and, establishes domain relevant proficiency levels and associated criteria that inform item development.

The Validation/Measurement Framework. Two sub–frameworks are included within this framework: the e–ECD Framework and the CP Framework. The e–ECD Framework explicitly adds multimodal evidence and learning into the framework (see Arieli–Attali et al., 2019). In e–ECD, test item configurations are designed with interactions in mind, and the raw rich data collected from test–takers is used to create construct–relevant features that can be used to model test–taker performance. For example, the collection of writing data can support the extraction of error features (such as grammar in writing and pronunciation in speech) as well as features that indicate the quality of a response (such as measures of lexical sophistication). In the context of modeling test–taker proficiency, the CP Framework combines psychometric methods and AI for modeling in assessment. Modeling methods include additional considerations, such as selection of representative corpora, process data, and other issues associated with algorithmic fairness. The psychometric principles related to reliability, validity, generalizability, and fairness guide the data collection and data analysis design, and help blend data–driven AI algorithms and psychometric models.

Test Security Framework. This framework uses AI for item security, test–taker identity, and test–taker data integrity (Cardwell et al., 2024; Isbell & Kremmel, 2020). HITL AI test security measures ensure that test users benefit from valid test scores. Test security framework uses a combination of statistics, computer vision, and NLP algorithms to analyze numerous behavioral and environmental factors that are captured during the test. When the algorithms raise flags indicating a potential security breach, human proctors intervene and evaluate the flag to determine if it is an actual breach. HITL AI test security measures provide assurance that the test score retains its validity with regard to test–taker identity and integrity.

6.5.1 Digitally–Informed Chain of Inferences

In contrast to the chain of inferences described in traditional validity–argument frameworks, the digitally–informed chain of inferences (DCI) makes explicit the digital considerations associated with inference assumptions. Figure 6.2 lists the DCI inference principles and assumptions; explanation and example digital considerations are provided below. The parenthetical next to the DCI inference name refers to the original inference names in Chapelle et al., (2008). See Burstein et al. (2022) for more details.

- The *construct definition* (domain description) inference principle asserts that the test score measures proficiency constructs for the target use domain. To do this, test items should be designed such that test–taker response data can be used to measure test–taker knowledge, skills, and abilities representative of the skills needed by the target use.
    - Example digital consideration: As part of item design, test developers should consider which AI capabilities can be used to accurately capture construct–relevant data from test–taker responses.
- The *scoring* (evaluation) inference principle asserts that automated evaluation methods used for scoring should accurately measure the skills required for target use.
    - Example digital consideration: With regard to measurement, AI methods should be able to detect construct–relevant features for producing test taker

observed scores that reflect proficiency. For instance, for language tests, measures for constructed–relevant writing items might include scores for grammatical accuracy and vocabulary (Yancey et al., 2023).

- The *generalization* inference principle asserts that observed performance measures are estimates of expected performance for parallel versions of the test, across automated/human raters and test administrations.
  - Example digital consideration: With regard to test security, AI–powered test security measures accurately identify the same test taker in a test–retest situation.
- The *transparency & explanation* (explanation) inference principle asserts that observed performance provides interpretable proficiency measures consistent with the required skills by target use.
  - Example digital consideration: With regard to test design, a digital consideration is that AI methods can generate construct–relevant, explainable, feature data from test–taker responses (e.g., constructed–response speaking and writing items).
- The *extrapolation* inference principle asserts that the test assesses skills consistent with the skills required for the target use.
  - Example digital consideration: For measurement, a digital consideration is that AI methods should accurately produce construct–relevant feature data from test–taker responses that can be objectively evaluated in relation to external measures (e.g., a constructed–response writing score is correlated to external writing measures).
- The *test score use* (utilization) inference asserts that the observed test performance is beneficial for stakeholders.
  - Example digital consideration: With regard to test design, a digital consideration is that stakeholders should be able to meaningfully interpret test scores generated using automated scoring methods (such as AI).

As illustrated, the overarching goal of Burstein et al.'s (2022) notion of the DCI is to ensure that the use of AI is addressed in the context of building the validity argument for a digital–first assessment. At its core, the DCI is human-centered: human values are embedded in the conception and development of the proposed digital considerations. Digital considerations will, of course, need to be maintained and updated to keep pace with the growth of technology and how it is being applied on an assessment.

6.6 Responsible AI Standards in Assessment

Responsible AI (RAI) standards and related processes aim to mitigate risks associated with AI that could impact the quality throughout an assessment ecosystem—specifically, test design, measurement, and security. As pointed out earlier with regard to challenges for human-centered AI, building RAI practices across the life-cycle of an assessment (i.e., the assessment ecosystem) meets the challenge posed by Amariles and Baquero (2023). Specifically, that the use of AI in an assessment must be addressed across the assessment ecosystem (life-cycle). Given the rise of AI for assessment, guidelines have proliferated that address the ethical use of AI on assessments (such as International Test Commission & Association of Test Publishers, 2022; OECD, 2023; The International Privacy Subcommittee of the ATP Security Committee, 2021; U.S. Department of Education, Office of Educational Technology, 2023). The ethical principles in assessment also resonate with those that have emerged from the larger literature on AI ethics that also specify concepts such as trust, responsibility, justice, and autonomy (see The National Institute for Standards and Technology's (NIST) Artificial Intelligence Risk Management Framework, 2023; Fjeld et al., 2020; Floridi & Cowls, 2022b; Jobin et al., 2019). Fjeld et al. (2020) identified *professional responsibility* as a key principle mentioned throughout responsible AI literature. This principle resonates with human–centered AI in that it is directly associated with human values and goals. Fjeld et al. (2020) characterize this principle by highlighting the importance of accuracy and scientific integrity (such as evaluating AI systems for assessments to ensure they provide accurate measures); responsible design (such as access to assessments); consideration of long–term effects, which considers the impact of the AI (such as the impact of a test score on an individual); and, multi–stakeholder collaboration, which recommends including a variety of relevant stakeholders (such as ensuring that humans who review test content represent a diverse group who are sensitive to the test–taker population). In the spirit of professional responsibility, it is important that the assessment community leverage RAI guidelines to develop RAI standards for AI–driven assessments. Consistent with this, AI researchers and computer scientists have highlighted varied risks associated with AI, especially those related to bias, which applies to assessment as well. They recommend systematic audits of AI (Mökander & Axente, 2023; Mökander & Floridi, 2021; Raji & Buolamwini, 2019; Raji et al., 2020).

6.6.1 *The DET Responsible AI Standards*

As part of professional responsibility, Burstein (2023) developed RAI standards for the DET. The standards were developed as an outcome of a systematic literature review of AI ethics, and multi-stakeholder collaboration with experts from applied linguistics, computational psychometrics, computer science, ethics, language assessment, law, and security. The DET's RAI standards offer public transparency about the guardrails being applied to the test that aim to mitigate risks associated with the use of AI. Further, the processes associated with the RAI standards provide a practical mechanism for auditing AI used throughout the assessment ecosystem for test design, measurement, and security. Examples of auditing practices are described earlier in the discussion about human-in-the-loop AI. As an extension of professional responsibility and in the spirit of fostering human trust and multi–stakeholder collaboration, the DET's RAI standards are publicly available and remain open for public comment. The DET RAI standards reflect themes specific to assessment. Common themes overlap with human–centered AI, that aim to ensure test validity, fairness, explainability, privacy and security, and accountability and transparency through four key ethical principles which are summarized below as they are discussed in Burstein (2023).

- The *Validity and Reliability* standard is crucial to ensure that the test is suitable for its intended purpose. The Validity standard involves evaluating construct relevance and accuracy, while the Reliability standard focuses on consistency;
- The *Fairness* standard promotes democratization and social justice through increased access, accommodations, and inclusion, representing test-taker demographics, and avoiding algorithms known to contain or generate bias;
- The *Privacy and Security* standard ensures that we (a) comply with relevant laws and regulations governing the collection and use of test taker data; (b) ensure test taker privacy and (c) to ensure secure test administration; and,
- The *Accountability and Transparency* standard aims to gain trust from stakeholders. This is essential for proper governance of AI used on the test.

While the DET RAI standards were designed for an assessment, for the most part, the standard's ethical principles are agnostic to a particular discipline or industry (see Burstein, 2023 for more details about the DET RAI standards).

6.7 Conclusions

Human values have been the traditional driver for assessment research and development. As technology continues to be integrated into assessments, it is critical to maintain these values in tandem with technological developments. AI use, in particular, is becoming increasingly more prominent in digital assessment. Given the potential risks (such as bias) and benefits (such as equitable access) of AI use on assessments, and especially high–stakes assessments, the insertion of human values into processes related to AI use is essential.

This chapter uses a human–centered AI lens to explore how human values (encoded in ethical principles) are integrated into the use of AI for assessment. It illustrates how ethical principles are incorporated as part of an assessment ecosystem through (a) validity arguments that consider AI use, (b) HITL AI practice, and (c) responsible AI standards and practices.

We have based the exposition presented here on our work with the Duolingo English Test, since, as a digital-first test, it uses AI extensively. We believe that these learnings and the example practices discussed are generalizable to other tests and other domains that may consider using AI. Given the novelty of the use of AI throughout the DET, we do expect that these approaches will evolve over time to address the rapid development of technology and the challenges yet to be discovered in the application of these nascent, powerful tools. As such, this chapter represents a snapshot in time.

References

Amariles, D. R., & Baquero, P. M. (2023). Promises and limits of law for a human-centric artificial intelligence. *Computer Law & Security Review, 48*, 105795.

American Educational Research Association, American Psychological Association, & National Council on Measurement in Education. (2014). *Standards for educational & psychological testing*. American Educational Research Association. https://www.testingstandards.net/uploads/7/6/6/4/76643089/standards_2014edition.pdf

Arieli-Attali, M., Ward, S., Thomas, J., Deonovic, B., & Von Davier, A. A. (2019). The expanded evidence-centered design (e-ECD) for learning and assessment systems: A framework for incorporating learning goals and processes within assessment design. *Frontiers in psychology, 10*, 853.

Auernhammer, J. (2020). Human–centered AI: The role of Human–Centered Design research in the development of AI. In S. Boess, M. Cheung, & R. Cain (Eds.), *Synergy–DRS International Conference 2020* (pp. 1315–1333). https://doi.org/10.21606/drs.2020.282

Belzak, W., Naismith, B., Burstein, J. (2023). Ensuring fairness of human– and AI–generated test items. In N. Wang, G. Rebolledo–Mendez, V. Dimitrova, N. Matsuda, & O. C. Santos (Eds.), *Communications in computer and information science. Artificial intelligence in education: Posters and late breaking results, workshops and tutorials, industry and innovation tracks, practitioners, doctoral consortium and blue sky*. Springer. pp. 701–707

Buolamwini, J., & Gebru, T. (2018). Gender shades: Intersectional accuracy disparities in commercial gender classification. In *Conference on fairness, accountability and transparency*. PMLR, pp. 77–91.

Burstein, J., LaFlair, G. T., Kunnan, A. J., & von Davier, A. A. (2022). *A theoretical assessment ecosystem for a digital–first assessment—The Duolingo English Test* [Research report]. Duolingo English Test. https://duolingo-papers.s3.amazonaws.com/other/det-assessment-ecosystem-mpr.pdf

Burstein, J. (2023). *Responsible AI standards*. Duolingo English Test. https://duolingo-papers.s3.amazonaws.com/other/DET+Responsible+AI+033123.pdf

Cardwell, R., Naismith, B., LaFlair, G. T., & Nydick, S. (2024) *Duolingo English Test: Technical manual* [Research report]. Duolingo English Test. http://duolingo-papers.s3.amazonaws.com/other/technical_manual.pdf

Cardwell, R., Liao, M., Belzak, W., & LaFlair, G. T. (2023a). *Incorporating test security into the validity argument of a remotely–proctored English test* [Conference Session]. 44th Language Testing Research Colloquium (LTRC), New York, NY. https://ltrc2023.weebly.com/conference-schedule.html

Chapelle, C., Enright, M., & Jamieson, J. (2008). *Building a validity argument for the Test of English as a Foreign Language*. Routledge.

Crevier, D. (1993). *AI: The tumultuous history of the search for artificial intelligence*. Basic Books.

Dotan, R., Blili-Hamelin, B., Madhavan, R., Matthews, J., & Scarpino, J. (2024). Evolving AI risk management: A maturity model based on the NIST AI risk management framework. arXiv preprint arXiv:2401.15229.

Fjeld, J., Achten, N., Hilligoss, H., Nagy, A., & Srikumar, M. (2020). Principled artificial intelligence: Mapping consensus in ethical and rights–based approaches to principles for AI [Research report]. *Berkman Klein Center for Internet & Society at Harvard University*. https://doi.org/10.2139/ssrn.3518482

Floridi, L., Holweg, M., Taddeo, M., Amaya Silva, J., Mökander, J., & Wen, Y. (2022a). CapAI-A procedure for conducting conformity assessment of AI systems in line with the EU artificial intelligence act. Available at SSRN 4064091.

Floridi, L., & Cowls, J. (2022). A unified framework of five principles for AI in society. In S. Carta (Ed.), *Machine learning and the city: Applications in architecture and urban design* (pp. 535–545). John Wiley & Sons Ltd. https://doi.org/10.1002/9781119815075.ch45

Foltynek, T., Bjelobaba, S., Glendinning, I., Khan, Z. R., Santos, R., Pavletic, P., & Kravjar, J. (2023). ENAI recommendations on the ethical use of Artificial Intelligence in education. *International Journal for Educational Integrity, 19*(1).

Future of Life Institute. (2023). *Pause giant AI experiments: An open letter*. https://futureoflife.org/open-letter/pause-giant-ai-experiments/

Garibay, O., Winslow, B., Andolina, S., Antona, M., Bodenschatz, A., Coursaris, C., & Xu, W. (2023). Six human-centered artificial intelligence grand challenges. *International Journal of Human-Computer Interaction, 39*(3), 391–437.

Greene, K. K., Theofanos, M. F., Watson, C., Andrews, A., & Barron, E. (2024). Avoiding past mistakes in unethical human subjects research: Moving from artificial intelligence principles to practice. *Computer, 57*(2), 53–63.

Grudin, J. (2009). AI and HCI: Two fields divided by a common focus. *AI Magazine, 30*(4), 48–57. https://doi.org/10.1609/aimag.v30i4.2271

Holmes, W., Persson, J., Chounta, I. A., Wasson, B., & Dimitrova, V. (2022). *Artificial intelligence and education: A critical view through the lens of human rights, democracy and the rule of law*. Council of Europe. https://rm.coe.int/artificial-intelligence-and-education-a-critical-view-through-the-lens/1680a886bd

Huggins-Manley, C., Booth, B. M., & DeMelo, S. K. (2022). Toward argument-based fairness with an application to AI-enhanced educational assessments. *Journal of Educational Measurement., 59*, 362–388. https://doi.org/10.1111/jedm.12334

International Test Commission & Association of Test Publishers. (2022). *Guidelines for technology–based assessment*. Association of Test Publishers.

Isbell, D. R., & Kremmel, B. (2020). Test review: Current options in at–home language proficiency tests for making high–stakes decisions. *Language Testing, 37*(4), 600–619. https://doi.org/10.1177/0265532220943483

Jobin, A., Ienca, M., & Vayena, E. (2019). The global landscape of AI ethics guidelines. *Nature Machine Intelligence, 1*, 389–399. https://doi.org/10.1038/s42256-019-0088-2

Kane, M. T. (1992). An argument–based approach to validity. *Psychological Bulletin, 112*(3), 527–535. https://doi.org/10.1037/0033-2909.112.3.527

Kane, M. T. (2011). Book review: Language assessment in practice: Developing language assessments and justifying their use in the real world. *Language Testing, 28*(4), 581–587. https://doi.org/10.1177/0265532211400870

Kane, M. T. (2013). Validating the interpretations and uses of test scores. *Journal of Educational Measurement, 50*, 1–73. https://doi.org/10.1111/jedm.12000

Langenfeld, T., Burstein, J., & von Davier, A.A. (2022). Digital-First Learning and Assessment Systems for the 21st Century. In Frontiers in Education. vol. 7 https://doi.org/10.3389/feduc.2022.857604

LaRosa, E., & Danks, D. (2018). Impacts on trust of healthcare AI. In *Proceedings of the 2018 AAAI/ACM Conference on AI, Ethics, and Society* (pp. 210–215). https://doi.org/10.1145/3278721.3278771

Levene, A. (2023). *Artificial intelligence and authorship.* Committee on Publication Ethics (COPE).
Madiega, M. (2023). *Artificial intelligence act* [Briefing]. European Parliamentary Research Service. https://www.europarl.europa.eu/RegData/etudes/BRIE/2021/698792/EPRS_BRI(2021)698792_EN.pdf
Markoff, J. (2005). What the dormouse said: How the sixties counterculture shaped the personal computer industry: Penguin Publishing Group.
Mislevy, R. J. (2018). *Sociocognitive foundations of educational measurement*. Routledge.
Mökander, J., & Axente, M. (2023). Ethics–based auditing of automated decision–making systems: Intervention points and policy implications. *AI & Society, 38*, 153–171. https://doi.org/10.1007/s00146-021-01286-x
Mökander, J., & Floridi, L. (2021). Ethics–based auditing to develop trustworthy AI. *Minds and Machines, 31*(2), 323–327. https://doi.org/10.1007/s11023-021-09557-8
Mosqueira-Rey, E., Hernández-Pereira, E., Alonso-Ríos, D., et al. (2023). Human-in-the-loop machine learning: A state of the art. *Artificial Intelligence Review, 56*, 3005–3054. https://doi.org/10.1007/s10462-022-10246-w
National Institute of Standards and Technology (2023). *Artificial intelligence risk management framework (AI RMF 1.0)*. U.S. Department of Commerce. https://doi.org/10.6028/NIST.AI.100-1
Nicoletti, L., & Bass, D. (2023). *Humans are biased. Generative AI is worse.* Bloomberg Technology + Equality. https://www.bloomberg.com/graphics/2023-generative-ai-bias/
OECD (2023). *Advancing accountability in AI: Governing and managing risks throughout the lifecycle for trustworthy AI* (No. 349). OECD Publishing. https://doi.org/10.1787/2448f04b-en
Raji, I. D., & Buolamwini, J. (2019). Actionable auditing: Investigating the impact of publicly naming biased performance results of commercial AI products. In *Proceedings of the 2019 AAAI/ACM Conference on AI, Ethics, and Society* (pp. 429–435).
Raji, I. D., Smart, A., White, R. N., Mitchell, M., Gebru, T., Hutchinson, B., Smith–Loud, J., Theron, D., & Barnes, P. (2020). Closing the AI accountability gap: Defining an end–to–end framework for internal algorithmic auditing. In *Proceedings of the 2020 Conference on Fairness, Accountability, and Transparency (FAT*'20), 27–30, Barcelona, Spain.* (pp. 33–44). ACM. https://doi.org/10.1145/3351095.3372873
Shneiderman, B. (2020). Human–centered artificial intelligence: Three fresh ideas. *AIS Transactions on Human-Computer Interaction, 12*(3), 109–124. https://doi.org/10.17705/1thci.00131
The Belmont report: Ethical principles and guidelines for the protection of human subjects of research," The Commission, Bethesda, MD, USA, (1979). [Online]. Available: https://www.hhs.gov/ohrp/regulations-and-policy/belmont-report/read-the-belmont-report/index.html
The IEEE Global Initiative on Ethics of Autonomous and Intelligent Systems. (2017). *Ethically aligned design: A vision for prioritizing human well–being with autonomous and intelligent systems* (Version 2). IEEE. https://standards.ieee.org/wp-content/uploads/import/documents/other/ead_v2.pdf
The International Privacy Subcommittee of the ATP Security Committee. (2021). *Artificial intelligence and the testing industry: A primer*. Association of Test Publishers.
The White House. (2023). *FACT SHEET: Biden-Harris administration secures voluntary commitments from leading artificial intelligence companies to manage the risks posed by AI* [Fact sheet]. https://www.whitehouse.gov/briefing-room/statements-releases/2023/07/21/fact-sheet-biden-harris-administration-secures-voluntary-commitments-from-leading-artificial-intelligence-companies-to-manage-the-risks-posed-by-ai/
U.S. Department of Education, Office of Educational Technology (2023). Artificial intelligence and future of teaching and learning: Insights and recommendations [Report]. https://www2.ed.gov/documents/ai-report/ai-report.pdf
von Davier, A. A. (2017). Computational psychometrics in support of collaborative educational assessments. *Journal of Educational Measurement, 54*, 3–11. https://doi.org/10.1111/jedm.12129

von Davier, A. A., Deonovic, B., Yudelson, M., Polyak, S. T., & Woo, A. (2019). Computational psychometrics approach to holistic learning and assessment systems. *Frontiers in Education, 4*, 69. https://doi.org/10.3389/feduc.2019.00069

von Davier, A. A., Mislevy, R. J., & Hao, J. (Eds.). (2021). *Computational Psychometrics: New Methodologies for a New Generation of Digital Learning and Assessment*. Springer.

Weir, C. J. (2005). *Language testing and validation: An evidence–based approach.* Palgrave McMillan.

Williamson, D. M., Xi, X., & Breyer, F. J. (2012). A framework for evaluation and use of automated scoring. *Educational Measurement: Issues and Practice., 31*(1), 2–13. https://doi.org/10.1111/j.1745-3992.2011.00223.x

Yancey, K. P., Laflair, G., Verardi, A., & Burstein, J. (2023). Rating short L2 essays on the CEFR scale with GPT-4. In Kochmar, E., Burstein, J., Horbach, A., Laarmann–Quante, R., Madnani, N., Tack, A., Yaneva, V., Yuan, Z., & Zesch, T. (Eds.), *Proceedings of the 18th Workshop on Innovative Use of NLP for Building Educational Applications (BEA 2023)* (pp. 576–584). https://doi.org/10.18653/v1/2023.bea-1.49

Chapter 7
The Role of AI Language Assistants in Dialogic Education for Collective Intelligence

Imogen Casebourne and Rupert Wegerif

Abstract This chapter combines evidence from empirical research studies with arguments drawn from philosophy to explore how we conceptualise the role of AI language assistants like ChatGPT in education. We begin with the challenge to existing models of education posed by AI's ability to pass examinations. We examine again the critique of the idea of AI from Dreyfus and from Searle and the critique of the value of writing from Socrates, to suggest that there may have been much too much focus on the skill of academic writing in education at the expense of the skill of dialogue, a skill which is more fundamental to intellectual development. We then look at the potential of AI for teaching through dialogue and for teaching dialogue itself in the form of dialogic thinking. We ask what it means for a person to enter into dialogue with a large language model. We conclude that dialogic education mediated by dialogues with large-language models is itself a form of collective intelligence which leads us to articulate a vision of individual education as learning how to participate in AI mediated collective intelligence.

Keywords Dialogue · Exploratory dialogue · Critical thinking · Creativity · Generative AI · Hybrid models · Tacit knowledge · Futures of education · Bias · AI ethics

---

I. Casebourne (✉)
DEFI, Hughes Hall, University of Cambridge, Cambridge, UK
e-mail: ic407@hughes.cam.ac.uk

R. Wegerif
Faculty of Education, University of Cambridge, Cambridge, UK
e-mail: rw583@cam.ac.uk

P. Ilic et al. (eds.), *Artificial Intelligence in Education: The Intersection of Technology and Pedagogy*, Intelligent Systems Reference Library 261,
https://doi.org/10.1007/978-3-031-71232-6_7

7.1 Introduction

The release of ChatGPT to the general public in November 2022, led to headlines suggesting that it could pass a variety of exams (Ali et al., 2023, Varanasi, 2023), in addition to reports of children using it to do their homework (Casebourne, 2023). In this chapter, we ask what the ability of generative AI based on large language models (LLMs) to write essays and pass exams means for current models of education. We conclude that it suggests a need to move to a more dialogic understanding of education.

In 2022 ChatGPT gave students access to a form of generative AI that was able to produce written essays that summarise existing knowledge on a topic, as well, if not better than many students (Sharples, 2022; Farazouli et al. 2023). However, the *process* of producing these essays is now completely different. The assumption of the existing assessment systems of education is that the student structures and writes the essay whereas in the new AI enhanced approach to essay writing, the main process is evolving to become the student offering a series of prompts to the LLM.

One immediate response from universities was to attempt to reinforce the status quo by upgrading existing plagiarism tools, setting AI to catch AI (Cotton, Cotton, & Shipway, 2023). However, almost as quickly, reports started to emerge suggesting that these new detection tools were generating both false positives and false negatives (Edwards, 2023). In particular they were more likely to identify writing by non-native speakers as having been generated by AI when this was not the case (Liang et al. 2023). However, rather than seeking ways in which AI might be used to maintain the print based approach to learning and assessment that evolved with mass schooling over the past 150 years, we argue that we could instead embrace its potential to facilitate new ways of learning and assessing knowledge that are in some ways also a return to older more dialogic approaches.

One way of framing this is to see large language models (LLMs) as a new form of collective intelligence (CI). CI describes the amplified capacity that collaboration enables, making it possible to engage with a larger and more diverse array of information and ideas than could be achieved by a single individual. Observable across various species, it encompasses small group collaboration and dialogue, crowdsourcing, swarm problem-solving, and stigmergy, which refers to the traces left by one individual which act to stimulate others to continue with an open ended task (Berditchevskaia et al., 2022; Mulgan, 2018). In humans CI extends beyond problem-solving to include problem-framing, priority selection, and decision-making. In the case of dialogue, it has been argued that an individual can represent an ongoing dialogue of many voices, and importantly, that technology can serve as an additional voice or perspective in the dialogue (Hermans, 2019; Wegerif & Major, 2023). LLMs literally invite us to dialogue and in doing so, offer an immediate entry into a series of ongoing dialogues in a way that has not previously been possible.

7.2 How Did We Get Here? A Brief History of AI

There are multiple overlapping definitions of artificial intelligence (AI) but in this chapter, we are especially interested in the relevance to education of a historic split in approaches to AI which Boden traces in her brief introduction (2016): a distinction between "symbolic" or GOFAI (good old fashioned AI) on the one hand and cybernetics and neural networks on the other, a distinction which Holmes and Tuomi, in considering AI and education (2022) present as being between knowledge based and data driven AI.

The first approach, the GOFAI or knowledge-based approach, broadly involves distilling a set of rules and information on a topic from what is known by human experts and programming these into a system, which is then able to operate by or advise on those rules. However, a problem with this, as was also encountered in efforts in vocational education to create detailed specifications of procedures and competencies for job roles, is that some of what human experts *know* may be tacit, that is, not easily accessed, explained or codified (Nonaka & von Krogh, 2009; Polanyi & Sen, 2009). Additionally, it becomes very cumbersome to specify every combination or sub-rule or circumstance. An early critique of AI articulated by Dreyfus (1972) focused on these types of limitations and influenced to an extent the direction of the *other* strand of AI, what Holmes and Tuomi term the *data driven strand*, which eventually led to the development of Large Language Models (LLMs), such as ChatGPT as well as computerised image generation such as MidJourney.

This second approach broadly involves the use of neural networks, computational structures loosely modelled on the human brain, which are given large data sets from which to "learn" concepts for themselves. This machine learning may be *supervised*, in which case the data is pre-labelled, sometimes by poorly paid people working in data centres (Gray & Suri, 2019) and the program learns to correctly identify new examples as belonging to a specific category (as an apple not a pear, for example). Alternatively, it may be *unsupervised*, in which case the data is *not* pre-labelled and the programme finds structure and assigns categories without guidance. A feature of this overall approach, in contrast to GOFAI, is that as with the human experts who proved unable to articulate all of their tacit knowledge or explain every element of their decision-making processes, with these models, it is often not clear how they arrive at the results they achieve (Russell & Norvig, 2021). That is to say, it is understood at a high level how such systems operate, but unlike with GOFAI systems, the associations formed, and decisions made are not typically accessible to human observers.

This raises an important conceptual question. Should decision-makers be primarily interested in and focus on the *outputs* of AI, or should the *process* by which AI achieves its results also be of concern? The original and famous test of computer intelligence proposed by Alan Turing, partly as a joke, focuses on what we can understand as output, or rather on the experience of engaging with AI. In brief, it suggests that if a conversation with a computer program is indistinguishable from a conversation with a human, we could view that program as being intelligent

(Epstein et al., 2009). Long before LLMs like ChatGPT were able to provide text that *is* often indistinguishable from that written by humans (Xiao et al., 2022), other philosophers objected that the *process* by which AI achieves its results and outputs should be viewed as equally important. For example, Dreyfus (1972) pointed out that AI's ways of knowing were rooted in the abstract, rather than in human experience.

Searle (1980) argued that if he was shut in a room, handed a set of symbols and given instructions on how to manipulate them in such a way that when he handed the results back to people outside the room, they gained the impression that he was conversing with them in Chinese, this wouldn't mean that he, in the room, actually understood Chinese. Searle's thought experiment has been criticised by Dennett (1991) among others, for mis-representing human consciousness and understanding. However, we agree with Searle that the facility with language that renders a LLM indistinguishable from a human conversational partner does not mean that it is intelligent or conscious in the way that humans are. We think current LLMs are *not* conscious—for a variety of reasons, for example, there is reason to believe that they don't have a representation of the world, just of language (Bender et al. 2021) and of images and the way in which text and images are related and this is in part because they are not embodied and do not have qualia, subjective experiences of the world linked to understanding (e.g. Kauffman & Roli, 2023). While they may be familiar with much of what has been written about cats and can match this to images of cats, they have never touched a cat or breathed in the scent of a cat. These are two of a number of attributes commonly understood to be necessary for consciousness. Chalmers gives a good overview of several more reasons for believing that LLMs are currently not conscious (2023).

Arguments about human consciousness aside, an element of Searle's argument, that AI programs *might* achieve results indistinguishable from humans via a different *process* to that underpinning human achievements remains relevant today as researchers attempt to identify measures of intelligence equally applicable to humans, animals and computers, as well as hybrid and collective intelligence efforts (Hernández-Orallo, 2017). For example, Hernández-Orallo argues that in passing visual tests, AI might be exploiting patterns and biases in the dataset to achieve high scores, thus using a strategy completely different from that employed by humans to achieve the same results. As with essays generated by ChatGPT, the similarity of the results to what might be achieved by an intelligent person does not imply that the underlying process is the same.

While the outputs of LLMs often pass the Turing test in that they are indistinguishable from the responses of a human, they are not produced by human intelligence but by a statistical process which involves predicting the most likely next word given the data set (corpus of text) that they have been trained on (Hoffmann et al. 2022). This is a process that can produce the results it does without requiring self-awareness or intentionality or the sense of ourselves as reflective reasoning beings that people commonly experience and that we imagine when we think of a sentient conversational partner. This is not to argue that AI will never become sentient in any future form, just that current LLMs are not sentient. Shanahan and colleagues offer an in-depth discussion of how LLMs work for those interested in reading more (Shanahan, 2023; Srivastava et al. 2022).

Relatedly, concerns have been raised that if people cannot understand the processes by which AI programs achieve results it is impossible to be confident that these processes are not cementing or worsening bias. This has been of concern in the case of generative AI, which involves neural networks trained on vast amounts of data. For one thing, the training data may well in itself be biased (Bender et al., 2021) but in addition to possible bias in the datasets used to train AI (e.g. Srinivasan & Uchino, 2021), it is also possible that *algorithms,* that is to say the *processes* by which AI achieves results, may inadvertently exacerbate dataset bias or introduce entirely new bias (Baker & Hawn, 2022; Noble, 2018). The movement for explainable AI, achievable or not, is an attempt to ensure the possibility of scrutinising the processes by which AI produces its results (Rudin et al. 2021).

A focus on process is relevant to education more broadly, as well as to AI in education, as we will go on to argue. For one thing, while education is typically viewed (rightly we think) as being a process, young people today are primarily *assessed* by means of written artefacts such as essays which may not be good proxies for the thought processes they are meant to measure. Now that LLMs such as ChatGPT can, with some prompting, produce human quality essays, it is, perhaps, time for a rethink.

7.3 AI and Education

There have been attempts to use AI in the service of education almost since its inception (Doroudi, 2022). Holmes and Tuomi (2022) provide a recent taxonomy of AI intended to support education, grouped under three main categories: AI to support students, teachers, and educational administration.

In the first category they include intelligent tutoring systems (typically based on GOFAI approaches) that provide adaptive feedback and learning pathways based on an individual's performance. The rationale behind these is that they offer students tailored and immediate feedback faster than might be possible from a teacher responsible for a classroom of students with different levels of understanding and needs. A recent review indicates that such systems continue to largely be directed at subjects such as science, maths or electronics (Kurni et al., 2023) although there have been attempts to build them for music theory and the analysis of historical documents (Bull and Kay, 2010). There have also been less domain specific attempts to create artificial dialogue and tutoring partners, for example, AutoTutor which builds on analysis of the discourse moves used by effective tutors (Nye, Graesser, & Hu 2014).

Finally, AI has also been used in games and simulations intended to support students to practise skills or to assess student skill levels. An example is the use of AI agents as collaborative partners, where AI agents are intended to provide reliable, repeatable collaboration opportunities appropriate for assessment of individual skills or potential (Chopade et al. 2018; Rojas et al., 2022; Rosen, 2015).

At the same time, AI intended to support teachers offers student observation and dashboards (Knoop-van et al., 2023), as well as AI generated lesson plans and test items (Wodzak, 2022), and even AI marking of student work (Noakes, 2022). This raises the possibility of a scenario in which children use generative AI to write essays which another form of AI then marks!

What the above approaches to supporting learners and teachers have in common is that they bring AI to bear on pedagogy, seeking to replace or supplement teachers or change teaching. However, recent advances in generative AI have also raised questions about the curriculum of the future, amid speculation that AI might automate swathes of job tasks if not jobs (Briggs & Kodnani, 2023; Eloundou et al. 2023).

7.4 Education as an Ongoing Process of Dialogue and Reflection

Key elements of global education systems have focused for the past 150 years on developing a student's ability to reproduce accurate and comprehensive overviews of the latest expert opinion using the written word (Garber, 2010).

Socrates famously critiqued writing, arguing that it leads to superficial understanding (Waterfield & Plato, 2002). His argument can be viewed as being in an interesting way somewhat related to the discussion about how we should engage with AI. He suggested that an issue with written text is that it is *silent when questioned*—i.e. it is an artefact and can be misinterpreted and misunderstood in a way that is less likely in a living exploratory dialogue between people. For Socrates the thinking *process* was important and someone absorbing a text had not necessarily gone through the process of reflection and questioning that was his aim as an educator. Of course, without the written text of the Phaedrus it is unlikely that people would be discussing this argument today, and it is also the case that debates and dialogues have been conducted through the medium of letters and texts. However, Socrates' focus on the importance of questioning, challenging, and engaging in dialogue in order to foster true understanding remains relevant to the contemporary debate about the role of generative AI.

The modern concept of dialogic space in education refers to an interactive and shared space where participants engage in open, respectful, and meaningful dialogue, a space where different voices, perspectives, and ideas are invited and explored, fostering a sense of curiosity, critical thinking, empathy, and collaboration (Wegerif, 2007). It emerged from analysis by Wegerif and Mercer of classroom talk underpinning effective collaborative learning in small groups (1997).

The idea of identification with 'dialogic space' was introduced to understand the form of identification involved in 'exploratory talk,' which was seen as key being able to change one's mind and engage with others' ideas. Various theorists, including Bakhtin (1981, 1984, 1986) and Buber (1958), have also addressed aspects of dialogue and introduced versions of the idea of dialogic space. The term 'dialogic' can be understood as seeing or thinking things from at least two points of view at once, as opposed to 'monologic,' where only one correct point of view is acknowledged. Real dialogues occur when people are prepared to listen to and learn from each other, leading to shared thinking and the experience of 'thinking together' (Mercer & Littleton, 2007).

As Nye, Graesser and Hu (2014) point out, a Socratic dialogue might not teach you everything, for example it won't teach you to play the piano,[1] but Socratic dialogue as a process can help learners achieve a profound understanding in many areas, and might be supported by large language models such as ChatGPT.

7.5 AI as a Dialogic Educational Companion

As touched on in the earlier part of this chapter, students have been learning in ongoing interaction or dialog with AI for some time, but until recently this AI was largely in the form of intelligent tutoring systems, restricted to leading students on a relatively predefined path through a very specific and structured knowledge domain. Now, however, with LLMs such as ChatGPT it is possible to initiate almost any conversation on a topic of the student's choosing and the experience certainly *feels* more like an exploratory dialogue, which can evolve in a direction of the student's choosing.

This raises questions about what we should think about such conversation with LLMs. Firstly, what exactly is the student conversing with? It has previously been argued that in interactions involving a computer and the digital sphere, we should consider the interaction as being with the designer or the owner of the system (or the organisation that owns the system) (Souza & Sieckenius, 2005). This is intuitive to grasp when thinking about GOFAI based intelligent tutoring systems, where, as explained above, a careful effort has been made to extract and codify a set of rules, and to design a system to tutor students according to those rules. However, in the case of LLMs, the way in which a person can be said to be conversing with an organisation, or a designer, rather than an artificial entity, becomes less clear because the organisation does not know in advance and does not completely control what their LLM will "say" in response to any novel conversational move that a student might make.

As explained in the introduction, while the principles by which LLMs produce their results are understood on one level, exactly what associations and predictions an individual LLM may have codified following training is typically less clear. The organisations know more than the public about the datasets and data they have used

[1] Although learning a vocational skill also involves engaging in an interactive *process* over time, one in which competency may also be assessed over time, as with an apprenticeship model, (Billett, 2014).

to train their models, and therefore at a high level what types of biases and political standpoints we might expect to have been present in those datasets (Brown et al. 2020; Reisner, 2023; Wu et al. 2021). Additionally, they are in a position to add what are known as guardrails (Ahn and Chen 2023), adjustments intended to ensure that their systems do not, for example, insult their conversational partners, encourage them to harm themselves or coach them in how to harm others. In deploying guardrails, organisations make decisions about what is and is not acceptable, so to an extent, the organisation retains a degree of control over what conversations can occur. Indeed, recent research suggests that LLMs created by different organisations appear to have different political leanings, perhaps evidence of varying organisational decision-making processes (West, 2023). However, it remains the case that currently organisations have less control over what their LLM might "say" than they might perhaps prefer or has been the case with anything that has gone before. Headlines about LLMs making up fake references (Casebourne, 2023), or recommending poisonous recipes (McClure, 2023) do not result from an intentional strategy on the part of the organisations to have their LLMs mislead people in these ways.

So, if the student can not straightforwardly be said to be conversing with, for example, OpenAI, Anthropic, Meta or Alphabet (Google's parent company) when interacting with a LLM, what *are* they in dialogue with? Some have claimed that individual LLMs, in their apparent creativity and unpredictability, have become sentient (Luscombe, 2022) and that the student is therefore conversing with a new type of self-aware being that has recently come into existence. However, as discussed earlier, we argue against this view.

Instead, we argue that in entering into a dialog with an LLM a student is enabled to enter into a particular kind of dialogic space where they experience and interact with a variety of voices and standpoints drawn from and often summarising communications previously uttered in textual form on the Internet. This kind of dialogue has the potential to fulfil one of the major functions of education which is to draw students into understanding and participating in the long-term dialogues of culture (Oakeshott, 1989; Wegerif and Major, 2023).

Over the course of 2023, DEFI (the Digital Education Futures Initiative at Cambridge University) conducted a series of ongoing experiments with large language models, primarily ChatGPT. The first set of experiments involved an attempt to use it to help social science students understand statistics. There are many tools on the market aimed at helping with statistics, but in addition to the calculations, students sometimes struggle to understand the process of choosing which tests to use and also how to write up their results. Part of the problem may be that they are only exposed to one or two relevant examples of good practice before being asked to repeat that practice (e.g. van Peppen et al., 2021). We attempted to get first ChatGPT 3.5, then GPT4 to generate a series of test items with a grading scale, which we thought might provide a cost effective means for providing formative assessment and worked examples. This involved three attempts with each tool using the same initial prompts, with slight variations in subsequent prompts.

While initially promising, it quickly became apparent that the LLMs by themselves were prone to a series of errors. These ranged from failing to calculate the mean correctly to underestimating the size of the dataset needed to carry out a T-test. It was noticeable that the later models did not initially perform more reliably than the original GPT3.5 publicly released in November 2022. One issue appeared to be that while the LLMs had been fed a lot of *text* discussing matters of statistics, they had not specifically been trained how to perform the underlying calculations. Another issue may be that the higher the *creativity* setting in GPT, the more likely it is to produce a less obvious answer—helpful for creative writing, but in maths and statistics the 'less predictable' answer will often be a wrong answer. We repeated the experiments in early 2024. Three PhD students tested ChatGPT 3.5 and 4. As before, GPT3.5 was error prone, incorrectly calculating the standard deviation, getting the T-test calculation wrong, providing a scoring rubric not based on worked examples. However, GPT4 now appears to draw on a python library which enables it to correctly perform T-tests. Similarly, GPT4 plus the Wolfram Alpha plug in was able to perform tests correctly. This suggests that a hybrid of LLMs plus a system with a representation of mathematical rules, and not just of language, can overcome the initial issues.

Another set of experiments involved asking ChatGPT to respond to a series of questions as if it were a well known philosopher or academic. These were more promising. It was possible to request a range of academic viewpoints, and to prompt the model to develop the contrasting arguments. Where the model could search the internet (Bing and GPT4) it was able to provide genuine links and references (although these were not of high quality). It provided a starting point for further research into the philosophers in question. This type of approach has recently been extended elsewhere by a team that set out to train a LLM on the works of a specific philosopher (Schwitzgebel et al., 2023). The promise here is that it provides an entry point to dialog with the philosopher in question which is focused around the student's current understanding and interests.

A third set of experiments involved using generative AI to summarise text and act as one reviewer among others in a literature review process. Previous studies found that large language models could effectively summarise text and extract text related to specific areas of interest, or generate themes from transcripts (Lennon et al., 2021) and this offers an addition to AI powered search and network analysis tools such as Research Rabbit[2] and Elicit.[3] The Rayyon literature review tool[4] already offers similar functionality as a beta, and this could have a significant impact on future literature reviews, enabling more papers to be included in reviews. DEFI's Camtree project has been trialling generative AI functionality to help teachers generate abstracts for their research projects, helping them to publish their work in a form that is findable thus making it easier for others to build on the findings. This experiment is in its early stages, but future work will seek to find the optimal balance between assisting teachers to present their research in a way that makes it findable in the Camtree library and retaining the authenticity of the teacher's voice. This type of functionality offers promise in enabling more voices to contribute to and enrich evolving long

[2] ResearchRabbit.

[3] Elicit.org.

[4] https://www.rayyan.ai/.

term cultural dialogues. Inducting students into participation in long-term cultural dialogues is, we suggested earlier, the essence of education. This is not only the case in the more academic subjects like philosophy but also very much the case where the dialogue revolves around human activities and practices such as in the case of teacher education.

Finally, we tried feeding GPT 4 with text that we had written ourselves and asking it to suggest counter views and possible weaknesses in our arguments. Here, we argue, it was useful—suggesting areas to consider and address. In this, the LLM acted as a critical friend, offering alternative viewpoints. Again, this is best regarded as a starting point for further investigation, research, and reflection, but often the LLM offers genuinely useful observations. In this use case, the advantage is that while human teachers and peers may not have time to read and comment on every piece of writing and thought at an early stage, a LLM never gets tired or bored and can always be on hand to help.

In thinking about AI as a dialogic learning partner, it might be useful to distinguish two different types of knowledge. The first has been characterised as 'well-structured domains' of knowledge found in some areas of mathematics science and engineering where there are clearly defined and explicit sets of principles and rules that need to be transmitted (Jonassen, 1997). For this kind of knowledge generalised LLMs fall short and more explicit GOFAI based systems have been successful. The second, larger case, is where knowledge is uncertain because it is complex and evolving. Here education takes the form not of transmitting settled knowledge but of inducting students into ongoing dialogues between multiple voices. Understanding how to work with both kinds of knowledge requires being inducted into what Wegerif refers to as ' the dialogue so far' to understand the kind of questions being asked to which this knowledge is understood as offering answers (Phillipson & Wegerif, 2016; Wegerif, 2019). However, while education into well-structured domains of knowledge seems effective with tutors using traditional AI the larger process of inducting students into the ongoing dialogues of knowledge might be better served using tutors based on generative AI able to access the multiple voices and perspectives of the internet.

7.6 Discussion

Our exploration of the issues suggests that generative AI based on LLMs such as Chat GPT might offer the possibility of developing a distinctively new type of education. This could be characterised as circulating collective codified knowledge through unique individual minds to in turn expand the collective knowledge. This could be visualised as circling an 'outside', objective external knowledge in the form of texts found already on the Internet, around an 'inside', internal subjective consciousnesses that make sense of this knowledge in an always new context. LLM-based AI has the potential to support human learning but only when working together with human intelligence. By focusing education on growing unique minds that can creatively participate in dialogue, supported by AI's external knowledge gathering abilities, education could potentially become a virtuous circle combining the collective information gathering of AIs with the individual questions and insights of students; students who are thereby engaged into a kind of game of leapfrog with this new technology, using it to gain more knowledge and stimulate further insights.

Of course, as discussed earlier, this raises questions as to what the *outside* consists of, that is, which dialogues and voices are available to the large language model and the extent to which all cultures and ways of thinking are represented and made available to the learner. Currently LLMs have been found to be Anglo-Saxon centric (Kasneci et al., 2023), and contain many other instances of bias, as discussed earlier. The reproduction of pre-existing cultural biases has always been a question of interest and debate in education and the design of curricula, but a danger with LLMs is that in the hands of a few organisations, they might act to centralise specific biases and reduce the richness of global dialogue, if the training data and algorithms are not themselves subject to scrutiny and debate. We therefore call for increased scrutiny and debate of training datasets and increased diversity of individuals involved in producing and authorising training data and algorithms, while at the same time we argue that LLMs nonetheless offer a powerful and novel resource for education.

7.7 Conclusion

AI agents based on LLMs have potential for teaching through dialogue and for teaching dialogue itself in the form of dialogic thinking. We have offered various experiments using AI language models, including using them to suggest counter views and possible weaknesses in arguments, to respond to questions as if they were a well-known philosopher or academic, and to summarise text and act as one reviewer among others in a literature review process. Our experiments suggest that AI-mediated dialogic education can be seen as a novel form of collective intelligence with the potential to significantly enhance educational experiences, in line with a vision of individual education as learning how to participate in AI-mediated collective intelligence.

While we acknowledge current issues and challenges related to the adoption of LLMs in and for education, we believe that the potential benefits are significant and warrant further exploration. We hope that this chapter provides a useful starting point for educators, researchers, and policymakers interested in the role of AI in education.

References

Ahn, M. J., & Chen, Y. C.. (2023). 'Building Guardrails for ChatGPT', February. https://policycommons.net/artifacts/4140623/building-guardrails-for-chatgpt/4949379/.

Ali, R., Tang, O. Y., Connolly, I. D., Zadnik Sullivan, P. L., Shin, J. H., Fridley, J. S.,& Asaad, W.F. et al. (2023). 'Performance of ChatGPT and GPT-4 on neurosurgery written board examinations'. https://doi.org/10.1101/2023.03.25.23287743.

Baker, R. S., & Hawn, A. (2022). Algorithmic bias in education. *International Journal of Artificial Intelligence in Education, 32*(4), 1052–1092. https://doi.org/10.1007/s40593-021-00285-9

Bakhtin, M. (1981). *Discourse in the Novel*. University of Texas Press.

Bakhtin, M. M. (1984). *Problems of Dostoevsky's Poetics*. University of Minnesota Press.

Bakhtin, M. M. (1986). *Speech Genres and Other Late Essays*. University of Texas.

Bender, E. M., Gebru, T., McMillan-Major, A., & Shmitchell, S. (2021). On the dangers of stochastic parrots: Can language models be too big?. In *Proceedings of the 2021 ACM Conference on Fairness, Accountability, and Transparency*, pp. 610–623. https://doi.org/10.1145/3442188.3445922

Berditchevskaia, A., Maliaraki, E., & Stathoulopoulos, K. (2022). A descriptive analysis of collective intelligence publications since 2000, and the emerging influence of artificial intelligence. *Collective Intelligence, 1*(1), 26339137221107924. https://doi.org/10.1177/26339137221107924

Billett, S. (2014). *Mimetic learning at work: Learning in the circumstances of practice*. SpringerLink. https://link.springer.com/book/https://doi.org/10.1007/978-3-319-09277-5.

Boden, M. A. (2016). *Artificial Intelligence: A Very Short Introduction* (Kindle). OUP Oxford.

Briggs and Kodnani. 2023. 'The Potentially Large Effects of Artificial Intelligence on Economic Growth'. Goldman Sachs. 2023. https://www.gspublishing.com/content/research/en/reports/2023/03/27/d64e052b-0f6e-45d7-967b-d7be35fabd16.html.

Brown, T. B., Mann, B., Ryder, N., Subbiah, M., Kaplan, J., Dhariwal, P., Neelakantan, A., et al. (2020). 'Language models are few-shot learners'. arXiv. http://arxiv.org/abs/2005.14165.

Buber, M. (1958). *I and Thou* (2nd ed.). T. and T. Clark.

Bull, S., & Kay, J. (2010). 'Open learner models'. In *Advances in Intelligent Tutoring Systems*, In Nkambou, R., Mizoguchi, R., & Bourdeau, J. (Eds.) Studies in Computational Intelligence. Berlin, Heidelberg: Springer. pp. 301–22 https://doi.org/10.1007/978-3-642-14363-2_15.

Casebourne, I. (2023). 'Should we trust ChatGPT?' *DEFI* (blog). 12 January 2023. https://www.deficambridge.org/should-we-trust-chatgpt/.

Chalmers, D. J. (2023). 'Could a large language model be conscious?' *Boston Review*, 9 August 2023. https://www.bostonreview.net/articles/could-a-large-language-model-be-conscious/.

Chopade, P., Khan, S. M., Edwards, D., & von Davier, A. (2018). 'Machine learning for efficient assessment and prediction of human performance in collaborative learning environments'. In *2018 IEEE International Symposium on Technologies for Homeland Security (HST)*, pp. 1–6. https://doi.org/10.1109/THS.2018.8574203.

Cotton, D. R. E., Cotton, P. A., & Shipway, J. R. (2023). Chatting and cheating: Ensuring academic integrity in the era of ChatGPT. *Innovations in Education and Teaching International, 0*(0), 1–12. https://doi.org/10.1080/14703297.2023.2190148

De Souza, C. S. (2005). *The semiotic engineering of human-computer interaction*. The MIT Press. https://doi.org/10.7551/mitpress/6175.001.0001

Dennett, D. C. (1991). *Consciousness Explained*. Penguin.

Doroudi, S. (2022). The intertwined histories of artificial intelligence and education. *International Journal of Artificial Intelligence in Education*. https://doi.org/10.1007/s40593-022-00313-2

Dreyfus, H. (1972). *What Computers Can't Do: The Limits of Artificial Intelligence*. Harper & Row.

Edwards, B. (2023, July 14). Why AI detectors think the US Constitution was written by AI. https://arstechnica.com/information-technology/2023/07/why-ai-detectors-think-the-us-constitution-was-written-by-ai/

Eloundou, T., Manning, S., Mishkin, P., & Rock, D. (2023). 'GPTs are GPTs: An early look at the labor market impact potential of large language models'. https://doi.org/10.48550/arXiv.2303.10130.

Epstein, R., Roberts, G., & Beber, G. (Eds.). (2009). *Parsing the Turing Test: Philosophical and Methodological Issues in the Quest for the Thinking Computer*. Springer, Netherlands. https://doi.org/10.1007/978-1-4020-6710-5

Farazouli, A., Cerratto-Pargman, T., Bolander-Laksov, K., & McGrath, C. (2023). Hello GPT! Goodbye home examination? An exploratory study of AI chatbots impact on university teachers' assessment practices. *Assessment & Evaluation in Higher Education, 0*(0), 1–13. https://doi.org/10.1080/02602938.2023.2241676

Garber, M. (2010). 'The Gutenberg parenthesis: Thomas Pettitt on parallels between the pre-print era and our own internet age'. *Nieman Lab* (blog). https://www.niemanlab.org/2010/04/the-gutenberg-parenthesis-thomas-pettitt-on-parallels-between-the-pre-print-era-and-our-own-internet-age/.

Gray, M. L., & Suri, S. (2019). *Ghost Work: How to Stop Silicon Valley from Building a New Global Underclass* (Illustrated). Harper Business.

Hermans, H. J. (2019). Dialogical self theory in a boundary-crossing society. In *Moral and Spiritual Leadership in an Age of Plural Moralities* (pp. 27–47). Routledge.

Hernández-Orallo, J. (2017). *The measure of all minds: Evaluating natural and artificial intelligence* (1st ed.). Cambridge University Press. https://doi.org/10.1017/9781316594179

Hoffmann, J., Borgeaud, S., Mensch, A., Buchatskaya, E., Cai, T., Rutherford, E., & Casas, D.D.L., et al. (2022). 'Training compute-optimal large language models'. arXiv. http://arxiv.org/abs/2203.15556.

Holmes, W., & Tuomi, I. (2022). State of the art and practice in AI in education. *European Journal of Education, 57*(4), 542–570. https://doi.org/10.1111/ejed.12533

Jonassen, D. H. (1997). Instructional design models for well-structured and ill-structured problem-solving learning outcomes. *Educational Technology Research and Development, 45*(1), 65–94.

Kasneci, E., Sessler, K., Küchemann, S., Bannert, M., Dementieva, D., Fischer, F., Gasser, U., et al. (2023). ChatGPT for Good? On opportunities and challenges of large language models for education. *Learning and Individual Differences, 103*, 102274. https://doi.org/10.1016/j.lindif.2023.102274

Kauffman, S. A., & Roli, A. (2023). What Is consciousness? Artificial intelligence, real intelligence, quantum mind and qualia. *Biological Journal of the Linnean Society, 139*(4), 530–538. https://doi.org/10.1093/biolinnean/blac092

Knoop-van Campen, C. A., Wise, A., & Molenaar, I. (2023). 'The equalizing effect of teacher dashboards on feedback in K-12 classrooms'. Interactive Learning Environments 31 (6): 3447–63. https://www.tandfonline.com/doi/epdf/https://doi.org/10.1080/10494820.2021.1931346?needAccess=true&role=button.

Kurni, M., Mohammed, M. S., & Srinivasa, K. G. (2023). 'Intelligent tutoring systems'. In *A Beginner's guide to introduce artifical intelligence in teaching and learning*. https://link.springer.com/chapter/https://doi.org/10.1007/978-3-031-32653-0_2.

Lennon, R. P., Fraleigh, R., Van Scoy, L. J., Keshaviah, A., Hu, X. C., Snyder, B. L., Miller, E. L., Calo, W. A., Zgierska, A. E., & Griffin, C. (2021). Developing and testing an automated qualitative assistant (AQUA) to support qualitative analysis. *Family Medicine and Community Health, 9*(Suppl 1), e001287. https://doi.org/10.1136/fmch-2021-001287

Luscombe, R. (2022). 'Google engineer put on leave after saying AI Chatbot has become sentient'. *The Guardian*, 12 June 2022, sec. Technology. https://www.theguardian.com/technology/2022/jun/12/google-engineer-ai-bot-sentient-blake-lemoine.

McClure, T. (2023). 'Supermarket AI meal planner app suggests recipe that would create chlorine gas'. *The Guardian*, 10 August 2023, sec. World news. https://www.theguardian.com/world/2023/aug/10/pak-n-save-savey-meal-bot-ai-app-malfunction-recipes.

Mercer, N., & Littleton, K. (2007). *Dialogue and the Development of Children's Thinking: A Sociocultural Approach* (1st ed.). Routledge. https://doi.org/10.4324/9780203946657

Mulgan, G. (2018). *Big mind: How collective intelligence can change our world* (Illustrated edition). Princeton University Press.

Noakes, J. (2022). ‘Is artificial intelligence the future of essay marking?’ School Management Plus: School & Education News Worldwide. 2022. https://www.schoolmanagementplus.com/assessment/is-artificial-intelligence-the-future-of-essay-marking/.

Noble, S. U. (2018). *Algorithms of Oppression: How Search Engines Reinforce Racism*. NYU Press. https://doi.org/10.2307/j.ctt1pwt9w5

Nonaka, I., & von Krogh, G. (2009). Tacit knowledge and knowledge conversion: Controversy and advancement in organizational knowledge creation theory. *Organization Science, 20*, 635–652. https://doi.org/10.1287/orsc.1080.0412

Nye, B. D., Graesser, A. C., & Hu, X. (2014). AutoTutor and family: A review of 17 Years of natural language tutoring. *International Journal of Artificial Intelligence in Education, 24*(4), 427–469. https://doi.org/10.1007/s40593-014-0029-5

Oakeshott, M. (1989). *The voice of liberal learning: Michael Oakeshott on education*. Liberty Fund.

Phillipson, N., & Wegerif, R. (2016). *Dialogic education: Mastering core concepts through thinking together*. Taylor & Francis.

Polanyi, M., & Sen, A. (2009). *The Tacit Dimension*. University of Chicago Press.

Reisner, A. (2023). ‘Revealed: The authors whose pirated books are powering generative AI’. The Atlantic. 19 August 2023. https://www.theatlantic.com/technology/archive/2023/08/books3-ai-meta-llama-pirated-books/675063/.

Rojas, M., Sáez, C., Baier, J., Nussbaum, M., Guerrero, O., & Rodríguez, M. F. (2022). Using automated planning to provide feedback during collaborative problem-solving. *International Journal of Artificial Intelligence in Education*. https://doi.org/10.1007/s40593-022-00321-2

Rosen, Y. (2015). Computer-based assessment of collaborative problem solving: Exploring the feasibility of human-to-agent approach. *International Journal of Artificial Intelligence in Education, 25*(3), 380–406. https://doi.org/10.1007/s40593-015-0042-3

Rudin, C., Chen, C., Chen, Z., Huang, H., Semenova, L., & Zhong, C. (2021). ‘Interpretable machine learning: Fundamental principles and 10 grand challenges’. arXiv. http://arxiv.org/abs/2103.11251.

Russell, S. J., & Norvig, P. (2021). *Artificial Intelligence: A Modern Approach* (4th ed.). Pearson.

Schwitzgebel, E., Schwitzgebel, D., & Strasser, A. (2023). Creating a large language model of a philosopher. *Mind & Language, 39*, 237–259. https://doi.org/10.1111/mila.12466

Searle, J. R. (1980). ‘Minds, brains and programmes’.

Shanahan, M. (2023). ‘Talking about large language models’. arXiv. http://arxiv.org/abs/2212.03551.

Sharples, M. (2022, May 17). New AI tools that can write student essays require educators to rethink teaching and assessment. *Impact of Social Sciences*. https://blogs.lse.ac.uk/impactofsocialsciences/2022/05/17/new-ai-tools-thatcan-write-student-essays-require-educators-to-rethink-teaching-and-assessment/

Srinivasan, R., & Uchino, K. (2021). ‘Biases in generative art -- A causal look from the lens of art history’. arXiv. http://arxiv.org/abs/2010.13266.

Srivastava, A., Rastogi, A., Rao, A., Shoeb, A. A. M., Abid, A., Fisch, A., Brown, AR., et al. (2022). ‘Beyond the imitation game: Quantifying and extrapolating the capabilities of language models’. arXiv. http://arxiv.org/abs/2206.04615.

Van Peppen, L. M., Verkoeijen, P. P., Kolenbrander, S. V., Heijltjes, A. E., Janssen, E. M., & van Gog, T. (2021). Learning to avoid biased reasoning: Effects of interleaved practice and worked examples. *Journal of Cognitive Psychology, 33*(3), 304–326. https://doi.org/10.1080/20445911.2021.1890092

Varanasi, L. n.d. ‘OpenAI just announced GPT-4, an Updated Chatbot that can pass everything from a bar exam to AP biology. Here’s a List of Difficult Exams Both AI Versions Have Passed.’ Business Insider. Accessed 17 April 2023. https://www.businessinsider.com/list-here-are-the-exams-chatgpt-has-passed-so-far-2023-1.

Waterfield, R. & Plato. (2002). *Phaedrus*. Kindle. Oxford: OUP Oxford. https://www.amazon.co.uk/Phaedrus-Oxford-Worlds-Classics-Plato/dp/0199554021/ref=sr_1_1?adgrpid=1176478340032996&hvadid=73530099343471&hvbmt=bp&hvdev=c&hvlocphy=69090&hvnetw=o&hvqmt=p&hvtargid=kwd-73530030524219%3Aloc-188&hydadcr=10777_2102412&keywords=plato%27s+phaedrus&qid=1693414264&sr=8-1.

Wegerif, R. (2007). *Dialogic, Education and Technology: Expanding the Space of Learning*. Springer.

Wegerif, R. (2019). *Towards a dialogic theory of education for the internet age* (pp. 14–26). Routledge.

Wegerif, R., & Major, L. (2023). *The Theory of Educational Technology: A dialogic foundation for design*. Routledge.

Wegerif, R., & Mercer, N. (1997). A Dialogical Framework for Investigating Talk. In R. Wegerif & P. Scrimshaw (Eds.), *Computers and Talk in the Primary Classroom* (pp. 49–65). Multilingual Matters.

West, D. M. (2023). 'Comparing google bard with OpenAI's ChatGPT on political bias, facts, and morality'. 23 March 2023. https://www.brookings.edu/blog/techtank/2023/03/23/comparing-google-bard-with-openais-chatgpt-on-political-bias-facts-and-morality/.

Wodzak, S. (2022). 'Can a standardized test actually write itself?' *Duolingo Blog* (blog). 6 April 2022. https://blog.duolingo.com/test-creation-machine-learning/.

Wu, J., Ouyang, L., Ziegler, D. M., Stiennon, N., Lowe, R., Leike, J., & Christiano, P. (2021). 'Recursively summarizing books with human feedback'. arXiv. http://arxiv.org/abs/2109.10862.

Xiao, Y., Chatterjee, S., & Gehringer, E. (2022). A new era of plagiarism the danger of cheating using AI'. In *2022 20th International Conference on Information Technology Based Higher Education and Training (ITHET)*, pp. 1–6. https://doi.org/10.1109/ITHET56107.2022.10031827.

Chapter 8
AI Powered Adaptive Formative Assessment: Validity and Reliability Evaluation

Yaw Bimpeh

Abstract This work presents an AI-powered formative assessment system for secondary school mathematics that uses learning objectives, cognitive mapping, and many factors—including detailed measures of competency, metacognition, time—to accurately measure each learner's weak/strong areas to determine where to place more attention. It provides timely feedback to students, enabling them to monitor their progress and identify areas of improvement. Feedback includes explanations, hints, and targeted remediation resources based on individual learning needs. Using empirical studies, the validity, reliability, and effectiveness of the AI-powered formative assessment system are discussed and evaluated. The results showed that the adaptive system engages students and give consistent information about students' knowledge and ability.

Keywords Adaptive formative learning · AI · Diagnostic · Assessment · Effectiveness · Math learning

8.1 Introduction

For more than three decades, computer scientists and cognitive scientists have been developing adaptive learning systems to mimic the interactions of human tutoring (Merrill et al., 1992). Adaptive learning is the technology that is most strategically influencing higher education, according to Calhoun Williams (2019). For example, these systems can be used to deliver content, ask questions, assign tasks, provide hints, or encourage learners to modify their attitudes (Ma et al., 2014). The fundamental design of all adaptive learning systems is a "closed loop". It collects data from the learner and uses it to assess progress, recommend learning activities, and provide personalised feedback.

Y. Bimpeh (✉)
AQA Education, Devas Street, Manchester M15 6EX, UK
e-mail: ybimpeh@aqa.org.uk

P. Ilic et al. (eds.), *Artificial Intelligence in Education: The Intersection of Technology and Pedagogy*, Intelligent Systems Reference Library 261,
https://doi.org/10.1007/978-3-031-71232-6_8

The fundamental ideas behind AI-powered formative assessment system are as follows: (1) A learner's ability, metacognition, time, and self-awareness; and (2) AI algorithms chooses the item from the concept map that maximises the informative values for the given examinee's ability and metacognition from the item bank.

Adaptive learning systems use various learning algorithms, such as item response theories, artificial intelligence, and machine learning to personalize the learning experience (e.g., Mavroudi et al., 2017; van der Linden, 2016). The systems can enhance student learning, according to research (e.g., Van Lehn, 2011, Jones, 2018). Most studies in the systematic review (86% i.e., 32 studies) conducted by Xie et al., (2019) on the impacts of adaptive learning on learning outcomes showed positive results. According to Bomash and Kish (2015) adaptive learning significantly increased student performance compared to non-adaptive learning.

Developing adaptive content that aligns with learning objectives and is engaging for students is a challenge. Previous research demonstrates that conceptual, metacognitive, and strategic hard scaffolds can be incorporated into both conventional and technologically aided learning contexts (e.g., Bulu & Pedersen, 2010; Chen, 2010).

This research explores designing formative assessment that aligns with the desired learning outcomes using learning objectives, cognitive mapping, The aim of the adaptive test is to assess how well students grasp and apply fundamental arithmetic knowledge and abilities. The test items are grouped into 3 sub-concepts that represent three general arithmetic content areas:

pre-requisite for understanding, arithmetic conceptual knowledge, and applications of basic arithmetic concept.

The purpose of this study is to present a framework for an adaptive formative assessment system that incorporates a relationship between metacognition and student performance together with numerous other data points to help adapt the content that the learner receives on the learning journey to mastery.

The system that is the subject of this study uses learning objectives, cognitive mapping, and many factors—including detailed measures of competency, metacognition, time, and self-awareness—to accurately measure each learner's weak/strong areas to determine where to place more attention. It probes each learner's current knowledge to uncover gaps and offer adaptive learning solutions focusing only on what each person does not know.

The platform gives personalized and adaptive feedback to students. Feedback includes explanations, hints, and targeted remediation resources based on individual learning needs. It also provides teachers with real-time analytics and reports on student performance. This information can help teachers tailor their instruction and interventions to meet individual student needs.

The research evaluates the validity and effectiveness of the AI-powered formative assessment system in improving learning outcomes using empirical evidence. We address the following research questions:

- Does the adaptive /formative test differentiate students of differing ability levels?
- Does the adaptive /formative test provide consistent and dependable information about students in terms of each student's knowledge and skills?

- Does the adaptive /formative test demonstrate strong concurrent validity with the Key stage 3 and 4 mathematics competency tests?
- Does the empirical evidence and theoretical rationale support the accuracy of inferences based on the diagnostic /formative test?

8.2 Platform

This is an extensive question bank which uses adaptive-learning technology. The algorithm tracks how a person perform on the concept map fundamental ideas before delivering questions that are specifically designed to fill in any knowledge gaps. The algorithm determines the optimal sequencing of additional questions by taking into account the learner's answer to a question, self-reported confidence on a four-point scale (1- "I know it", 2 - "I think so", 3 - "not sure", and 4 - "no idea"), the learner's overall confidence in the topic, the time the learner takes to indicate an answer, and how others have answered.

Scores, timings, responses, metacognition, and the structure and priorities of the learning objectives are all examples of measures used. This measurement is taken for each individual learner, and both the learning path and the learning analysis are tailored to each learner's needs.

The platform also utilises math probe, which can be used when solving math problems. It allows the learners to enter their work towards the solution (see Fig. 1 for the general user interface). The underlying math engine uses AI to evaluate the response according to mathematical truths.

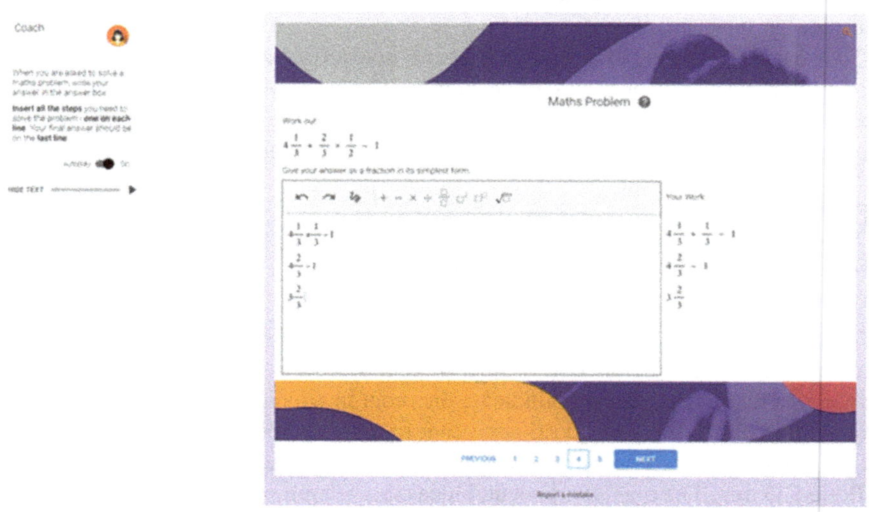

Fig. 1 AI-powered adaptive testing platform

The platform attempts to establish the milestones that individuals have attained in their learning. This comprises determining what they know, understand, and can do at the time of test. This data can then be used to decide the next steps in teaching and learning, as well as to track individual growth over time.

8.2.1 AI Algorithms Used

The adaptive learning platform employs a range of AI algorithms to customize learning experiences and improve educational results. Through machine learning methods, the platform examines learner data and dynamically adapts the complexity and speed of content delivery. These algorithms constantly evaluate learner performance, learning objectives, and skill levels to create personalized learning paths and recommendations.

Additionally, the platform integrates natural language processing (NLP) technologies to process and understand text-based data, including learner responses and feedback. NLP algorithms enable the platform to interpret learner inputs, provide relevant feedback, and create interactive learning experiences. NLP enhances the platform's ability to interact with learners in a meaningful and personalized way.

Furthermore, the platform utilizes content-based filtering to suggest items based on both their characteristics and the learner's previous interactions with related items. By analysing the learner's past performance on assessments and other learning activities, the algorithm identifies strengths, weaknesses, and areas for improvement. This data is then used to customize recommendations to meet the learner's specific needs. In addition, the platform evaluates the relevance of learning materials to the learner's current learning objectives.

AI-driven data analytics tools are also employed to extract actionable insights from learner data and platform usage metrics. These analytical capabilities empower educators and administrators to oversee learner progression, gauge engagement levels, and use data-driven insights to enhance teaching and learning experiences.

8.3 Test Materials and Concept Map

The concept map in Fig. 2 depicts all the concepts that a student must understand to grasp the arithmetic by the end of Key Stage 3. Each key concept is represented by a node on the map. The arrows indicate the order in which they must be understood. Each node has a series of questions meant to assess a specific aspect of that node. The questions are extremely detailed, and a student may answer as many as 60 in the allotted 25 min. The most fundamental ideas that are pre-requisite for understanding are shown in green on the map. The green nodes reflect concepts that all students should understand by the time they begin General Certificate of Secondary Education

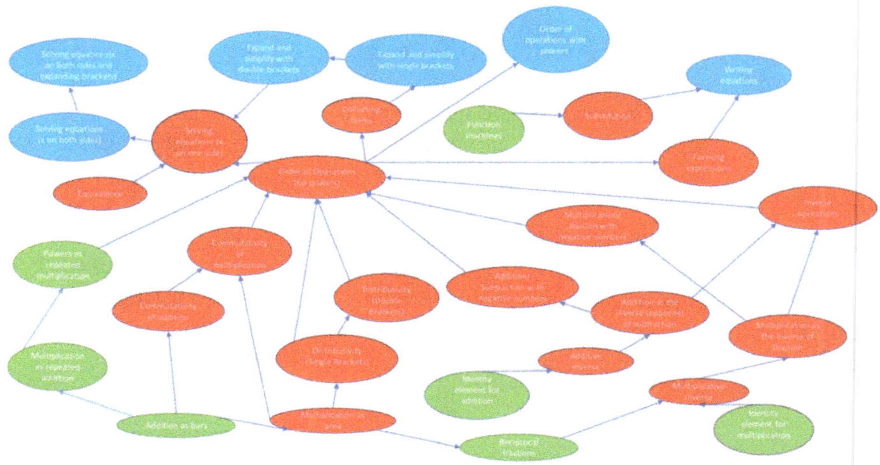

Fig. 2 Arithmetic concept map

(GCSE). The red nodes represent the arithmetic conceptual knowledge at Key Stage 3.

The red nodes are the key ideas that students really need to master to make satisfactory progress at General Certificate of Secondary Education (GCSE). The blue nodes represent more advanced ideas that will be developed at GCSE.

The map's linkages are used to programme the adaptive algorithm. This, combined with student responses, is used to determine which questions from the question bank to ask. This helps to develop a picture of what a student knows and does not know.

8.4 Methods

8.4.1 Participants

A sample of the secondary schools in England, United Kingdom, were used for this study. The trial began with the recruitment of 1504 students, ages 14 to 15, from various schools. Two cohorts of students were created. Cohort 1 has the remediation features of platform enabled and for cohort 2 the remediation function was disabled. Following the test, survey questions were given to the study participants to gather their feedback on their testing experience. A mixed method was used, consisting of (1) the quantitative and qualitative analysis of test performance, as well as user experience, engagement, and satisfaction with the adaptive learning platform and (2) thematic analysis of semi-structured interviews.

An empirical study was conducted to investigate the construct validity of a set of assessment of objectives developed by the subject matter experts, embodied in items in the adaptive test. Using structural equation modelling in the framework put forward by Fornell and Larcker (1981), an evaluation of the construct validity of AI-powered formative assessment systems was conducted. The confirmatory factor analysis, which is the first component, relates the assessment objective to test items that are assumed to measure that assessment objective; considering the measurement error. The covariance of the assessment objectives is examined in the second component. We can determine the validity of assessment objectives, the degree of variance explained by a construct versus the amount due to measurement error, construct reliability, and the relationships between various constructs by using this method.

We also used Rasch-based analysis for providing evidence for the six aspects of Messick's validity for the adaptive test in the framework of evaluation used by Wolfe and Smith (2007) and Beglar (2010).

8.4.2 Data Sources and Evidence

Students' item response data from the diagnostic tests and data from their interactions with the platform were collected and analysed. Feedback from teachers and students on each of the key maths concepts was also obtained to help determine whether the platform is appropriate for its intended application and will be helpful to teachers. The effectiveness and usability of the platform were improved using teacher and student survey data as well as qualitative comments.

8.4.3 Analysis and Results

In total, 1504 students participated in this study. Their mean time spent per question, average number of questions attempted scores on the adaptive test are displayed in Table 1, and their mean scores on the sub concepts are in Table 4 (see the Appendix).

Table 1 Summary of the diagnostic test

	Cohort 1 (size = 621)	Cohort 2 (size = 883)
Number of Items used in the diagnostic test trial	359	371
Average time spent per question	34 s	34 s
Average number of questions attempted per person	51	49.76
Average time spent per test	28.63 min	28 min
Internal Consistency of Learning objectives	0.75	0.82

Table 2 Results of convergent* validity of constructs

Results of convergent validity of constructs			
	Constructs	Construct reliability (> 0.7)	AVE (>0.5)
Arithmetic	Green	0.72	0.28
	Red	0.78	0.27
	Blue	0.75	0.37

* Convergent validity: CR > = 0.7 and AVE > 0.5

8.4.4 Assessing Internal Consistency Reliability

The internal consistency of the measurement constructs was examined through calculating Cronbach's alpha (>0.70; Cronbach, 1951), the test–retest reliability, and the construct reliability (>0.70; Hair et al., 2017) analysis (see Tables 1 and 2) to assess the correlations of individual item scores. The rationale behind using both metric is as follows: the Cronbach alpha of items is underweighted, which can make it a less precise measure of reliability though, contrarily, construct reliability (CR) of items is weighted based on each item's loading, which makes it a more precise reliability measure (Hair et al., 2017).

8.4.5 Assessing Convergent Validity of Constructs

Convergent validity was assessed to confirm that the items together measure the same latent constructs (Henseler et al., 2015) through loadings, CR, and average variance extracted (AVE) (e.g., Ali et al., 2016).

Table 2 shows the values of construct reliability (CR) and average variance extracted (AVE). The AVE represents the average proportion of variance in the items that is explained by the assessment objective (i.e., grand mean value of the squared loadings of the items associated with the assessment objective). According to Table 2 the CR values are above the satisfactory threshold of 0.7 for all assessment objectives. The average variances extracted are all below the acceptable level 50%. To test for convergent validity, the items factor loadings, AVE, and CR should be checked using Fornell and Larcker (1981) criterion. All assessment objectives have a satisfactory level of construct reliability (more than 0.7); (Green = 0.72, Red = 0.78 and Blue = 0.75) for arithmetic assessment. However, all the assessment objectives have AVE below the acceptable level (0.5), Green = 0.28, Red = 0.27 and Blue = 0.37.

The recommended value of factor loadings (> 0.50, Hair et al., 2017) and it was also confirmed that the constructs explain at least 30% of the variance of their corresponding items (AVE; Hair et al., 2017). The values of the AVE for green, red, and blue constructs were less than 0.50 (see Table 2). However, we decided to accept that construct as valid, as a construct with an AVE less than 0.50 but the CR more than 0.70 can still be considered to have adequate convergent validity; (Fornell & Larcker, 1981).

The path diagram in Figs. 3 and 4 illustrate visually the relationships among the Arithmetic learning objectives and the items. There are created using lavaan R package for structural equation modelling. The arrows leading from the assessment objectives (Red, Blu and Grn) to the items illustrate the casual effect of the learning objective on the items. The values on the arrows show the factor loadings. These values convey the extent to which the response to the item is attributable to the common cause shared by all the items reflecting the assessment objective. High factor loadings on an assessment objective indicate the associated items have much in common, which is captured by the assessment objective. The factor loading is the same as the item discrimination index when there are no cross-loading items. The arrows between Red, Blu and Grn represent the inter-construct correlations.

The values on the left side of the items in the path diagram represent the proportion of variance in the items that is not explained by the learning objective. These values are referred to as the variance of the measurement errors (i.e., residual variance). For example, the proportion of variance in the A22 (see Fig. 1) variance, that is not explained by the blue bubble is 50%. This is the variance unique to the item itself (i.e., measurement error).

TRUE

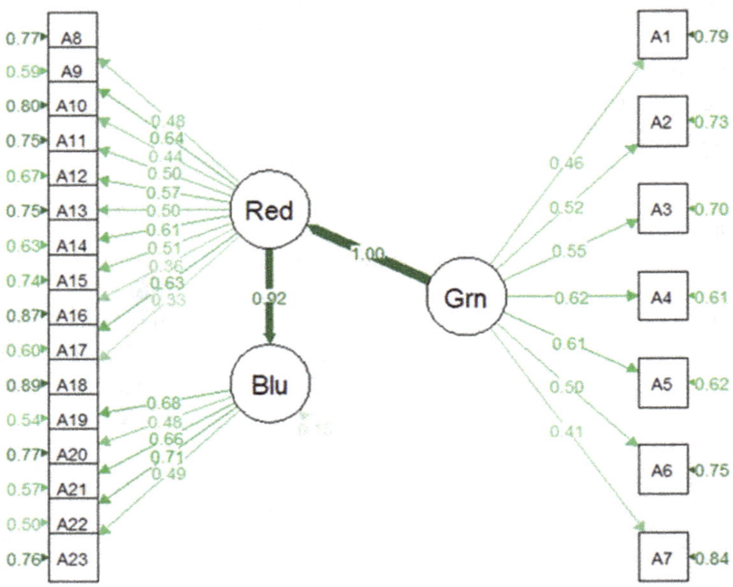

Fig. 3 Path diagram illustrating the structural equation model of three assessment objectives (Grn, Red and Blu) and their items-Cohort 1

TRUE

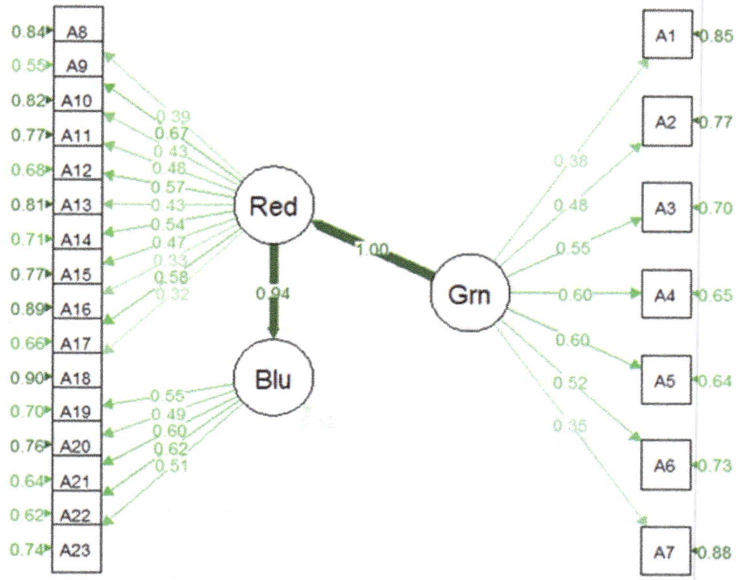

Fig. 4 Path diagram illustrating the structural equation model of three assessment objectives (Grn, Red and Blu) and their items-Cohort 2

Table 3 in the appendix shows details of the goodness of fit for the notes and concept map model. Hu and Bentler (1999) recommended that the assessments of model fit should be based on a joint evaluation of several fit indices. Our study shows that the goodness of fit indicators (GFI) = 0.985, CFI (Comparative Fit Index) = 0.962 and TLI = 0.958 satisfy the thresholds of 0.90, the standard deemed vital for model fit. Furthermore, the root mean square error of approximation (RMSEA = 0.030) indicated a good fit of the hypothesized model.

8.4.6 Content Aspect of Construct Validity

The results of the assessments are displayed on the item-person maps (Figs. 5 and 6). These show the linear relationship between the 1,504 test-takers' Rasch calibrations and the arithmetic tests. Using the evaluation framework by Wolfe and Smith (2007) for the content aspect of construct validity we examine the person-item map gaps and redundancy along the vertical line, mismatch between item and person means, infit and outfit statistics.

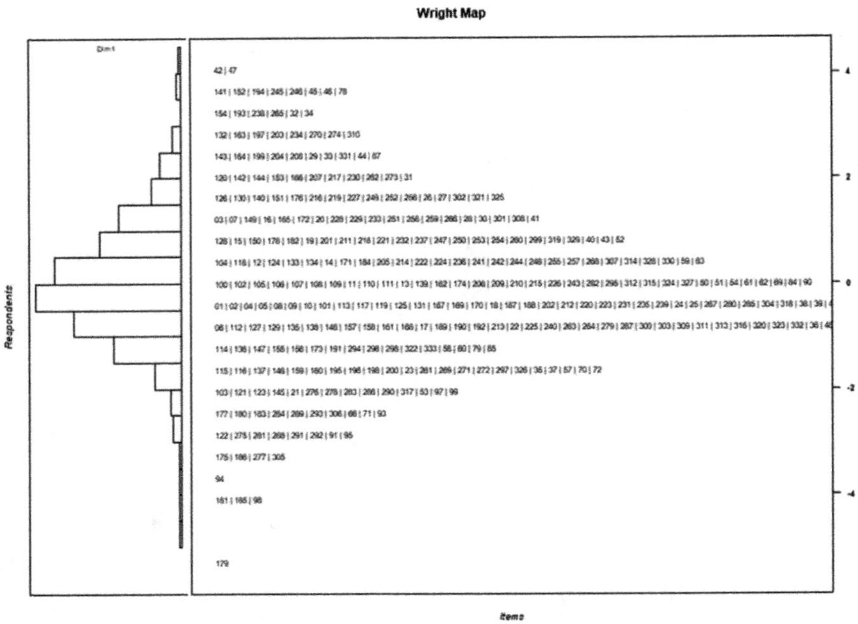

Fig. 5 Construct Map: Learning objective- person response (cohort 1)

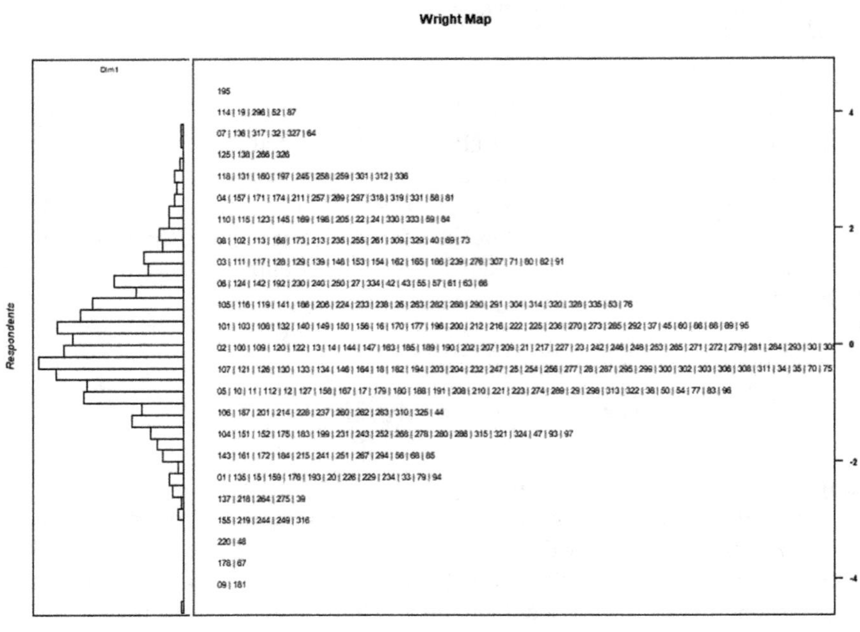

Fig. 6 Construct Map: Learning objective- person response (cohort 2)

The Item-person maps provide a convenient summary of item statistics for the diagnostic test. On the vertical axis, the figure locates items according to item difficulty. Easier items are located at the bottom of the Wright map, with more difficult items at the top. Items which are closer to the deterministic Guttman model are located to the left; items which misfit the model are located to the right.

By looking at "fit" data (such as MNSQ Item Outfit, MNSQ Item Infit) for each test item, it was possible to identify items that do not contribute to useful measurement. Items that did not fit the test clearly were either reviewed or taken out and replaced with new ones. The mean-square values for both infit and outfit in the Rash model fit statistics fall within the acceptable range (i.e., ranges from 0.7 to 1.3; Linacre, 2007). The adaptive test has acceptable mean values for the infit and outfit mean-square of 0.87 and 0.91, respectively. The map suggests good representativeness. This has beneficial implications for the consequential aspect of validity. The person-item map offers numerous justifications for each of Messick's (1989) six facets of construct validity (Beglar, 2010). The map shows gaps or overlaps in item difficulty or ability, as well as the distribution of items and abilities. It enables one to evaluate the degree of alignment between a student's proficiency and the difficulty of the learning objectives.

For example: For cohort 1, the most challenging learning objectives are "Solve equations of the form as fractions ($x + b = c$ with x is negative)" and "Solve equations of the form $a(bx + c) = d\ (ex + f)$ (x is negative)"; these are labelled 42 and 47 in the construct map (see Fig. 4). Overall, the test-takers response patterns somehow reflect, the degree to which test-takers are engaging with the test.

8.4.7 Differential Functioning of a Diagnostic Test

We further investigated the diagnostic test's effectiveness using a sample of 183 students from the cohort who had information about their maths abilities, higher (79 students) and lower abilities in maths (104 students). The diagnostic test's capacity to distinguish between students of varying abilities is the question at hand.

Draba's (1977) criteria was used in the current study to assess differential item functioning (DIF). That is an item is flagged for DIF if the difference in item difficulty estimates for the two groups (I.e. higher and lower achieving students) is greater than 0.5 logits. Some of the items have a DIF contrast value greater than 0.5 logits as shown in Fig. 7. Although a considerable number of items' difficulty measures varied, the DIF size for 10 nodes out of 23 the nodes evaluated by Draba's (1977) criteria, was significant.

The direction and size of the changes in item difficulty between the higher and lower subgroups are shown in Fig. 7. Each bar in the plot shows the variation in difficulty between subgroups for an individual arithmetic sub-concept. Bars that point to the left of the plot indicate that the item was easier to answer correctly by higher ability group, and bars that point to the right of the plot indicate that the item was easier for lower ability group to give correct response to. The values of + 0.5

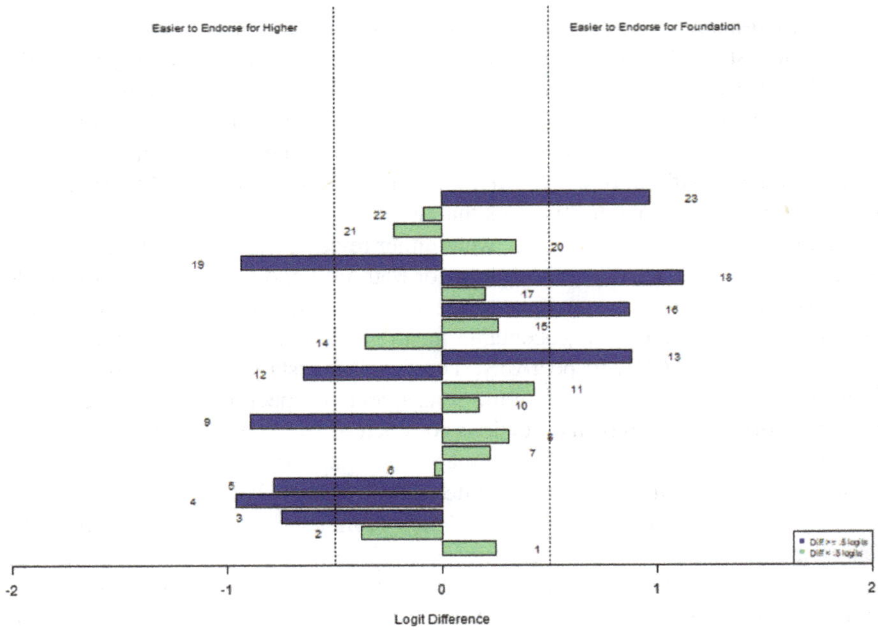

Fig. 7 Differential nodes functioning by lower and higher ability students

and − 0.5 logits are presented as dashed vertical lines as an indication of significant differences in item difficulty between subgroups. The empirical evidence suggests that diagnostic test can distinguish between test-takers' abilities.

8.5 User Experience Feedback

After the test, the study participants were given survey questions to answer in order together their thoughts on the testing experience. A mixed method was used, consisting of two main parts: (1) a quantitative and qualitative study of test performance coupled with user experience, engagement, and happiness with the adaptive learning platform; and (2) a thematic analysis of semi-structured interviews. Key findings from the student survey are outlined in Figs. 8, 9, 10, 11 and 12.

Regarding their overall test-taking experience, most respondents (78%) tended to rate it favourably. About 40% of participants said the test was difficult in one way or another.

Approximately 73% of participants expressed agreement with the platform's assessment of their strengths or competencies. A test like this, according to 75% of respondents, would be beneficial for their learning.

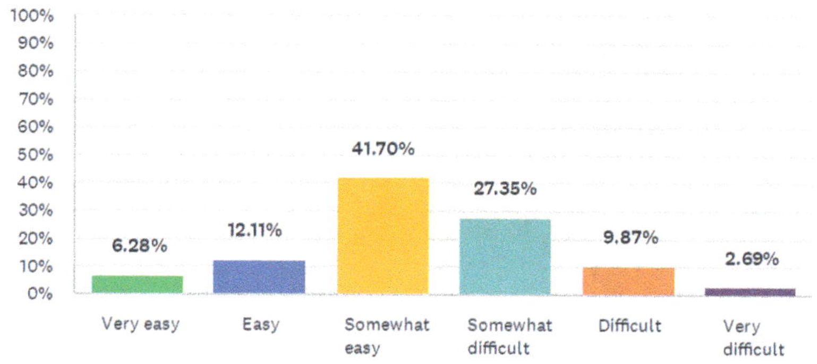

Fig. 8 How easy or difficult was the test for you?

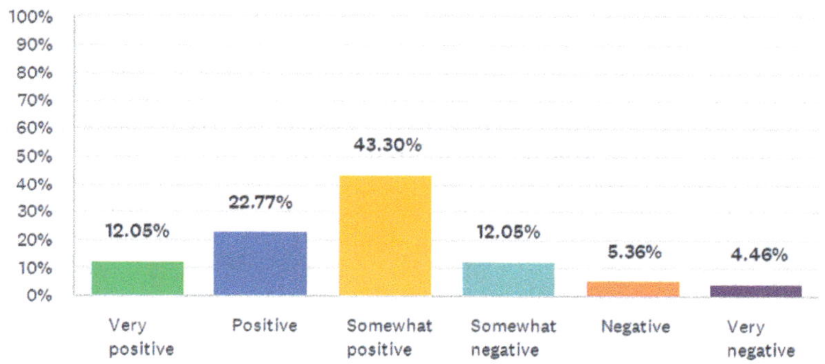

Fig. 9 Overall, how would you rate your experience of taking this test onscreen?

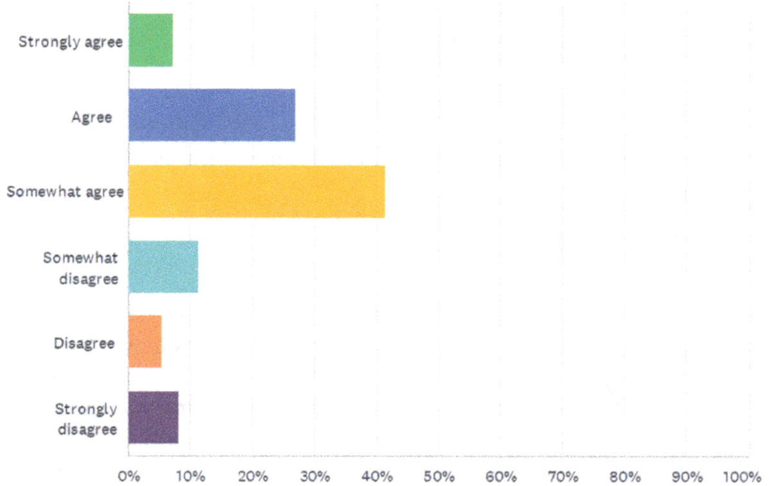

Fig. 10 When you saw the results about how you performed on this test, how much did you agree with the areas of strength or strong skills that the platform has identified?

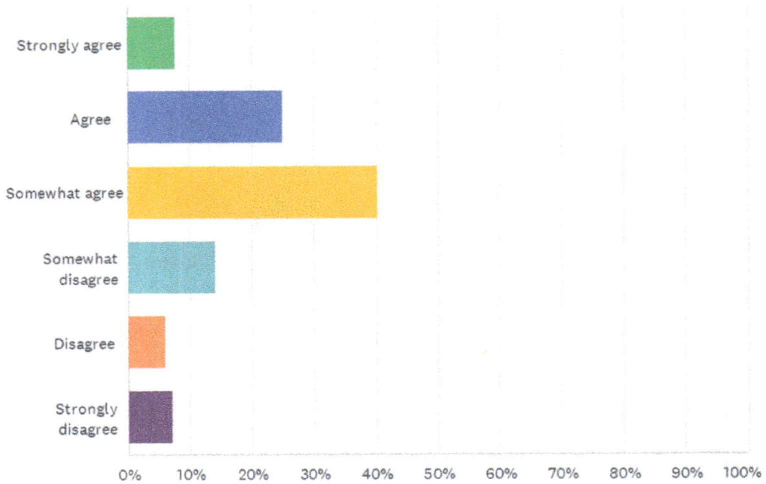

Fig. 11 When you saw the results about how you performed on this test, how much did you agree with the areas of weakness or weak skills that the platform has identified?

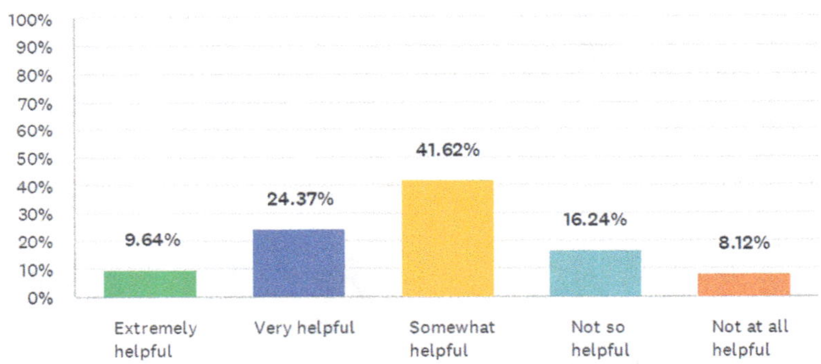

Fig. 12 Do you think that a test like this would be helpful for your learning?

8.6 Discussion and conclusions

In this study, we looked at AI-powered formative assessment systems that claim to offer personalized and adaptive feedback to students. Feedback may include explanations, hints, and targeted remediation resources based on individual learning needs. It also supports teachers by providing them with real-time analytics and reports on student performance. This information can help teachers tailor their instruction and interventions to meet individual student needs. The result is a testing experience that is more efficient at gauging students' progress over time and more informative for both students and teachers.

The evaluation of the construct validity of AI-powered formative assessment systems was developed using structural equation modelling in the framework proposed by Fornell and Larcker (1981). The first component is the so-called confirmatory factor analysis that relates the assessment objective to test items that are assumed to measure that assessment objective; considering the measurement error. The second component examines the covariance of the assessment objectives. Using this method, we able to answer questions such as; are the items valid and reliable indicators for the assessment, the validity of assessment objectives, the level of variance that a construct explains versus the level due to measurement error, construct reliability and how different constructs are related.

Applying the structural equation method and the concept map for secondary school arithmetic outlined in Fig. 1, this paper evaluated the assessment objective validity. Several relevant goodness of fit statistics was used to check whether the empirical data sufficiently fit the theoretical model. There was a strong support for the hypothesized relationship among the assessment objectives. Thus, the content for mathematics of arithmetic (Red bubbles) is dependent on having the prerequisite knowledge and understanding (Green bubbles); and being able to apply the content for mathematics of Arithmetic (Blue bubbles). The results show that all the three assessment objectives met the construct reliability minimum threshold of 0.7. Specifically, the following findings were revealed.

The empirical evidence indicates that all assessment objectives have a satisfactory level of construct reliability. The average proportion of variance in the items that is explained by the assessment objectives also ranged from 27 to 37% (below the acceptable threshold of 50%). We decided to accept that construct as valid, as a construct with an AVE less than 0.50 but the CR more than 0.70 can still be considered to have adequate convergent validity; Fornell & Larcker, 1981).

The results provide sound evidence that the diagnostic tests give consistent information about students' knowledge and ability. The testing process is more helpful for both students and teachers, and it is more effective at tracking students' process over time. The AVE values still require improvement, thus further work is needed.

When asked about their overall test-taking experience, most respondents (78%) inclined to give it a positive rating. Approximately 40% of participants acknowledged that the test was challenging in some way. About 73% of respondents said they agreed with the platform's identification of the respondents' strong points or competencies. A test like this, according to 75% of respondents, would be beneficial for their learning.

Using machine learning and AI technologies, the platform enhances its adaptive learning system to provide personalized, efficient, and effective learning experiences covering a range of mathematical topics. Whether it is numerical fluency or algebra, the platform utilizes AI-driven capabilities to continuously tailor its approach to meet learners' needs, enhance learning outcomes, and foster innovation in education.

One of the key strengths of AI-powered formative assessment systems is their ability to tailor learning experiences to individual learners. By analysing each learner's responses, behaviour, and progress, it can identify learners' strengths and weaknesses. This personalized method facilitates adjustments in learning materials and teaching methods to address the unique requirements of each learner, thereby enhancing their engagement and understanding.

Table 3 Assessment of structure model fit for arithmetic concept map

Assessment of structure model fit							
Model fit indices	GFI	AGFI	TLI	CFI	NNFI	SRMSR	RMSEA
Recommended threshold	≧0.8	≧0.8	≧0.9	≧0.9	≧0.9	≦0.08	≦0.05
Obtained arithmetic	0.985	0.985	0.958	0.962	0.958	0.033	0.030
Supported	Yes	Yes	Yes	Yes	Yes	Yes	Yes

AI-powered formative assessment systems differ from traditional assessments in that they allow for ongoing monitoring of learner progress instead of occurring at set intervals. By integrating frequent assessments into learning activities, these systems collect detailed data on learners' performance and understanding in real-time. This continuous flow of information drives adaptive learning by facilitating immediate adjustments to instruction, such as providing extra assistance or presenting more challenging tasks based on individual performance.

The extensive data produced by AI-driven formative assessment systems offers educators and instructional designers valuable understanding of learners' learning paths and challenging areas. Through analysing this data, educators can pinpoint common patterns, trends, and misconceptions among learners, facilitating precise interventions and improvements in instruction. These insights empower educators to make informed choices regarding curriculum design, teaching methods, and resource distribution, all aimed at enhancing learning outcomes.

Timely and constructive feedback plays a crucial role in aiding learning and steering learners toward mastery. AI-driven formative assessment systems stand out in providing prompt and practical feedback to learners, tailored to their responses. This feedback loop encompasses corrective guidance for misconceptions, acknowledgment for correct answers, and recommendations for enhancement, fostering metacognition, self-regulation, and mastery learning. Furthermore, by scrutinizing learners' responses and interactions, these systems can dynamically adjust feedback to suit individual learning requirements and preferences.

Building on the insights gained from formative assessment data, AI-powered systems can dynamically adjust the delivery of learning content to suit each learner's ability level and learning pace. This adaptive approach to content delivery prevents learners from feeling overwhelmed by material that exceeds their comprehension or being hindered by content they have already mastered. Through the provision of appropriately challenging tasks and tailored support, adaptive learning environments cultivate feelings of competence and independence, motivating learners to persevere and achieve success.

Appendix

Table 4 Nodes performance statistics (Cohort 1: Baseline of 621 students)

Nodes	Description	Number of students moving through the node	Mean score (%)	Standard deviation	Correlation between node and scale
A1	Powers as repeated multiplication	621	84.5	29.3	0.45
A2	Multiplication as repeated addition	606	83.9	26.5	0.52
A3	Addition as bars	581	72.3	28.6	0.55
A4	Fractions as bars	604	74.6	23.7	0.62
A5	Reciprocal fractions	584	55.3	41.4	0.62
A6	Place value	584	69.0	36.0	0.51
A7	Distributivity	620	44.9	39.7	0.40
A8	Inverse operations	617	70.7	37.0	0.48
A9	Addition as the inverse of subtraction	589	73.3	32.5	0.65
A10	Additive inverse	614	59.5	41.8	0.45
A11	Multiplication as the inverse of division	615	53.3	42.6	0.50
A12	Multiplicative inverse	609	67.0	36.0	0.57
A13	Identity elements	615	89.3	23.3	0.50
A14	Multiplication as area	602	63.2	31.9	0.61
A15	Commutativity	621	71.1	35.5	0.51
A16	Order of operations	613	45.1	41.7	0.37
A17	Equivalence	598	80.3	29.0	0.63
A18	Multiplying and dividing by powers	620	56.9	34.8	0.33
A19	Powers and roots	609	52.4	40.2	0.66
A20	Multistep calculations	576	45.3	38.4	0.39
A21	Rearranging expressions	613	42.9	43.4	0.61
A22	Collecting terms	561	52.6	40.3	0.57
A23	Solving equations	620	63.7	40.3	0.45

References

Beglar, D. (2010). A Rasch-based validation of the vocabulary size test. *Language Testing, 27*, 101–118.

Bomash, I., & Kish, C. (2015). *The improvement index: Evaluating academic gains in college students using adaptive lessons*. Knewton.

Bulu, S., & Pedersen, S. (2010). Scaffolding middle school students' content knowledge and ill-structured problem solving in a problem-based hypermedia learning environment. *Educational Technology Research & Development, 58*(5), 507–529.

Calhoun Williams, K. (2019). Prepare for AI's new adaptive learning impacts on K-12 education. Gartner. https://www.gartner.com/en/documents/3947160

Chen, C. H. (2010). Promoting college students' knowledge acquisition and ill-structured problem-solving skills: Web-based integration and procedure prompts. *Computers & Education, 55*(1), 292–303

Cizek, G. J., Rosenberg, S. L., & Koons, H. H. (2008). Sources of validity evidence for educational and psychological tests. *Educational and Psychological Measurement, 68*, 397–412.

Draba, R.E. (1977). The identification and interpretation of item bias. Research Memorandum 26.

Fornell, C., & Larcker, D. F. (1981). Evaluating structural equation models with unobservable variables and measurement error. *Journal of Marketing Research, 18*(1), 39–50.

Hair, J. F., Hult, G. T. M., Ringle, C. M., & Sarstedt, M. (2017). *A Primer on Partial Least Squares Structural Equation Modeling (PLS-SEM)* (2nd ed.). Sage Publications

Haoran, X., Hui-Chun, C., Gwo-Jen, H., & Chun-Chieh, W. (2019). Trends and development in technology-enhanced adaptive/personalized learning: A systematic review of journal publications from 2007 to 2017. *Computers & Education., 140*, 103599.

Henseler, J., Ringle, C. M., & Sarstedt, M. (2015). A new criterion for assessing discriminant validity in variance-based structural equation modeling. *Journal of the Academy of Marketing Science, 43*(1), 115–135. https://doi.org/10.1007/s11747-014-0403-8

Hu, L. T., & Bentler, P. M. (1999). Cutoff criteria for fit indexes in covariance structure analysis: Conventional criteria versus new alternatives. *Structural Equation Modeling: A Multidisciplinary Journal, 6*(1), 1–55. https://doi.org/10.1080/10705519909540118

Linacre, J. M. (2007). A user's guide to WINSTEPS-MINISTEP: Rasch-model computer programs. Chicago, IL: winsteps.com.

Ma, W., Adesope, O. O., Nesbit, J. C., & Liu, Q. (2014). Intelligent tutoring systems and learning outcomes: A meta-analysis. *Journal of Educational Psychology, 106*(4), 901–918.

Mavroudi, A., Giannakos, M., & Krogstie, J. (2017). Supporting adaptive learning pathways through the use of learning analytics: Developments, challenges, and future opportunities. *Interactive Learning Environments, 26*(2), 206–220. https://doi.org/10.1080/10494820.2017.1292531

Merrill, D. C., Reiser, B. J., Ranney, M., & Trafton, J. G. (1992). Effective tutoring techniques: A comparison of human tutors and intelligent tutoring systems. *Journal of the Learning Sciences, 2*(3), 277–305. https://doi.org/10.1207/s15327809jls0203_2

Messick, S. (1989). Validity. In R. L. Linn (Ed.), *Educational Measurement* (3rd ed., pp. 13–103). American Council on Education/Macmillan.

Van Lehn, K. (2011). The relative effectiveness of human tutoring, intelligent tutoring systems, and other tutoring systems. *Educational Psychologist, 46*(4), 197–221.

Wolfe, E. W., & Smith, E. V. (2007). Instrument development tools and activities for measure validation using Rasch models: Part II-validation activities. *Journal of Applied Measurement, 8*(2), 204–234.

Xie, H., Chu, H. C., Hwang, G. J., & Wang, C. C. (2019). Trends and development in technology-enhanced adaptive/personalized learning: A systematic review of journal publications from 2007 to 2017. *Computers & Education, 140*, 103599. https://doi.org/10.1016/j.compedu.2019.103599

Chapter 9
Decimal Point: A Decade of Learning Science Findings with a Digital Learning Game

Bruce M. McLaren

Abstract The McLearn Lab at Carnegie Mellon University (CMU) first designed and developed the artificial intelligence (AI) in education learning game, *Decimal Point*, in 2013 and 2014 to support middle school children learning decimals and decimal operations. Over a period of 10 years, the McLearn Lab has run a series of classroom experiments with the game, involving over 1,500 elementary and middle school students. In these studies, we have explored a variety of game-based learning and learning science principles and issues, such as whether the game leads to better learning—demonstrated learning gains from a pretest to a posttest and/or a delayed posttest—than a more traditional online instructional approach; whether giving students more agency leads to more learning and enjoyment; whether students benefit from hints and error messages provided during game play; and what types of prompted self-explanation lead to the best learning and enjoyment outcomes. A fascinating finding also emerged during the variety of experiments we conducted: the game consistently led to a gender effect in which girls learned more from the game than boys. In this chapter I will discuss the current state of digital learning games, how we designed and developed *Decimal Point*, the technology it is built upon—including AI techniques—and the key results of the various experiments we've conducted over the years. I conclude by discussing the important game-based learning take-aways from our studies, what we have learned about using a digital learning game as a research platform for exploring learning science principles and issues; and exciting future directions for this line of research.

Keywords Digital learning games · Research platform · Middle school mathematics · Artificial Intelligence

---

B. M. McLaren (✉)
Carnegie Mellon University, Pittsburgh, USA
e-mail: bmclaren@cs.cmu.edu

P. Ilic et al. (eds.), *Artificial Intelligence in Education: The Intersection of Technology and Pedagogy*, Intelligent Systems Reference Library 261,
https://doi.org/10.1007/978-3-031-71232-6_9

9.1 Introduction

Digital learning games are omnipresent and embraced by many educators and K-12 schools in the U.S. and around the world. Countless schools use digital learning games as a regular part of instruction (Juraschka, 2019). Such games span many topic areas, including math, science, language, and social science. Some examples of commonly used learning games include *Math Blaster* (https://en.wikipedia.org/wiki/Math_Blaster!), one of the oldest learning games, first publicly distributed in 1983. *Math Blaster* offers skill-building games in basic math for first- to sixth-grade students. *Legends of Learning* (https://www.legendsoflearning.com/) is a commercial organization that offers more than 2,000 math and science games for grades K-8 across more than 350 learning objectives. *Free Rice* (https://freerice.com/home) is a website with a trivia learning game, spanning a variety of content, such as English vocabulary, grammar, geography, and literature, that is designed to help kids learn and, at the same time, support donations of rice and other goods to third world and developing countries. A few of the many other learning games that are commonly seen and used in K-12 classrooms (and beyond) include *Cool Math 4 Kids* (Math), *Math Playground* (Math), *Brain Pop* (Science), *Oregon Trail* (History/Social Studies), and *Duolingo* (Language).

Taking notice of the increasing use of digital learning games during the many classroom studies my lab, the McLearn Lab, conducted with intelligent tutors in middle and high school math and science classrooms between 2006 and 2013 (Adams et al., 2012, 2013; Aleven et al., 2010; McLaren et al., 2008, 2011a, 2011b, 2014a, 2014b; Roll et al., 2011; Walker et al., 2007), I became interested in exploring the learning benefits of digital learning games. It was clear to me that teachers believe in the educational value of learning games and most of these games are highly engaging to students. However, the question I asked myself was: Do digital learning games really help students learn?

In short, the enthusiasm about and proliferation of digital learning games made me curious about their efficacy. What I found was that the evidence was very limited at the time of my initial classroom observations. There was some evidence that games can lead to more engagement and learning than conventional instructional technology, but evidence across subjects, and in particular in the domain of mathematics, was lacking (Honey & Hilton, 2011; Mayer, 2014; O'Neil & Perez, 2008; Tobias & Fletcher, 2011). In 2014, Richard E. Mayer, one of my collaborators, published a book that carefully evaluated the scientific evidence that learning games provide more learning benefits than traditional instructional approaches (Mayer, 2014—so-called *media comparison* studies). At that time there had only been five rigorous studies—those that used a controlled experiment with a comparison and measured objective learning outcomes (versus enjoyment or other subjective measures)—of digital learning games in mathematics. Of those, only three showed learning benefits for the games, with a negligible effect size of 0.03.[1]

These results—or more specifically, the lack of conclusive results—piqued my interest and led me to start a digital learning games research agenda, beginning in 2013. Given that I had already done extensive research with intelligent tutoring systems in math classrooms, it was largely a matter of switching the instructional mechanism from intelligent tutors to digital learning games for math. We focused on middle school math—and, in particular, decimals and decimal operations—as we had in prior studies with tutoring systems (Adams et al., 2012, 2013; Isotani et al., 2010; McLaren et al., 2012). My lab carefully reviewed the literature on learning games for math (Chang et al., 2012; Mayer, 2014; Van Eck & Dempsey, 2002) and game design (Schell, 2008), and began to design and develop a new math learning game. This chapter discusses that journey, the game we created, what we discovered in studies with the game, and where we, as a learning games community, are headed.

The first and primary emphasis of the chapter is an overview and discussion of the many and varied studies the McLearn Lab has conducted with *Decimal Point*, the learning game my lab designed and developed. Besides the game-based learning studies my lab has done with *Decimal Point*, our decade of experimentation has also led to some insights about using educational technology—and in particular a learning game—as a platform for exploring learning science issues and principles. That is the second emphasis of the chapter.

Because this is a long chapter, and not all readers are likely to be interested in all aspects of my lab's work with *Decimal Point*, here are some recommendations on how to read the chapter. For readers interested in learning only a bit about digital learning games, the *Decimal Point* game in particular, and the general results we've found from our many classroom studies over the past decade, I recommend reading Sect. 9.2 ("Background on Digital Learning Games"), Sect. 9.3 ("*Decimal Point*: A Digital Learning Game for Middle School Mathematics"), and Table 9.1 at the beginning of Sect. 9.4 ("Experiments with the *Decimal Point* Learning Game"), which summarizes all of the studies we've conducted. For the reader seeking a bit more depth, perspective, and understanding of the impact of the *Decimal Point* results, I suggest also reading the final three sections—5 ("Key Take-Aways: Digital Learning Game Findings"), 6 ("Key Take-Aways: Use of a Digital Learning Game as a Research Platform"), and 7 ("Conclusions")—which highlight the key take-aways about the studies and use of the game as a research platform, as well as conclusions and future directions. Finally, for readers interested in digging into details about the many results my lab has gotten with the game, I recommend reading the entire chapter, including the lengthy Sect. 9.4, which summarizes all of the classroom studies we've conducted.

[1] In the intervening years there have been many more studies of learning games in STEM subjects, and in particular mathematics, with a higher effect size in game to non-game comparisons (Hussein et al., 2022; Wang et al., 2022).

9.2 Background on Digital Learning Games

A substantial segment of the global population actively participates in digital gaming, a trend that spans various age groups. According to a report from TrueList (2023), approximately 3.26 billion people worldwide play video games with 41% of the world population estimated to be playing or have played a video game. The NPD Group, a leading market research organization, reports that video or digital gaming attracts 73% of children aged two and above (NPD, 2019). Gaming is most popular among the 18–34 age group in the United States, accounting for 36% of gamers, yet, at the same time, 24% of gamers are under the age of 18 (PlayToday, 2023), the population targeted in our work.

In summary, these findings underscore the widespread impact of digital games on recreational activities, particularly among younger people. Hence, there exists a clear rationale for leveraging games as tools to support learning, given their already widespread popularity among people in general and children in particular. Moreover, the evidence strongly suggests that digital games possess a high degree of engagement and motivation for children, which is evident in their sustained and prolonged play (Johnston, 2021). The challenge lies in seamlessly integrating instructional content into gameplay to facilitate effective learning outcomes.

In fact, there is a natural tension in digital learning games between engagement and learning. Table 9.1, taken from a Mayer and Johnson (2010) game-based learning paper, shows the pitfalls and potential of digital learning games crossed against game features and instructional features. This, to my knowledge, is one of the best depictions of the trade-offs between engagement and enjoyment, on one hand, and the educational efficacy of digital games, on the other hand.

In short, game features can be engaging and motivating but also potentially distracting, thus diminishing learning. On the other hand, instructional features, designed with a primary focus on promoting learning, can lead to student learning but can also be boring, thereby reducing motivation. Thus, the dynamic interplay between engaging game features and purposeful instructional elements presents an ongoing challenge in learning game design, demanding thoughtful design to maximize the benefits while mitigating potential drawbacks.

Table 9.1 The trade-offs between game and instructional features (Used by permission)

Potential and pitfalls of game features and instructional features in computer games for learning		
	Game features	**Instructional features**
Potential	Game features can promote motivation to learn (increasing generative processing)	Instructional features can promote learning (increasing essential and generative processing)
Pitfalls	Game features can diminish learning (increasing extraneous processing)	Instructional features can diminish motivation to learn decreasing generative processing)

Source Adapted from Mayer and Johnson (2010)

There are a variety of theories often cited for the benefits of learning games. For instance, flow theory—which posits that people can become so engaged that time passes quickly and concentration and enjoyment are deeply felt—is often cited as a reason for the benefits of games (Czikszentmihalyi, 1975, 1990; Johnston, 2021). Flow induces focused concentration and total absorption in an activity, which may in turn support better learning by enhancing engagement and persistence. Another oft-cited theorist and proponent of game-based learning, James Gee, has put forth 36 key principles of learning with games, including an "Active Learning" principle and a "Committed Learning" principle (Gee, 2003, 2007).

One of the earliest theorists of learning with games was Malone (1981), who discussed how games often trigger intrinsic motivation, employing game features such as fantasy, curiosity, and challenge. He emphasized that the immersive nature of games taps into individuals' intrinsic desires for autonomy, competence, and relatedness, aligning with Deci and Ryan's self-determination theory (Deci & Ryan, 1985; Ryan et al., 2006). Malone's insights add depth to the understanding of how games fulfill psychological needs, making them powerful tools not only for education but also for personal development. Other relevant theories are Piaget's view that play is integral to a child's cognitive development. Piaget's theory, outlined in his seminal work "Play, Dreams, and Imitation in Childhood" (1962), posits that play is not just a recreational activity but an essential component of a child's intellectual growth. Vygotsky's perspective that a child's motivation to play is related to the "Zone of Proximal Development" (ZPD—Vygotsky, 1978) further highlights the interconnectedness of play, cognitive growth and learning. These developmental theories underscore the importance of games in fostering intellectual and social skills during crucial stages of childhood. Moreover, the role of emotions in engagement and game-based learning is explored by Loderer and colleagues (2019). Their work explores the intricate connection between emotional experiences and effective learning within a gaming context, casting light on how games can be powerful tools not only for transmitting knowledge but also for shaping positive emotional responses that enhance the learning process. These diverse theories provide a comprehensive framework for appreciating the benefits of incorporating games into educational settings.

Given these learning theories about flow, motivation, and emotion as a foundation, educational technology researchers, including myself, have investigated various ways to inject the learning of traditional academic subjects, and the theory behind that learning, into digital games (see e.g., Benton et al., 2021; Cheng et al., 2017; Hooshyar et al., 2021; Lomas et al., 2013; McLaren & Nguyen, 2023; McNamara et al., 2010; Shute et al., 2019). For instance, Habgood and Ainsworth (2011) explored how to leverage intrinsic motivation (Deci, 1975) in a game context to create what has been called intrinsic integration—tightly integrating instructional content with game mechanics (Kafai, 1996). Meta-analyses of digital learning games in recent years have reported positive learning results (Clark et al., 2016; Hussein et al., 2022; Mayer, 2019; Wouters & van Oostendorp, 2017). For instance, Clark et al. (2016), in a review of 69 rigorous, empirical studies (filtered from over 1,000 studies reported in published papers), found that digital learning games were associated with a 0.33 standard deviation improvement in learning over non-game comparison conditions.

In addition, motivational and affective benefits of digital learning games have also been supported in some meta-analyses. For instance, Sitzmann (2011) found self-efficacy was higher when learning with games (average $d = 0.52$).

9.3 *Decimal Point*: A Digital Learning Game for Middle School Mathematics

The McLearn lab, along with CMU colleague Jodi Forlizzi, designed and developed the *Decimal Point* learning game, which operates using an amusement park metaphor (Forlizzi et al, 2014; McLaren et al, 2017a). We used playtesting design concepts (Walsh, 2009; Yáñez-Gómez et al., 2017) to conceptualize and design the game (Forlizzi et al., 2014). For instance, we used a co-design process in which students acted as producers, rather than consumers, in the early stages of our design work. The co-design sessions involved 32 sixth grade children over multiple sessions. Those sessions also prompted input from students on both known and established games, as well as presenting the students with preliminary game concepts we had devised for them to review. Some of the key ideas that emerged from our sessions with these children were:

- Students mentioned 54 different games, with their top choices being Minecraft, Angry Birds, Temple Run[2];
- The students particularly liked games with familiar, real-world metaphors; and
- Students liked an obstacle course concept best.

In general, the feedback we collected led us to the idea of an amusement park game (i.e., a familiar metaphor) with a series of "mini-games" (i.e., similar to obstacle courses, with multiple, different challenges).

In the *Decimal Point* game, students "travel" through a theme park playing a variety of mini-games that help them learn (and reinforce their knowledge of) decimal concepts and operations, such as place value, comparing decimal magnitude, placing decimals on a number line and adding decimals. In the base version of the game, students follow the dashed line of the amusement park map, playing mini-games in sequence, as shown in Fig. 9.1. For each mini-game, the student is prompted to play that game twice, each time with a different specific decimal problem. Across 24 mini-games, students play a total of 48 mini-game problems throughout the entire amusement park. A group of fantasy, non-player characters (NPCs) encourage students to play, congratulate them when they correctly solve problems, and provide feedback when they make mistakes (see right side of Fig. 9.1).

There are a wide variety of mini-games within *Decimal Point*, each designed to support students in learning one of following five decimal operations:

[2] Of course, this survey was conducted in 2014 when there were many different games than there are today. It would be interesting to see how this may have changed in the intervening years.

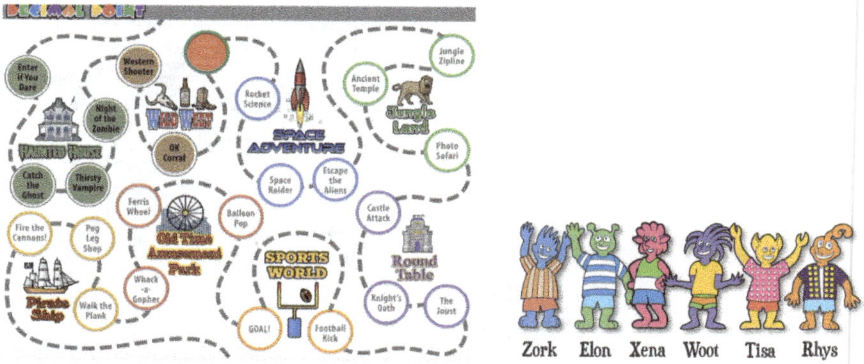

Fig. 9.1 The *Decimal Point* learning game and fantasy characters that are part of the game

- Addition of decimal numbers (*addition mini-games*—Fig. 9.2 shows an example)
- Placing decimals in less-than and greater-than "buckets" (*bucket mini-games*—Fig. 9.3)
- Placing decimals on a number line (*number line mini-games*—Fig. 9.4),
- Sorting decimals in less-than or greater-than order (*sorting mini-games*—Fig. 9.5),
- Completing sequences of decimals, given the first three numbers in a sequence (*sequence mini-games*—Fig. 9.6).

The game is embedded within a narrative designed to contextualize the math work, and the NPCs serve as guides and cheerleaders throughout the game. The narrative of *Decimal Point* provides elements of fantasy (Malone, 1981; Malone & Lepper, 1987), which is important in supporting student engagement in the early phases of interest and domain development, when students do not have enough knowledge to

Fig. 9.2 An addition mini-game

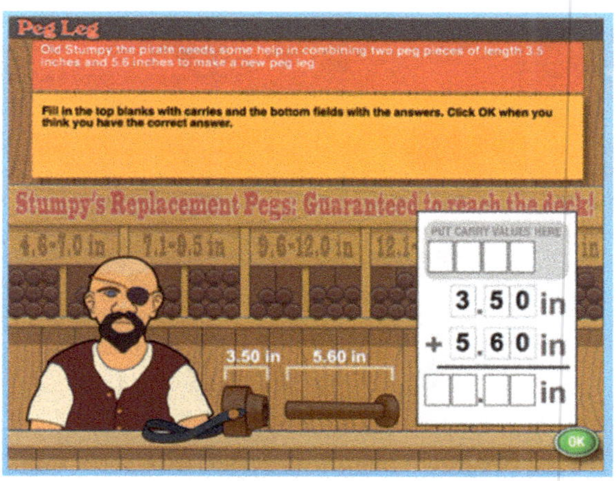

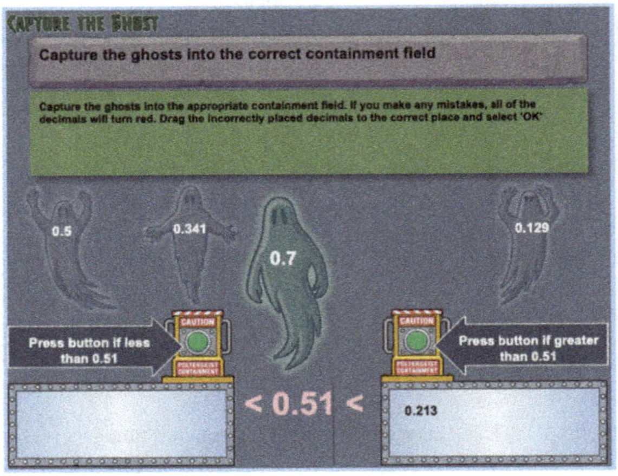

Fig. 9.3 A bucket mini-game

Fig. 9.4 A number line mini-game

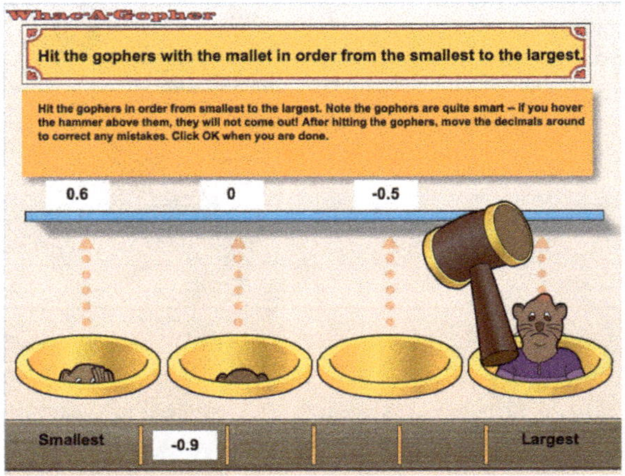

Fig. 9.5 A sorting mini-game

Fig. 9.6 A sequence mini-game

have developed intrinsic interest in a topic (Hidi & Renninger, 2006). This narrative also provides context for the utility of the mathematics that students are learning—and for why they are performing various problem-solving activities—a strategy that is important since it allows students to connect new knowledge to established and current knowledge (Noël et al., 2008). This connection of new to old knowledge may, in fact, be particularly important for overcoming misconceptions, which is critical for learning (Bransford et al., 2000).

All of the 24 mini-games and 48 problems in *Decimal Point* are built using learning science principles. For instance, each mini-game problem targets an established decimal misconception, such as "longer decimals are larger" (Glasgow et al, 2000; Isotani et al., 2010; Irwin, 2001; Stacey et al., 2001). This misconception occurs when kids think that, for instance, a decimal such as 0.213 is greater than 0.51 simply because the former decimal is longer (see, for instance, what the student in Fig. 9.3 has done). This misconception likely occurs because kids learn decimals after whole numbers, a domain in which the "longer is larger" heuristic works. Self-explanation, one of the most robust and widely studied learning science principles (Chi et al., 1989, 1994; Darling-Hammond et al., 2020; Rittle-Johnson & Loehr, 2017; Wylie & Chi, 2014), is also prominently used in the game. After solving the problem correctly in each of the mini-games, students are prompted to self-explain their solution. For instance, see Fig. 9.7, which shows the self-explanation step for the sorting mini-game in Fig. 9.5. The default prompted self-explanation step is a multiple-choice question, which has been argued to be minimally disruptive to gameplay while still providing the learning benefits of self-explanation (Johnson & Mayer, 2010); however, one of our studies explored other forms of prompted self-explanation, such as focused and open-ended self-explanations (McLaren et al., 2022a).

The mini-games of *Decimal Point* are essentially self-contained "intelligent tutors," built with the Cognitive Tutor Authoring Tools (CTAT), a widely used authoring tool developed by Aleven et al. (2016). The game runs on the Internet (originally implemented in Flash, and later ported to HTML/JavaScript) and takes

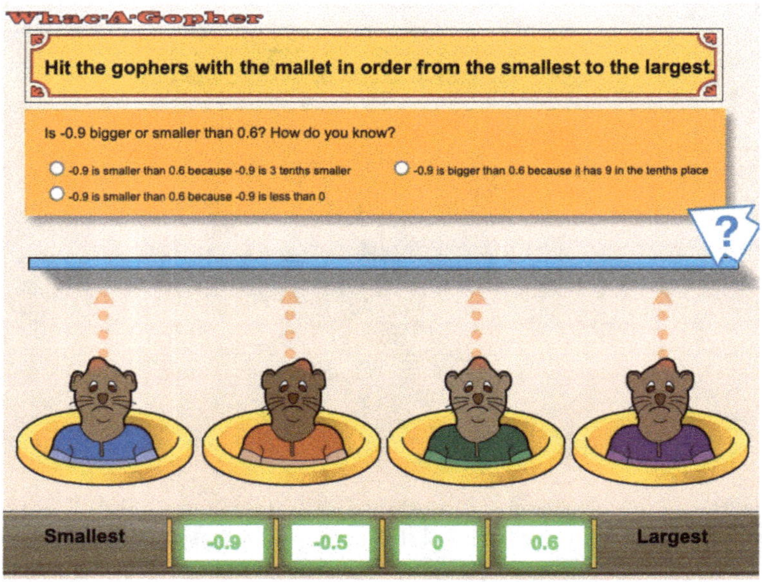

Fig. 9.7 Prompted self-explanation after the sorting mini-game of Fig. 9.5

advantage of tools that have been developed at Carnegie Mellon University, such as the TutorShop (Aleven et al., 2009) and DataShop (Koedinger et al., 2010). The TutorShop is how we deploy *Decimal Point* on the Internet, while the DataShop is how we capture and analyze student data. We designed and developed reusable, aggregated sets of CTAT components for *Decimal Point;* these support game mechanics that are shared across the mini-games. This approach supports shared interactions and stylistic elements across the mini-games and provides a consistency in presentation and interaction.

Artificial Intelligence techniques were used in the development of *Decimal Point* and in the studies we have run with the game. For instance, we implemented adaptive learning (Aleven et al., 2017) and Bayesian Knowledge Tracing (BKT—Corbett & Anderson, 1994) in the game for one study (Hou et al., 2020, 2022). We've also used educational data mining techniques to build detectors of cognitive, behavioral, and affective aspects of learning to analyze game play (Baker et al., 2024; Mogessie et al., 2020; Richey et al., 2021); and we recently used GPT (Ye et al., 2023) to experiment with AI-generated feedback to students (Nguyen et al., 2023b). We also recently explored using a new AI-based knowledge tracing algorithm—Deep Knowledge Tracing—and found that it performed better than BKT (Baker et al., 2023). In general, infusing AI in the context and analysis of learning games is a burgeoning and highly promising area of research (McLaren & Nguyen, 2023). I will return to this theme in the conclusions section of this chapter.

Three decimal tests were used for all studies. These tests are also designed to target common decimal misconceptions, predominantly the same ones targeted in the game, and measure near, medium, and far transfer learning. The tests have been tweaked slightly throughout the years, but generally have had between 42 and 45 items, many with multiple parts, worth a total of 52 to 61 points on each test. There are three forms of the test —A, B, and C—which are isomorphic to one another and were positionally counterbalanced in all studies, such that approximately 1/3 of the students in each condition received Test A as the pretest, 1/3 received Test B as the pretest, and 1/3 received Test C as the pretest; likewise for the posttest and delayed posttest. Some examples of test problems are: "Complete the following sequence: 0.3, 0.6, 0.9, ___, ____."; "Place 0.34 on a number line that already has 0.1, 0.3, and 0.4 on it"; "Order the following decimals, smallest to largest: 0.721, 0.3, 0.42." The test problems were largely derived from math education research and studies, with an emphasis on probing misconceptions (Brueckner, 1928; Glasgow et al., 2000; Graeber & Tirosh, 1988; Hiebert & Wearne, 1985; Irwin, 2001; Putt, 1995; Resnick et al., 1989; Sackur-Grisvard & Léonard, 1985; Stacey et al., 2001). Finally, through the various studies described in this chapter we have been able to statistically validate that the three tests are equivalent to one another.

9.4 Experiments with the *Decimal Point* Learning Game

Our overarching interest in experimenting with *Decimal Point* has always been to explore a wide and varied set of learning science-related questions in connection with game-based learning. Over the decade we have experimented with *Decimal Point*, in which a total of 1,542 students have completed our studies (See Table 9.2), we have pursued a variety of research questions, including:

- Does *Decimal Point* lead to better learning[3] and more enjoyment than a more conventional (i.e., non-game) instructional approach?
- Do female students benefit more, less, or the same as compared to male students playing the game?
- Do students learn more or less—and enjoy the game more or less—when they are given more agency in playing *Decimal Point*?
- Do students learn more or less—and enjoy the game more or less—if they are presented with a learning-focused or enjoyment-focused version of the game?
- Do students benefit from hints and error messages provided in the context of the game?
- How does the instructional context—in particular, the classroom versus remote learning—impact playing of the game and learning?
- What types of self-explanation prompts in the context of *Decimal Point* lead to the best learning and enjoyment outcomes?

---

[3] All mentions of 'better learning', 'more learning' or 'learning benefits' with respect to *Decimal Point* throughout this chapter means: learning gains from a pretest to a posttest and/or a delayed posttest.

- Can GPT correctly grade students' focused and open-ended self-explanations and provide correct and instructionally helpful feedback?
- Could mindfulness inductions provided in conjunction with the game enhance learning outcomes?

In effect, we have used *Decimal Point* as a *research platform* for exploring learning science questions and principles, as they relate to learning games. The game has offered us a unique opportunity to explore the nuances of learning science questions and principles. Through variations of the game and a variety of classroom studies, we have been able to probe into the underlying dynamics that define the intersection of game-based learning, education, and educational psychology.

Decimal Point has allowed us to scrutinize the effectiveness of different learning science methodologies, instructional designs, and game mechanics. Through our variations of and studies with *Decimal Point*, we have gained valuable insights into the cognitive processes, motivational factors, and emotional dimensions that contribute to the success (and failure) of learning games as educational tools. The game's customizability—which was part of the design from the start—has enabled us to test hypotheses, analyze data, and derive meaningful conclusions about the ways in which learning science principles can be applied to optimize learning with games.

The studies we have conducted over the past decade, and the key results, are summarized in Table 9.2 and discussed in the following sections.

In all of our studies we worked with a subset of the 10 public elementary and middle schools we regularly work with in a medium-size U.S. city, Pittsburgh, Pennsylvania, U.S.A. The 10 schools are distributed between urban, suburban, and rural areas. In all cases we conducted our studies in school during actual class time—except for Study 4, which was conducted during lockdown periods of the 2020–2022 worldwide pandemic, and Study 5, which was conducted in hybrid fashion due to the pandemic—over a period of approximately 6 days (5 days first week, 1 day the following week for a delayed posttest). All of the studies replaced regular instruction with our materials and online instruction. Although not mentioned in the description of every study, it is also important to note that students learned significantly from pretest to posttest and from pretest to delayed posttest in *all* conditions across *all* studies. Thus, we only report on the comparative learning benefits between conditions in each study in what follows.

Table 9.2 Studies with the *Decimal Point* learning game

Study	Description	Key research question	Key results	*N*	Relevant papers
Study 1	Comparing a learning game with a non-game digital tutor	Does *Decimal Point* lead to more learning and more enjoyment than a more conventional computer-based instructional approach (i.e., a tutor)?	*Decimal Point* led to more learning and enjoyment than the *Decimal Tutor,* a tutoring system with the same academic content. This was also the first study that uncovered a gender effect, in which girls learned more from the game (but not the tutor) than boys	153	McLaren et al., (2017a), McLaren et al., (2017b), Richey et al., (2021), Baker et al., (2023), (2024), Nguyen et al., (2022a)
Study 2	Comparing different levels of student agency in a learning game	Do students learn more or less when they are given more agency in playing *Decimal Point*?	Students did not learn more given more agency. Most students followed the canonical sequence of mini-games, rather than exercising agency	158	Nguyen et al., (2018), (2019); (2022a), Baker et al., (2023), (2024)
Study 2a (follow-up to Study 2)	Comparing students who were subject to indirect control with those not subject to indirect control in a learning game	How does the inclusion of indirect control impact students' exercise of agency in a digital learning game?	Indirect control did not lead to learning differences, but students varied in how they tried the mini-games, with some approaches leading to more enjoyment	238	Harpstead et al., (2019); Baker et al., (2023), (2024); Nguyen et al., (2019), (2022a)
Study 3	Comparing a Learning-Focused game-based learning approach to an Enjoyment-Focused approach	Do students learn more or less—and enjoy the game more or less—if they are presented with a learning-focused or enjoyment-focused version of *Decimal Point*?	Learning and enjoyment did not vary across conditions, but learning-focused students did more repeated practice, while enjoyment-focused students did more exploration	159	Hou et al., (2020), (2022), Nguyen et al., (2022a)

(continued)

Table 9.2 (continued)

Study	Description	Key research question	Key results	*N*	Relevant papers
Study 4	Comparing the use of hints and feedback in a learning game to not having hints and feedback. Also, comparing game learning in class versus at home	Do students benefit from hints and error messages provided in the context of the *Decimal Point* game? How does instructional context (i.e., classroom vs. remote) impact learning with the game?	Remote students learned more than Classroom students, but the remote drop-out rate was also very high. Surprisingly, *No-Hint* students did better in the classroom than *Hint* students on the delayed posttest	277	McLaren et al., (2022b)
Study 5	Comparing different forms of prompted self-explanation in a learning game	What form of self-explanation prompt in the context of *Decimal Point* lead to the best learning and enjoyment outcomes?	Students in the focused self-explanation condition learned more than students in the menu-based self-explanation condition at delayed posttest, with no other learning differences between conditions. Thus, it appears that focused self-explanations may be especially beneficial for retention and deeper learning	214	McLaren et al., (2022a, (2022c), Nguyen et al., (2023a), Ni et al., (2024)
Study 5a (follow-up to Study 5)	Analyzed the data from Study 5 to test whether recent advances in Large Language Models (LLMs), and in particular GPT, can support learning in a learning game	Can GPT provide instructionally meaningful feedback to incorrect student self-explanation answers in *Decimal Point*?	Results showed that GPT does very well in responding to and providing feedback for students' self-explanation errors; it also provided encouragement and flagged inappropriate language used by students, even though it was not prompted to do so. On the other hand it struggled with procedural math problems (e.g., placing points on a number line). In general, it appears teachers could gainfully use GPT, but they should stay in the loop in responding to student problems	–	Nguyen et al., (2023b)

(continued)

Table 9.2 (continued)

Study	Description	Key research question	Key results	N	Relevant papers
Study 6	Comparing a version of *Decimal Point* that includes prompted mindfulness, with one that prompts reading and jokes, and the base version	Can mindfulness inductions during *Decimal Point* gameplay lead to different behaviors and more learning?	Mindfulness inductions did not enhance learning or change students' gameplay behaviors. This suggests that mindfulness inductions are not beneficial for learning within digital learning games or that we may not have successfully induced mindfulness	166	Nguyen et al., (2022b), Bereczki (2024), in press; Ni et al., (2024)
Study 6a (follow-up to Study 6)	Replication of Study 6, in which we implemented a manipulation check for mindfulness to gain a better understanding of the effects of the intervention	Same basic question as Study 6, but also this: Did we manage to induce mindfulness in students in the mindfulness condition?	Once again, mindfulness inductions did not enhance learning or change students' gameplay behaviors. The only benefit detected was that students had more correct answers after listening to mindfulness reminders in the mindfulness condition as compared to listening to jokes in the story-enriched condition. The manipulation check result suggests that we did not successfully induce state mindfulness	177	Bereczki et al., (2024); Ni et al., (2024)

The N values represent the final number of students that were analyzed, i.e., those who completed all materials, in all studies. Note that in all studies some students were excluded from analyses for not having completed the materials or for having test scores too far under or over the mean

9.4.1 *Study 1: Learning Game Versus Conventional Learning Technology*

Our first and most fundamental study with the *Decimal Point* game was to test whether the game would lead to better engagement, enjoyment, and learning results than an equivalent, more conventional tutoring technology (McLaren et al., 2017a). As mentioned earlier, prior to this study there had only been a handful of rigorous game-based learning studies in mathematics and only three that showed learning benefits for the games, with a very small effect size of 0.03. Our research question for this first study was:

> *Study 1 RQ*: Does *Decimal Point* lead to more learning and more enjoyment than a more conventional computer-based instructional approach (i.e., a tutor)?

Study 1 was conducted during the fall of 2015 and involved students either playing and learning with the *Decimal Point* game or using a more conventional learning technology, the *Decimal Tutor*. The game and tutor share the same underlying instructional architecture and decimal content (i.e., 48 decimal problems). The decimal problems are the same across the pre-defined sequence of items in *Decimal Point* and the *Decimal Tutor*, with one example of a matching problem shown in Fig. 9.8.

One-hundred and fifty-three (153—87 female, 66 male) 6th grade students participated in and completed the initial *Decimal Point* study, from eleven classes at two middle schools. A total of two-hundred and thirteen (213) students began the study, with sixty (60) students eliminated from analyses either because they didn't finish all of the materials (52) or for having gain scores that were 2.5 standard deviations above or below the mean learning gain. Due to the potential distraction and demotivation that might have occurred with students sitting next to one another but working with very different materials, we assigned students by classroom to one of the two instructional conditions. We asked teachers to characterize classes as low, medium, and high performing and then equally distributed these across the two conditions, i.e., a quasi-random condition assignment. Seventy (70) students played and learned with

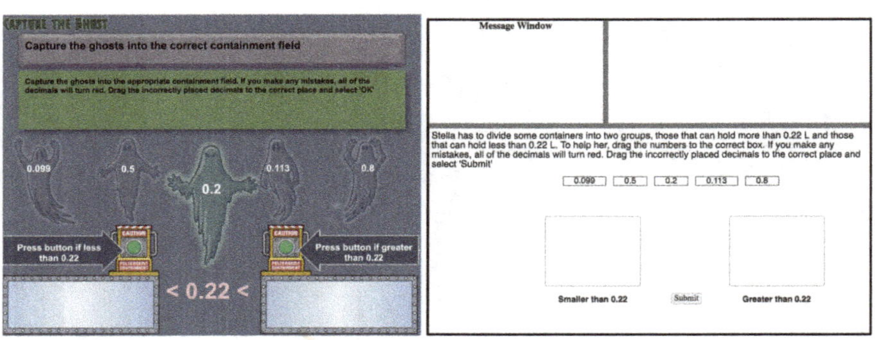

Fig. 9.8 Screenshots of the "Capture the Ghost" mini-game in *Decimal Point* (left) and the corresponding problem presented in the *Decimal Tutor* (right)

Decimal Point, while eighty-three (83) students learned the same content by using a *Decimal Tutor*. Materials included three decimal tests (pretest, posttest, delayed posttest—the tests described previously), the *Decimal Point* game as the experimental condition, the *Decimal Tutor* as the control, and two questionnaires (demographic, evaluation).

The learning results were as follows. The Game group learned more than the Tutor group, with relatively high effect sizes, on the immediate posttest ($p < 0.001$, d = 0.65 for adjusted means) and the delayed posttest ($p < 0.001$, d = 0.59 for adjusted means). In addition, the Tutor group made significantly more errors while working with the tutor (M = 273.4) than the Game group made while playing *Decimal Point* (M = 175.0). This is at least some indication that students playing the *Decimal Point* game were more engaged than those using the *Decimal Tutor* (i.e., The larger number of errors with *Decimal Tutor* likely suggests that students were guessing more frequently with the tutor). Finally, the Game group appeared to enjoy their experience more than the Tutor group, according to the evaluation questionnaire, with students expressing a significantly higher "liking" of the game than the Tutor group liked the tutor. Additional support for that finding is that the Game group expressed that the game interface was significantly easier to use than the Tutor group expressed about the Tutor. Also, the Game group expressed significantly more positive feelings about mathematics after playing than the Tutor group.

We subsequently conducted a post-hoc analysis of the data from Study 1 in which we investigated the differential impact of learning with *Decimal Point* on boys and girls (McLaren et al., 2017b). Given the established gender gap in middle school math education (Breda et al., 2018; Wai et al., 2010), where female students report higher anxiety (Huang et al., 2019; Namkung et al., 2019) and lower engagement (Else-Quest et al., 2013), we were interested in whether our game might help to address that gap. The key finding in the follow-on analysis was that female students learned more than male students from the game. This established a thread throughout all of our *Decimal Point* studies where we repeatedly investigated whether the game benefited girls or boys more. This theme is taken up and discussed in greater detail in a later section of this chapter: "The Gender Effect: A Replication Across Multiple Studies."

9.4.2 Study 2: Student Agency Versus System Control in a Learning Game

Since the beginning of our work and experiments with *Decimal Point*, we have been interested in identifying the game features that have the biggest impact on both enjoyment and learning. Agency, allowing players to make their own decisions about how to play, is one game feature that could impact both enjoyment and learning. A game might give players high agency—the ability to make many, if not all, decisions on what to do next and how to play—or low agency—where players are more restricted in what they can do, often focusing players on learning objectives they might otherwise miss if left to their own devices. Student agency is often seen as related to engagement and, consequently, learning and fun (Ryan et al., 2006). Agency is also related to self-regulated learning (SRL—Zimmerman, 2008), which depending on a student's SRL abilities, could either be helpful or harmful to learning.

Thus, the second key question we pursued in this line of research was:

> *Study 2 RQ*: Do students learn more or less—and enjoy the game more or less—when they are given more agency in playing *Decimal Point*? (Nguyen et al., 2018)

We were inspired to pursue this issue by a study of agency in the context of the *Crystal Island* learning game (Sawyer et al., 2017). *Crystal Island* is a learning game in the area of microbiology in which students try to discover the origins of an infectious disease on a remote island by interacting with key non-player characters (NPCs, e.g., a nurse, a doctor) and objects on the island. Sawyer and colleagues compared three conditions of learning*: high agency*: Students could move freely and explored throughout the island, with no guidance; *low agency*: Students investigated the infectious disease by being guided to talk to characters in a fixed order; *no agency*: Students watched a video of an expert solving the problem, essentially a worked example (Atkinson et al., 2000; Renkl, 2014; Wittwer & Renkl, 2010). They found that the low agency students attempted more incorrect submissions but at the same time learned more than the other two conditions. Interestingly, their study suggests that limiting agency can *improve* learning performance but can also lead to undesirable student behaviors, such as a propensity for guessing. Other studies have provided agency to students by allowing them to customize game features, such as icons and names in a fantasy-based arithmetic tutor (Cordova & Lepper, 1996) or customizing in-game currency, which could be spent on either personalizing the student interface or extra play in a game to improve reading comprehension skills (Snow et al., 2015). While these studies led to increased engagement and learning, student agency was essentially focused on game mechanics, and not instructional features, thus giving students a sense of control but limiting the possibility of students making poor pedagogical decisions.

Study 2 was conducted during the fall of 2017 and involved 158 5th and 6th grade students (77 female, 81 male) from two schools. Students were randomly assigned to either a high-agency (HA—79 students) or low-agency condition (LA—79 students). Thirty-nine (39) additional students were eliminated from analyses either because they didn't finish all of the materials (32) or for having gain scores that were 2.5 standard deviations above or below the mean learning gain (7). Materials included the previously discussed pretest, posttest, delayed posttest, along with two questionnaires (demographic, evaluation). The two conditions that were compared in this study—*high agency* and *low agency*—are illustrated in Fig. 9.9.

In the *high agency* condition, students were presented with a dashboard that displays the 5 different categories of mini games (e.g., Addition, Number Line), as well as the specific mini-games within each category (see the left side of the screenshot on the left of Fig. 9.9). Mini-games that had been played were marked

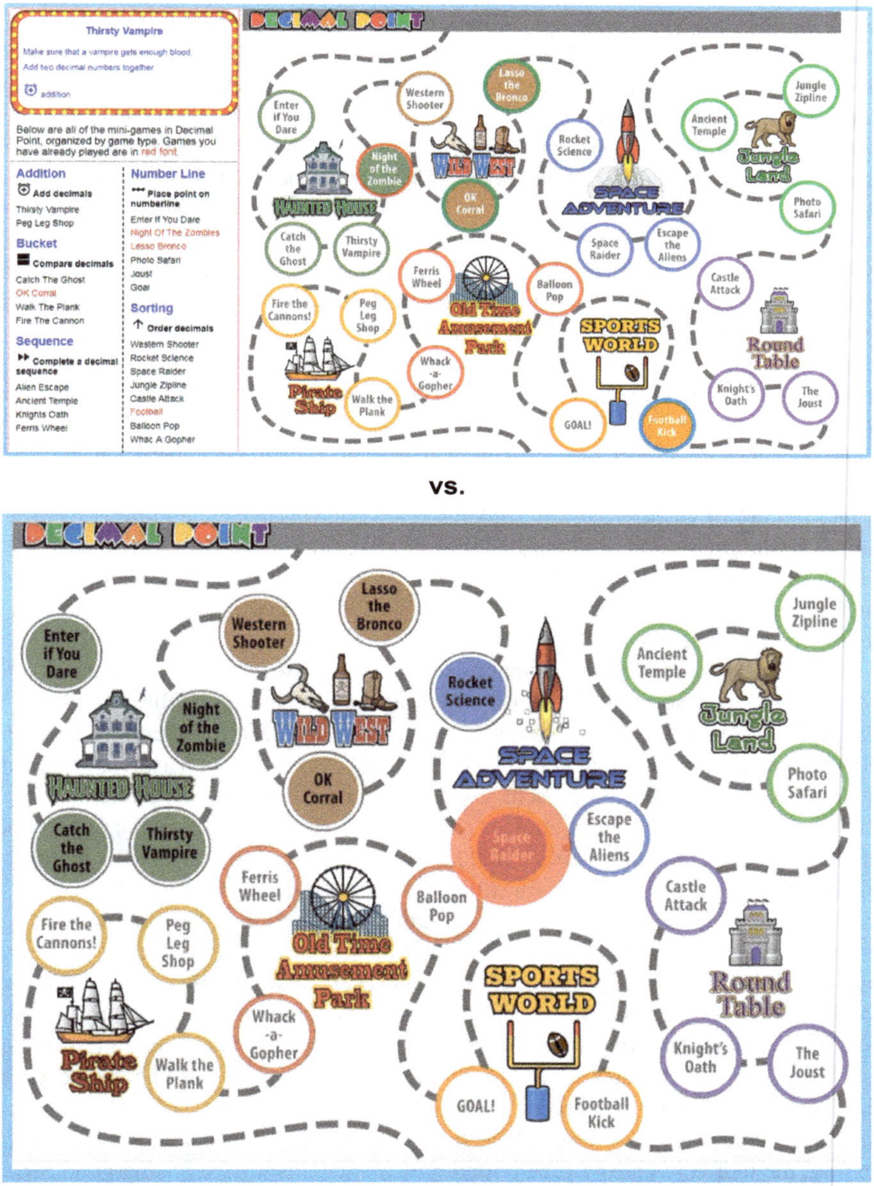

Fig. 9.9 Screenshots of the dashboard that guides the *Decimal Point* high agency condition (left) and the predefined sequence of the *Decimal Point* low agency condition, which is identical to the Game condition of Study 1 (right)

in red on the dashboard with icons filled in on the map. By mousing over the mini-game icons, students were able to learn about how each mini-game is played and what decimal skill it targets. Students could then select whatever mini-game they want to play by clicking on the corresponding mini-game icon. Students could play between 24 and 72 mini-game problems, according to their own desire. More specifically, the students could stop playing *Decimal Point* once they had finished at least 24 mini-game problems, at which point they would be presented with a dialog that contains a "Stop Playing" button. They could also play *more* than the pre-defined set of 48 mini-game problems, up to a total of 72. At any time after tackling 24 mini-game problems, the student could click on the "Stop Playing" button and thus halt game play. The *low agency* condition was the original version of *Decimal Point* from Study 1, in which students must play through all 48 mini-game problems, following the sequence shown by the dashed line on the map, starting from the upper left of the map.

There were no significant enjoyment or learning differences between the *high agency* and *low agency* groups in Study 2. In conducting post-hoc analyses, it was found that 54 of 81 *high agency* students (68%) played the same number of mini-games as the *low agency* students. Eighteen (18) of 81 *high agency* students (22%) exactly followed the canonical sequence. Also, on average, *high agency* students' sequences differed by about 10.77 edits (SD = 8.83) from the canonical sequence, a relatively minor deviation.

In short, it appears that students generally did not exercise much agency and consequently did not benefit from the high agency intervention. But why did this happen? We had several hypotheses about these results. First, students were given choices (autonomy) but may not have felt in control (agency). In particular, being in a classroom, with a teacher present, could have given many students the sense that they were not as free to make choices as we hoped. Second, perhaps the dotted line connecting all of the mini-games could have implicitly, yet unintentionally, communicated the sequence of mini-games the student should play (Schell, 2005, 2008). Finally, while we thought that students might exercise good self-regulated learning with their agency, clearly most students did not, a finding that would be predicted by some SRL research (Schunk & Zimmerman, 1998; Zimmerman, 2008). Perhaps the game environment made it even less likely that students would exercise good SRL than in other learning environments, i.e., they may have been more interested in enjoying their experience with the game than regulating their learning. The bottom line is that the hoped-for student agency and resultant benefits to enjoyment and learning did not occur. This could have been because of the teacher and/or classroom setting, the indirect control of the dotted line, or students not exercising good SRL.

9.4.2.1 Study 2a: The Impact of Indirect Control in a Learning Game

We chose to explore the second possibility, the dotted line guiding students, as the most logical next step in altering game features, as opposed to classroom or student factors. Jesse Schell has defined "indirect control" as subtle cues or design elements

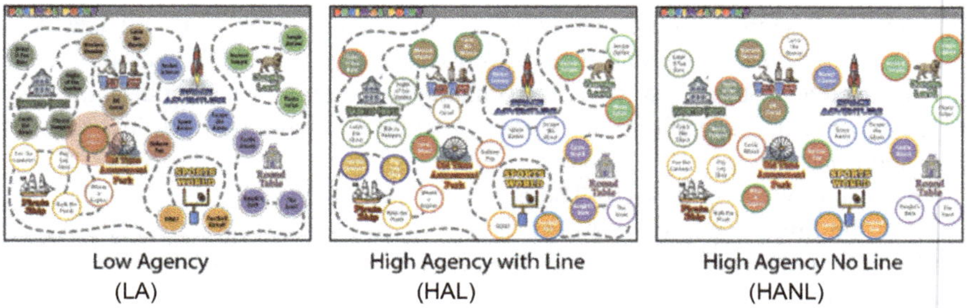

Fig. 9.10 Screenshots of the three conditions of Study 2a with *Decimal Point*

that can lead players toward certain—perhaps unwanted—behaviors (Schell, 2005, 2008). Indirect control can be exerted in a variety of ways, including by presenting players with game goals, specific interface elements, characters, or visual design. Given the results of Study 2—that students generally did not exercise agency (Nguyen et al., 2018)—we wanted to more deeply explore the issue of agency and indirect control in *Decimal Point* in a follow up study (Harpstead et al, 2019). I call this Study 2a here and our question was:

> *Study 2a RQ*: Does the design of the map in *Decimal Point* exert indirect control on players by communicating an implicit order of mini-games? And does this make a difference to learning and/or enjoyment?

To explore this research question, we conducted an expanded replication of the Nguyen et al (2018) study in the spring of 2018 in which we again compared the low agency and high agency conditions, as described for Study 2, but also added a third high agency condition—what we called High Agency No Line or HANL—that operated exactly as the prior high agency condition, but with no line shown on the map (see Fig. 9.10, right). Thus, we had one condition—HAL—that exerted indirect control and another—HANL—that did not.

Two-hundred thirty-eight (238) 5th and 6th grade students from two schools (130 females, 107 males, 1 declined to respond) participated in and completed Study 2a. Forty-nine (49) other students were eliminated because: (a) they didn't finish all of the materials (35), (b) they had login errors during at least one session (13), or (c) they spent an exceptionally long time in the instructional intervention (1). Students were randomly assigned to one of three conditions: low-agency (LA—88 students), high-agency with line (HAL - 78 students) or high-agency condition with no line (HANL—72 students). Materials included the previously discussed pretest, posttest, delayed posttest, along with the two questionnaires (demographic, evaluation). The three conditions that were compared in this study—*low agency (LA), high agency with line (HAL),* and *high agency with no line (HANL)*—are illustrated in Fig. 9.10. The text is very small in this figure, but the key attributes to note in each condition is that in the LA condition students would have to follow the dotted line sequence of mini-games, in the HAL condition students could randomly choose mini-games to play, but be guided by the visible dotted line, and in the HALN condition students could randomly choose mini-games to play but without the dotted line for guidance. Note that the HAL and HALN conditions also had a dashboard like the one on the left of Fig. 9.9 that would allow students to make their own choices about game play, including choosing specific mini-games to play and playing more or less than the LA condition.

The results were as follows. There were no significant differences in learning between the three conditions. However, because students in the HAL and HANL conditions could quit early—and they largely chose to do so—they learned the same amount in significantly less game playing time, i.e., they had greater learning efficiency.[4] In particular, HAL learning efficiency > LA learning efficiency ($p = 0.012$, $d = 0.45$) and HALN learning efficiency > LA learning efficiency ($p = 0.011$, $d = 0.41$). There was no learning efficiency difference between HAL and HALN. Students in HAL and HANL played significantly fewer mini-games than in the Low Agency condition, in which they had to play all of the mini-games. There were also no differences between the three groups in enjoyment. Finally, students in HANL deviated from the canonical sequence significantly more, as measured by the length-matched Dameru-Levenshtein edit distance of a student's mini-game sequence from the canonical sequence (Bard, 2007).

The basic take-aways from this study are that while agency did not improve learning it *did* improve learning efficiency. The results further suggest that indirect control can be limited through subtle game design decisions and that students can exercise agency that ultimately leads to learning more efficiently. This suggests that the game had sufficient support in place to scaffold students' self-regulated learning (a notion discussed in Sawyer et al., 2017).

To more carefully investigate the effects of the agency we provided to students, we conducted post-hoc analyses on the combined data from Studies 2 and 2a, as reported in Wang et al., 2019. In particular, we did a cluster analysis (Bauckhage, 2015) across the 160 students who were in the HAL and HANL conditions in these two studies. We clustered students' mini-game sequences by edit distance—the number of edit operations to turn one sequence into another. We found four distinct clusters of navigation behavior in the HAL and HANL conditions. *Canonical Sequence* students (89 students) stayed very close to the prescribed order of mini-games. *Initial Exploration* students (14) initially jumped around in playing mini-games but then followed the canonical sequence. *Half-on-Top* students (100) only played half the games, in particular the ones at the top of the amusement park map. *Half-on-Left* (32) students only played half the games, in particular the ones on the left of the amusement park map. There was no difference in learning across these clusters, but a key result is about differences in enjoyment. Specifically, we observed significant differences in enjoyment between some clusters, in particular, between *Half-on-Top* and *Half-on-Left*, in which HL > HT. In general, those who deviated more from the canonical order and switched more frequently between theme areas of the Decimal Point amusement

[4] Learning Efficiency was calculated as the z-score of a student's pre-post or pre-delayed test gains minus the z-score of the total amount of time they spent playing the mini-games (McLaren et al., 2008).

park reported higher enjoyment scores. While increasing enjoyment is important, it's also important, of course, to emphasize the instruction and learning aspects of game-based learning. More investigation into the amount of instructional content needed within the game to maximize learning efficiency was clearly necessary. This prompted us to pursue our next research question.

9.4.3 *Study 3: Learning Focus Versus Enjoyment Focus in a Learning Game*

In our next study—Study 3, conducted in the fall of 2019—our goal was to explore the trade-off of a "learning focused" version of the game with an "enjoyment focused" version of the game. That is, we wanted to answer the question:

> *Study 3 RQ*: Do students learn more or less—and enjoy the game more or less—if they are presented with a learning-focused or enjoyment-focused version of *Decimal Point*? (Hou et al., 2020, 2022)

In much of game-based learning the tension between game and instructional features is palpable, as earlier depicted in Table 9.2. A successful learning game skillfully straddles the boundary between engagement and learning. A challenging aspect of digital learning game design is that features that promote engagement in learning games may also disrupt the cognitive processes that are essential for learning. For instance, one study found an inverse relationship between engagement and the difficulty of the learning task (Lomas et al., 2013). Although Lomas and colleagues found that easier learning tasks were more engaging to students in the short term, easier activities also resulted in lower learning gains and less long-term engagement. Some studies have compared enjoyment and learning constructs in the *same* game play context (Erhel & Jamet, 2013; Wechselberger, 2013). In contrast, our intent in Study 3 was to separate learning focus and enjoyment focus by comparing enjoyment and learning across *different* game play contexts. In particular, we were interested in comparing one game context that explicitly emphasized the enjoyable aspects of the game and one that explicitly emphasized the instructional aspects of the game.

To conduct this study we designed three conditions that differed based on how students were presented game information and control through a dashboard attached to the main game map. These conditions are illustrated in Fig. 9.11 and are defined as follows.

- The *Learning-Focused* condition (Fig. 9.11, left) featured an open learner model (Bodily et al., 2018; Bull, 2020), where the knowledge components (i.e., skills) displayed to students were the five decimal skills targeted in *Decimal Point* (addition, bucket, number line, sorting, sequence—Figs. 9.2 through 9.6). The bars indicate the mastery probability of each skill, which is computed by Bayesian Knowledge Tracing (Corbett & Anderson, 1994). This condition also recommended 3 specific mini-games for students to pick next, chosen randomly from

Fig. 9.11 At the top are screenshots of the three conditions of Study 3 with *Decimal Point*. On the top left is the *Learning-Focused* dashboard, in the top middle is the *Enjoyment-Focused* dashboard, and on the top right is the *Control* dashboard. At the bottom is the Fun-O-Meter dialog (Read & MacFarlane, 2006) used in the *Enjoyment-Focused* condition to rate a mini-game

those related to the top 2 skills that students needed improvement on. Our intention with this condition was that it would encourage students to focus on and practice their skills that are lacking; however, they could also choose not to follow these recommendations.

- The *Enjoyment-Focused* condition (Fig. 9.11, middle) featured an analog to the open learner model by displaying the students' enjoyment level of each skill—skills that we renamed to appear more playful (e.g., Pattern Perfect, Mad Adder). Enjoyment data was collected in this condition by prompting students to rate every mini-game round that they finished, from 1 to 5 (see Fig. 9.11, bottom). The score of a skill was the average score of all mini-games belonging to that skill. Similar to the *Learning-Focused* condition, we also recommended 3 mini-games from the game types that the student liked the most so far.

- The *Control* condition (Fig. 9.11, right) simply displayed a list of all mini-games and marked the mini-games that had been played with the red text color. Thus, this design was neutral with respect to both learning and enjoyment. Another difference with the Control condition is that students had to finish all the mini-games once before they could replay more rounds. This is a feature that was present in prior studies of *Decimal Point*, so we wanted to preserve it in the *Control*.

In all three conditions of Study 3, students would have to play at least one-half of the content of the overall game—24 mini-games—but could also play additional mini-games, up to a maximum of 72 mini-games. We hypothesized that the *Learning-Focused* version of the game would lead to the best learning outcome, whereas the *Enjoyment-Focused* version of the game would lead to the best enjoyment outcome.

One-hundred and fifty-nine (159) 5th and 6th grade students (77 females, 82 males) from two schools participated. Thirty-five (35) other students were removed from our analyses due to not finishing all of the materials and two (2) other students were excluded due to their gain scores being 2.5 standard deviations from the mean. Materials included the previously discussed pretest, posttest, delayed posttest, along with the two questionnaires (demographic, evaluation). Each student was randomly assigned to one of the three conditions—*Learning-Focused* (55 students), *Enjoyment-Focused* (54 students) or *Control* (50 students).

Our results showed that there were no significant differences in learning outcomes between the three conditions. With pretest score as a covariate, an ANCOVA showed no significant condition differences in posttest scores, $F(2, 155) = 0.201, p = 0.818$, or delayed posttest scores, $F(2, 155) = 0.143, p = 0.867$. Thus, our first hypothesis that students in the *Learning-Focused* condition would learn the most from the game was not confirmed. Regarding enjoyment, we also found that there were no significant differences across conditions according to three enjoyment constructs (i.e., achievement emotion, game engagement, affective engagement); thus our second hypothesis was also not confirmed. We additionally conducted a number of post hoc analyses. For instance, we compared the number of mini-game rounds played in each condition. With respect to the number of mini-game rounds, we found that the total number of mini-games played was *Control* > *Learning-Focused* > *Enjoyment-Focused*. (Recall that students in all three conditions could choose to stop playing at any time after finishing the first 24 mini-game rounds.) With respect to mini-game replay rate, we found that the *Learning-Focused* condition had a higher replay rate than the *Enjoyment-Focused* condition.[5]

While we didn't see either a learning or enjoyment difference between conditions, our dashboards appeared to prompt students toward significantly different learning behaviors, in particular, the *Learning-Focused* students engaged in more repeated practice and the *Enjoyment-Focused* students did more exploration, behaviors that could have led to, respectively, more learning and more enjoyment. Yet a key question

[5] Mini-game replay rate was calculated by how often students would replay any mini-games.

that arises from these results is: Why were there no learning or enjoyment differences between conditions given these behaviors?

Regarding learning: since the *Learning-Focused* version of the game used the often-effective BKT algorithm and an open learning model, one might have expected that condition to have shown significantly more learning gains than the other two conditions. We have a couple conjectures as to why this did not happen. First, while the *Learning-Focused* condition clearly encouraged blocked practice (i.e., playing mini-games with the same underlying skill back-to-back) it could be, as has been shown in some prior research for different domains (Carvalho & Goldstone, 2015), that interleaved practice is equally as effective as blocked practice. This explanation seems especially likely given the limited number of skills emphasized in *Decimal Point*—essentially only five different skills. Second, although there were obvious differences in the game dashboards and choices presented to students between the conditions, they still spent the majority of their time playing the actual mini-games, which are identical across conditions. In other words, even with the choices students were allowed to make, they were exposed to similar instructional content across conditions.

Regarding enjoyment, it is important to note that our study was conducted in classrooms, where students had limited time each day to play the game, were subject to teacher and experimenter expectations, and were aware of the posttests still to come. Some prior studies have, in fact, shown that game play enjoyment can be lost in the classroom (Rice, 2007; Squire, 2005). Thus, the intended playful and more enjoyable nature of the *Enjoyment-Focused* condition may have been reduced for this reason. Alternatively—and similar to the second explanation for no learning differences—students in the *Enjoyment-Focused* condition may not have experienced more enjoyment because they still spent the majority of their time in the mini-games, which are identical across conditions. In short, both with respect to learning and enjoyment, the student experience and exposure to mini-games may have been more similar across conditions than we intended.

9.4.4 Study 4: Hints and Error Messages in a Learning Game

While hints and feedback may seem an obvious inclusion to a learning game, the research is divided on this point. On one hand, much of intelligent tutoring systems research has demonstrated the learning benefits of providing cognitive hints and feedback to students (VanLehn, 2006, 2011; Woolf, 2008; Xu et al., 2019). Timely, contextualized feedback (Ahmadi et al., 2023; Hattie & Timperley, 2007) could also be helpful to students' learning as they engage with digital learning games. On the other hand, it could be that hints and feedback might disrupt the hoped-for engagement (Bouvier et al., 2013) and flow (Czikszentmihalyi, 1990) of students during game-based learning, a key to learning with games. Some studies have, in fact, precisely uncovered this issue (Moyer-Packenham et al., 2019; O'Rourke et al., 2014). For instance, O'Rourke et al. (2014), in an experiment involving over 50,000 students with *Refraction*, a digital learning game to help students learn fractions, explored different hint types (concrete versus abstract) and hint presentation (by level versus reward). In a 2 × 2 comparison of hint type and hint presentation, plus a condition with no hints at all, they found that students in the no-hint condition learned more than students in any of the other conditions. Thus, we were intrigued with how hints and feedback would help or hurt students in the context of learning with *Decimal Point*, and Study 4 explored the question:

> *Study 4 RQ 1*: Do students benefit from hints and error messages provided in the context of the *Decimal Point* game?[6] (McLaren et al., 2022b).

In addition to the exploration of hints and feedback, the pandemic provided a rare opportunity to explore the use of learning games in the classroom versus learning games at home. While we were conducting Study 4, during the winter and spring of 2020 and after having already administered the study at two K-12 schools, the COVID pandemic forced students across the U.S. to learn from home. Thus, we conducted Study 4 at the final three schools with students playing the game online at home. While the pandemic was of course unfortunate for students in the U.S. and around the world, this change in the study context provided us with a unique possibility to contrast how hints and error messages worked in classrooms versus at home. Thus, a second research question we pursued in this study was:

> *Study 4 RQ 2:* How does instructional context (i.e., classroom vs. remote) impact learning with the game? (McLaren et al., 2022b)

To conduct this study we extended the original, low agency version of the game. In the *Hint* condition students played a version of the game that, in addition to correctness feedback, also provided on-demand hints and error messages for common student errors (i.e., when students made a common error, they received a message specifically addressing the error immediately after entering the incorrect response). The hints were developed together with a mathematics education specialist who participated on the project (Jon Star, Harvard University School of Education). Hints were context-sensitive and three levels in length: the first level reminding the student of their goal and the general procedure to solve the problem, the second taking the student through the mathematics procedure specifically applied to the current problem, and the third providing the student with the answer, also called the "bottom-out hint." In the *No-Hint* condition students played the original version of the game that provided no hints and only correctness feedback (i.e., turning correct answers green and incorrect answers red) within the individual mini-games. These conditions are depicted in Fig. 9.12. Some examples of hints and error messages in the *Hint* Condition are shown in Table 9.3.

The study effectively became a 2 × 2 design, crossing *Hint* and *No-Hint* with *Classroom* and *Remote* game play. The study was conducted in two phases. The

---

[6] Note that before this study, the *Decimal Point* learning game did not include hints, beyond providing correctness feedback (red and green highlighting), nor error messages focused on the common decimal misconceptions.

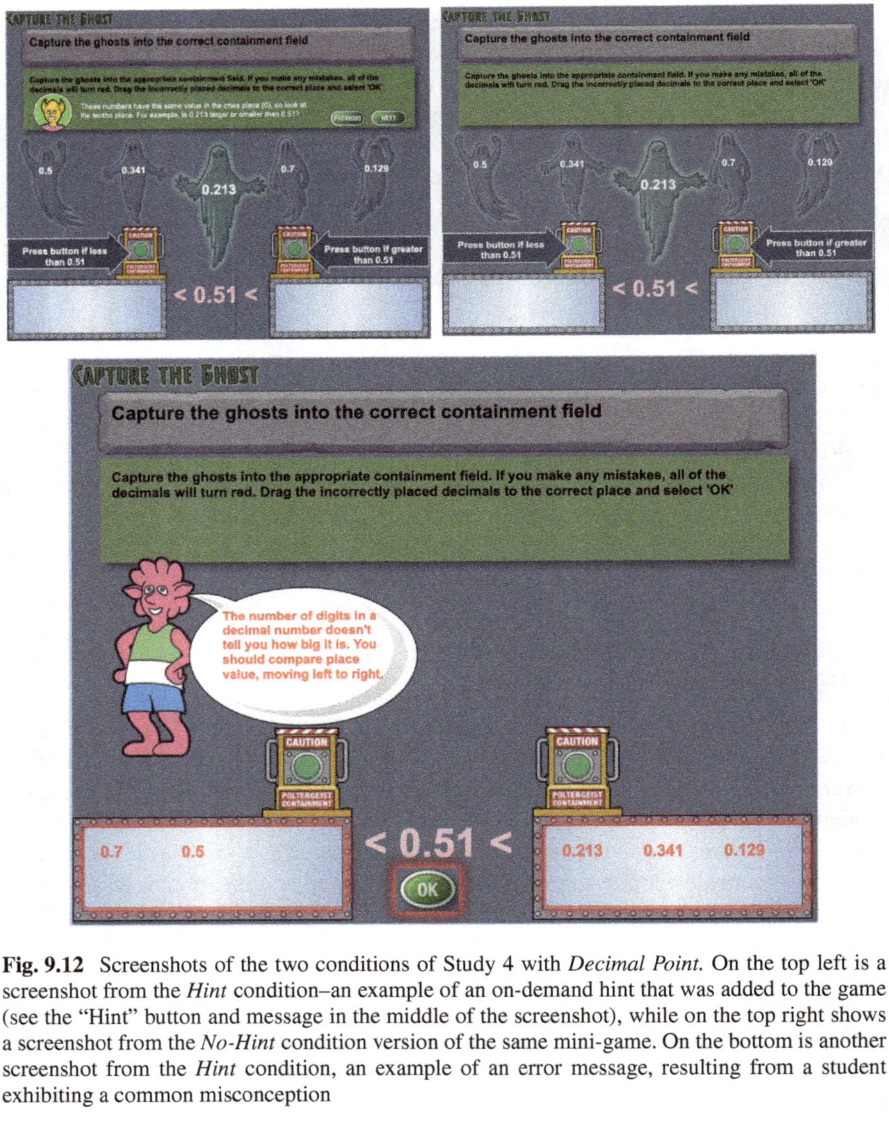

Fig. 9.12 Screenshots of the two conditions of Study 4 with *Decimal Point*. On the top left is a screenshot from the *Hint* condition–an example of an on-demand hint that was added to the game (see the "Hint" button and message in the middle of the screenshot), while on the top right shows a screenshot from the *No-Hint* condition version of the same mini-game. On the bottom is another screenshot from the *Hint* condition, an example of an error message, resulting from a student exhibiting a common misconception

first phase, conducted in the classroom pre-COVID at two schools, had a total of 151 5th and 6th grade students, sixty-seven (67—31 females, 36 males) assigned to the *Hint* condition and eighty-four (84—41 females, 43 males) assigned to the *No-Hint* condition. For this phase of the study, we assigned students to condition by class, due to concerns about students within a classroom observing one another's work and seeing differences in the game (i.e., students without hints noticing their classmates receiving hints, and students with hints might share them with classmates

Table 9.3 Example hint and error messages in *Decimal Point*

Mini-game problem type	Hint examples	Error message example
Sorting	*Level 1*: Compare digits in the same place values of the decimal numbers, moving from the leftmost digit to the rightmost *Level 2*: Since these numbers all have the same ones place (0), compare the tenths place. Which has the smallest tenths place? *Level 3*: 0.0234 has the smallest tenths place, followed by 0.14, 0.323, 0.4 *(These are the three hint levels provided when the student is given a sorting problem with the decimal numbers 0.14, 0.4, 0.0234, 0.323)*	Start by comparing the first digit to the right of the decimal point, even if the digit is 0 *(If the student is presented with sorting the decimal numbers 0.14, 0.0234, 0.323, 0.4)*
Number line	*Level 2*: If you divide the space between 0 and 1 into two pieces, 0.5 is at the end of the first piece. Is 0.456 smaller or larger than 0.5? *(Level 2 hint when the student is given a number line problem to place 0.456 on a number line running -1.0 to 1.0)*	0.456 is greater than 0, so it goes to the right of 0 *(If student clicks to the left of 0, to where the decimal number would be negative)*

not receiving hints). We asked teachers to characterize classes as low, medium, and high performers and then did quasi-random condition assignments so that we had close to the same number of classes of each level within each condition. Eighteen (18) students were excluded (8 in the *Hint* condition and 10 in the No-Hint condition) for failing to complete the materials. An additional student in the *Hint* condition was excluded for performing more than 3 standard deviations below the mean on the posttest and delayed posttest.

For the second phase of the study, when students were working from home due to COVID, three schools with a total of 126 6th grade students participated in the study remotely (64 female, 62 male), with sixty-four (64) students assigned to the *Hint* condition and sixty-two (62) students assigned to the *No-Hint* condition. For this phase, we randomly assigned students to condition, since there was no longer a concern about students seeing one another's work. Ninety-seven (97) students (51 in the *Hint* condition and 46 in the *No-Hint* condition) were excluded from analyses for failing to complete the materials in the allotted time. In summary, the numbers for each of the 2 × 2 conditions is shown in Table 9.4.

Table 9.4 Conditions in 2 × 2 study 4

Condition	N	Female	Male
Hint/classroom	67	31	36
No-hint/classroom	84	41	43
Hint/remote	64	33	31
No-hint/remote	62	31	31

The key results of Study 4 were as follows. Regarding completion rate, the different instructional settings led to significantly different completion rates: *Classroom* students completed the materials at a rate of 88.8%; *Remote* students completed at a rate of only 56.5%. Regarding learning, the *Remote* students learned significantly more than the *Classroom* students, likely due to the fact that in the *Remote* condition more of the students with lower prior knowledge (and/or students with less at-home support) failed to complete the materials. In addition, the two versions of the game, *Hint* and *No-Hint*, led to different *Classroom* versus *Remote* results. In particular, on the delayed posttest, students in the *No-Hint* condition did significantly better than the *Hint* condition in the classroom, while there was no significant difference between conditions at home. Another finding was that female students learned more in the classroom than male students, but the same effect did not occur remotely. We also conducted some post-hoc analyses, finding that students in the *Hint* group used significantly more hints in the *Classroom* than *Remotely*. In addition, higher prior knowledge students used hints more productively, with a significant negative correlation between hints and learning gains.

Some interesting conclusions emerge from these results. First, the different completion rates, as well as better test performance for *Remote* students, were likely due to more and better supervision and guidance in the classroom than at home. The students in the classroom ($N = 151$) were monitored by experimenters and teachers. On the other hand, students at home ($N = 126$), especially because this was at the beginning of the pandemic, may have been unmotivated and not pushed to work by their caretakers at home. The higher performing students working from home likely persevered more, completed the materials more frequently, and thus performed better. Second, why did students in the *No-Hint* condition do better in the classroom on the delayed posttest? While at first this may seem counter-intuitive, in light of the Interactive-Constructive-Active–Passive (ICAP) framework from Chi and Wylie (2014), this is perhaps not so surprising. *No-Hint* students may have worked harder, struggling harder to construct their knowledge, and thus learned more. In support of this, a learning curve analysis showed us that *No-Hint* students initially did worse than *Hint* students, but eventually caught up with their *Hint* counterparts. Finally, why did female students in the *Classroom* condition do better than male students, but not remotely? This was due to girls performing the same in both contexts, but, interestingly, boys did much better at home. This finding aligns with some prior research that girls tend to outperform boys in classroom settings (Dwyer & Johnson, 1997; Entwisle et al., 1997).

9.4.5 Study 5: Comparing Different Forms of Prompted Self-Explanation in a Learning Game

Another learning science principle that intrigued us with respect to game-based learning was self-explanation. Thus, for Study 5, conducted in the spring of 2021, we set about exploring the best approach to prompt self-explanation within *Decimal Point* (McLaren et al., 2022a; Nguyen et al., 2023a). Prompted self-explanation is a feature of instructional technology in which learners are induced to explain their work; it is one of the most robust of learning science principles, supported by decades of research (Chi et al., 1989, 1994; Darling-Hammond et al., 2020; Rittle-Johnson & Loehr, 2017; Wylie & Chi, 2014). Self-explanation supports learners in a number of ways; it helps them fill in gaps in their understanding, revise errors in their prior knowledge, and connect fragmented and disconnected knowledge (Chi et al., 1989; Nokes et al., 2011). When paired with problem-solving, prompted self-explanation can help learners connect problem-solving steps with principles and application conditions (Ainsworth & Burcham, 2007; Aleven et al., 2003). Prompted self-explanation has been shown to be effective in a variety of empirical studies, for instance, in prompting students to explain the principles behind steps in solving geometry problems in a cognitive tutor (Aleven & Koedinger, 2002) and prompting and coaching of self-explanations in a physics tutor (Conati & VanLehn, 2000).

Thus, a key question is:

> *Study 5 RQ*: What form of self-explanation prompt in the context of *Decimal Point* leads to the best learning and enjoyment outcomes?

A variety of approaches have been attempted within instructional technology. Wylie and Chi (2014) cast these various forms of prompted self-explanation along a continuum between unconstrained, on one extreme, and highly constrained self-explanations, on the other extreme. Unconstrained self-explanations allow learners to freely create their own explanations, while presenting the greatest cognitive challenge to learners (i.e., *open-ended self-explanations*). Highly constrained self-explanations, on the other hand, present the learner with a small set of options to choose from to self-explain and thus create the least cognitive challenge for learners (i.e., selecting self-explanations from a menu, *menu-based self-explanations*). Between the two extremes Wylie and Chi cite three other types of prompted self-explanation: *focused self-explanations,* which are constructive but focused in a specific way, such as prompting learners to identify relationships between mental models; *scaffolded self-explanations*, which provide support and/or feedback as learners construct explanations or fill in blanks of an explanation sentence; and *resource-based self-explanations*, in which explanations are selected by learners with the support of a resource, such as a glossary. Chi and Wylie's (2014) ICAP framework for cognitive engagement predicts that students will learn more from cognitively engaging tasks, meaning that *constructive* self-explanations, such as open-ended self-explanations and focused self-explanations, should be more effective for learning than active self-explanations, such as menu-based self-explanations.

Prompted self-explanation has been minimally explored in digital learning games, in which a study by Johnson and Mayer (2010) found that menu-based prompts led to better learning than open-ended prompts. This work was, in fact, the inspiration for us to explore the issue of self-explanation in the context of *Decimal Point*. In other work, Hsu and Tsai (2011) found that prompting learners to explain their errors from a menu of choices led to better learning gains than not prompting for error explanations.

Yet, not all studies have shown learning benefits through prompted self-explanation in digital learning games. In a study with *Newtonian Game Dynamics*, Adams and Clark (2014) compared menu-based self-explanation with explanatory feedback and a control condition with neither self-explanation nor explanatory feedback. They found no learning differences between the three conditions and, in fact, students in the menu-based self-explanation condition completed fewer game levels than the condition with no self-explanation or feedback.

Thus, in Study 5, we set out to experiment with different versions of prompted self-explanation after problem solving in the game (Fig. 9.13). We decided to experiment with three types of prompted self-explanations across the Wylie and Chi continuum from unconstrained to highly constrained, starting with *focused self-explanations*, in which students must create their own explanations, but with prompting text to focus their attention on a particular aspect of the problem they are explaining. For instance, in Fig. 9.13 at the bottom left, the student is prompted to self-explain just one comparison—9.2111 compared to 9.222—of the sorting problem of four decimal numbers. Next, we created a *scaffolded self-explanation* condition, which essentially presents students with sentence builders that provide all of the components of a correct self-explanation but prompts students to correctly piece together those components into a self-explanation response. Finally, *menu-based self-explanations*—the default self-explanation approach of *Decimal Point*—prompts students to select a self-explanation from a multiple-choice list of predefined options. Note that this approach is essentially what Johnson and Mayer (2010) showed led to the best learning in the context of their game, in contrast to the prediction of the ICAP theory (Chi & Wylie, 2014).

Study 5 involved 214 5th and 6th grade students (114 females; 99 males; 1 did not report) from 4 schools (1 rural, 2 suburban, and 1 urban), with students randomly assigned to condition. Seventy-five (75) were in the menu-based condition, 72 were in the scaffolded condition, and 67 were in the focused condition. An additional one hundred and forty-three (143) students were dropped due to (a) failing to complete part or all of the learning materials or any tests and (b) having participated in one of our studies the previous year. (Note that the relatively high attrition rate was due, at least in part, to running the study during the COVID-19 period. Some students participated in person, some at home, and some in a hybrid format.)

The results of Study 5 showed that students in the *focused self-explanation* group learned more on the delayed posttest than the *menu-based self-explanation* group. There were no other significant effects. Regarding time on task, the *menu-based self-explanation* group spent significantly less time than the *focused* or *scaffolded self-explanation* groups. This indicates that at least the menu-based approach takes less time, i.e., it is more efficient. The only significant effect of engagement was that students in the *focused self-explanation* group reported a significantly higher sense of player mastery.

So, in conclusion, the key finding of this study was that focused self-explanations led to better learning than menu-based self-explanation, without any loss of engagement. This result is in line with Chi and Wylie's ICAP theory (2014), but in contrast to the Johnson and Mayer study (2010) that found menu-based self-explanations led

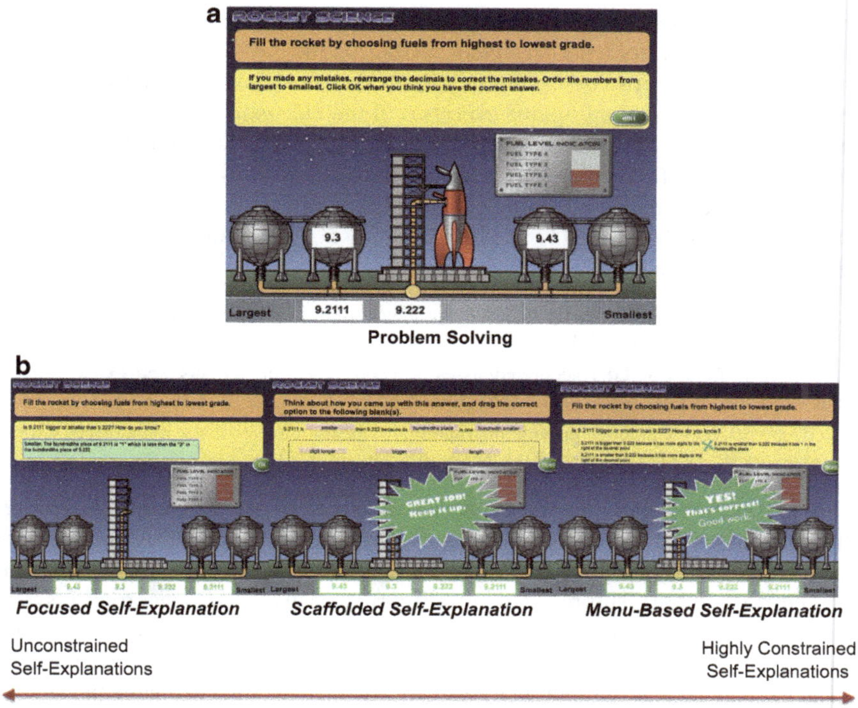

Fig. 9.13 Screenshots of the three conditions of Study 5 with *Decimal Point,* from unconstrained self-explanations to highly constrained self-explanations. On the top is an example problem solving step, within the *Rocket Science* mini-game of *Decimal Point*. On the bottom left is the subsequent prompted self-explanation step, a focused self-explanation. On the bottom middle is a scaffolded self-explanation. On the bottom right is a menu-based self-explanation, the default self-explanation approach of *Decimal Point*, used in all other studies described in this chapter

to better learning than open-ended self-explanations in a game context. This indicates that focused self-explanations used in the context of a digital learning game may be better for deeper, more conceptual learning than other forms of prompted self-explanation, without accompanying loss of game play engagement.

9.4.5.1 Study 5a: Using a Large-Language Model to Assess and Provide Feedback for Self-Explanations in a Learning Game

The recent emergence and advances with large language models (LLMs), and in particular ChatGPT (Ye et al., 2023), intrigued my lab and I, as it did many other researchers. When a commonly available version of ChatGPT appeared in November of 2022, we decided to do a post-hoc study of the data from Study 5 to explore whether

ChatGPT/GPT[7] could provide instructionally meaningful feedback to the focused self-explanations of students (Nguyen et al., 2023b). Given over 5,000 focused self-explanations from students in Study 5, we conducted analyses to assess GPT's capability to (1) solve the in-game exercises of the *Decimal Point* game, (2) determine the correctness of students' self-generated self-explanations, and (3) provide instructionally helpful feedback to incorrect self-explanations.

Study 5a was conducted completely off-line, using the 5,142 focused self-explanations collected from 117 students in Study 5. We had three specific research questions for this study:

> *Study 5a RQ1*: Can GPT correctly answer the problem-solving and self-explanation questions in the game *Decimal Point*? (i.e., Is GPT a good student in this domain?)
>
> *Study 5a RQ2*: Can GPT accurately assess the correctness of students' self-explanation answers? (i.e., Is GPT a good grader in this domain?)
>
> *Study 5a RQ3*: Can GPT provide instructionally meaningful feedback to incorrect self-explanations? (i.e., Is GPT a good teacher in this domain?)

Three coders manually graded as correct, incorrect, or off-topic all of the focused self-explanations from the 117 students, using an iterative process which included inter-rater reliability as a means of assessing coding agreement, as described in (Nguyen et al., 2023a). This resulted in 1000 correct answers, 4076 incorrect answers, and 66 off-topic answers that did not address the question. For the purpose of Study 5a's analysis, we treated off-topic answers as incorrect.

The general approach of our analyses was, for each decimal problem and student self-explanation, to send GPT the question and, in the case of the self-explanations, the student's response and a grading rubric. We developed a script to automatically send all of the prompts to GPT and then harvested all of its answers. We used GPT 3.5 for Study 5a, as that was the current version of GPT when we conducted the study.

For RQ1 we wanted to see how well GPT could solve the *Decimal Point* math problems and self-explanations. Since GPT gives a unique answer each time it is queried, we sent each math question and self-explanation prompt to GPT 10 times to assess how correctly and consistently it handled each. The correctness of GPT's responses to both the problem-solving and self-explanation items were assessed by a math expert on the research team, with the results shown in Fig. 9.14. As can be seen on the left side of Fig. 9.14, GPT was excellent at solving sorting and sequence problems, very good at solving bucket problems, but quite poor at both addition and number line problems. GPT had a much better overall performance for self-explanations than for the problem-solving activities (right side of Fig. 9.14). The only problem type where GPT's explanations were occasionally incorrect was sorting, where it sometimes slipped at assessing decimal place values.

[7] ChatGPT is the chat interface that enables sending data to and receiving data from the underlying GPT model. While ChatGPT is the commonly used term, in fact, we used the GPT API in this study; thus, from this point on I will only use the more precise term: "GPT."

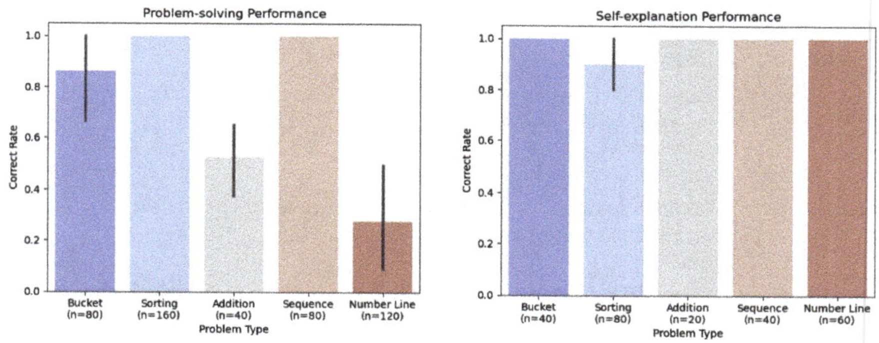

Fig. 9.14 Results of Study 5a's RQ1. GPT's problem solving (left) and self-explanation (right) performance

For RQ2 we wanted to see how well GPT could assess student self-explanations. To do this, we prompted GPT to provide a response of correct or incorrect, per self-explanation, given the self-explanation prompt, the student's self-explanation, and the grading rubric for self-explanations. GPT's performance compared to that of the human coders is shown in Table 9.5. Notice that there were a relatively small number of false negatives (lower left cell, in boldface font), but a much larger number of false positives (upper right cell, in boldface font). Most of these were due to GPT grading an incorrect answer as correct, suggesting that it did not follow the grading rubrics as closely as the human graders did. For instance, for bucket and sorting items, we found that the presence of comparison keywords such as "bigger" or "smaller" was sufficient to get a correct rating from GPT. For example, if the student just wrote "A is smaller than B because it is smaller"—clearly an example of fallacious circular reasoning—GPT would rate it as correct. Similar errors based on shallow keyword matching occurred across all problem types.

For RQ3 we wanted to know whether GPT could provide accurate and instructionally meaningful feedback to students. To generate feedback per incorrect student self-explanation we provided GPT with the self-explanation prompt, the rubric items specific to that self-explanation, and the student's self-explanation. We then coded GPT's feedback according to six relevant categories with results as shown in Table 9.6.

In summary, GPT did much better as a teacher—providing feedback to incorrect self-explanations—than it did as a student—solving and self-explaining the math

Table 9.5 Results of Study 5a's RQ2

	Human: Correct	Human: Incorrect
GPT: Correct	830	**1,118**
GPT: Incorrect	**170**	3,024

GPT's assessment of student self-explanation compared to the assessment of human coders

Table 9.6 Results of Study 5a's RQ3

Category	Description	Results
Accuracy	Does GPT distinguish between partially and fully correct self-explanations?	GPT assigned correct partial credit 75% of the time
Fluency	Is GPT's feedback grammatical and natural sounding?	GPT was 100% proficient in English
Regulation	Does GPT's feedback address all decimal misconceptions reflected in the self-explanation?	Chat GPT was very effective at identifying and addressing student misconceptions
Solution	Does GPT's feedback tell the student the correct answer?	GPT did not provide solutions to 794 incorrect and low-effort self-explanations (e.g., "idk," "by adding up")
Rationale	Does the feedback provide a rationale?	GPT demonstrated good understanding of 85% of self-explanations and provided a range of nuanced explanations (e.g., "your answer is not specific enough")
Encouragement	Does the feedback provide any form of encouragement?	GPT provided encouragement to 20% of the answers ("great job," "keep practicing") and detected 9 cases of inappropriate language used by students

Assessment of GPT's feedback on incorrect self-explanation responses according to six relevant categories

problems—or a grader—assessing the correctness of student self-explanations. In providing feedback to the incorrect student self-explanations, GPT's feedback was high quality and nuanced; it provided encouragement and flagged inappropriate language, even though it was not prompted to do so. It also did very well understanding student answers but provided incorrect feedback more frequently than a teacher likely would have. GPT did less well in solving math problems; it had difficulty with the nuances of math, such as carrying when performing addition and placing points on a number line. (Note that this shortcoming of GPT is now widely recognized, with at least some preliminary suggestions for how to correct it (Wolfram, 2023)). It also struggled a bit in assessing the correctness of student self-explanations, likely due to shallow keyword matching with the grading rubric, which led to many false positives. It appeared not to detect all of the nuances in the grading rubric. Overall, our assessment at the conclusion of Study 5a was that GPT, at least the version 3.5 current at the time of this study, is more suited for conceptual analyses (e.g., giving feedback to self-explanations) than procedural math questions. In short, at the time of Study 5a, GPT was still in a state where a teacher should remain in the loop, double checking answers before they are presented to students.

9.4.6 Study 6: Mindfulness Induction When Learning with an Online Game

Study 6, conducted in the fall of 2021, involved an investigation of the interaction of mindfulness—attending to the present moment with focus and without judgment—with game-based learning (Bereczki et al., 2024; Nguyen et al., 2022b). Mindfulness meditation has been shown to support self-regulated learning (Dunning et al., 2019; Takacs & Kassai, 2019), improve attention skills (Dunning et al., 2019, 2022; Takacs & Kassai, 2019), and reduce math anxiety (Samuel & Warner, 2021). On the other hand, the role of mindfulness in children's academic achievement and outcomes is less clear. There have been several studies that have assessed the efficacy of mindfulness-based interventions, but those studies have shown a non-significant, small average effect on learning (Maynard et al., 2017). More promising results have been found with older students and those with ADHD, but still the results are inconclusive (Güldal & Satan, 2020; Singh et al., 2018).

More specific to math skills, it has been shown that executive function—which entails working memory, inhibitory control, and cognitive flexibility—is key to learning math skills (Cragg & Gilmore, 2014). For instance, prior research has found that kindergarten-age children with higher executive function skills but lower math skills are more likely to catch up with their higher-performing peers by the 5th grade than those students with lower executive function skills (Ribner et al., 2017). Furthermore, mindfulness appears to play a role in supporting executive function (Dunning et al., 2019, 2022; Takacs & Kassai, 2019). Yet, even with the promising connection between mindfulness, executive function, and math learning, Study 6 is, to the best of our knowledge, the first to explore the benefits of employing mindfulness as a means for boosting learning in the context of a digital learning game for mathematics. The question we asked in this study is:

> *Study 6 RQ*: Can mindfulness inductions during *Decimal Point* gameplay lead to different behaviors and more learning?

To explore this issue, we created three *Decimal Point* conditions, *Mindfulness*, *Story*, and *Control*, as shown in Fig. 9.15. The order and content of the mini-games within *Decimal Point* during gameplay was identical across all three conditions. The key differences between conditions were as follows. In the *Mindfulness* and *Story* conditions students would listen to a five-minute audio session at the start of each day of the study, prior to playing and learning with *Decimal Point*. In the *Mindfulness* condition (Fig. 9.15, top), the audio content involved an alien character prompting students to be mindful by asking them to close their eyes, focus on their breath and sounds in the environment, and let go of passing thoughts (Vekety et al., 2022). In the *Story* condition (Fig. 9.15, middle), the audio content was age appropriate, emotionally neutral (i.e., not emotionally arousing or upsetting) science fiction stories that were unrelated to the learning content. This condition was created to control for time with respect to the mindfulness condition, but with material that was not designed to induce mindfulness. Both the *Mindfulness* and *Story* conditions also featured in-game, minute-long reminders that would appear when the student had made three consecutive errors in a mini-game. In the *Mindfulness* condition, students would be encouraged to slow down, close their eyes, and focus on their breath for a moment. In the *Story* condition, students would listen to a joke from an alien character. Each reminder would appear at most once every 10 min to avoid overwhelming the students. Finally, students in the *Control* condition (see Fig. 9.15, bottom) were not presented with any opening audio material before starting gameplay each day, nor did they receive reminders when they made errors.

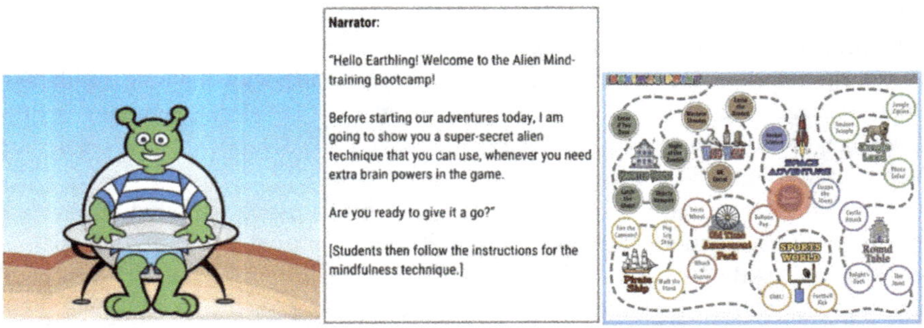

***Mindfulness* condition, each day begins with a mindfulness induction**

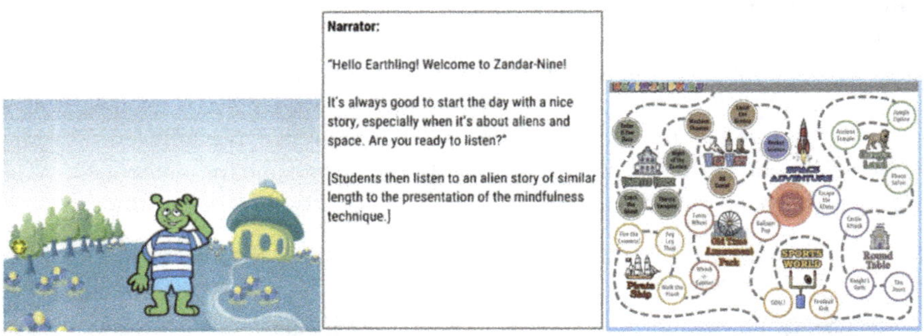

***Story* condition, each day begins with a science fiction story**

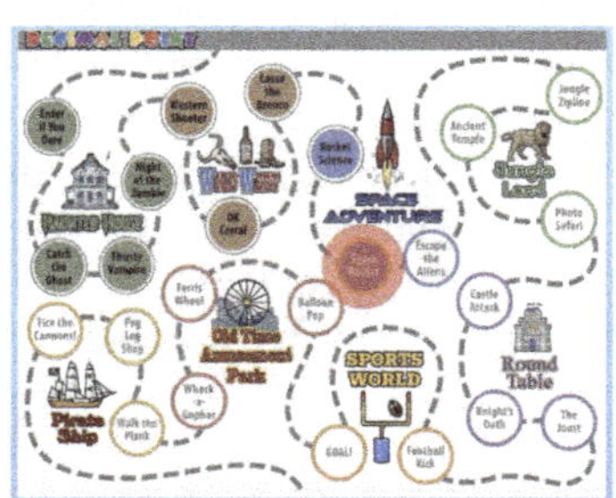

***Control* condition, students play the standard version of the *Decimal Point* game**

Fig. 9.15 Screenshots of the three conditions of Study 6 with *Decimal Point*

We hypothesized that students in the *Mindfulness* condition would learn the most, due to the expected additional benefits of mindfulness when added to game-based learning. We also hypothesized that students in the *Mindfulness* condition would take more time and make fewer errors during game play than the other two conditions.

The final analyzed set of participants in Study 6 included 166 students (90 females; 76 males) from three schools, with 53 students randomly assigned to the *Mindfulness* condition, 56 to the *Story* condition and 57 assigned to the *Control* condition. A total of 77 students were excluded from our analyses because they did not complete all of the materials.[8] Note, importantly, that students were randomly assigned to condition, meaning that every class would have a mix of students in all three conditions. (We return to this point in the discussion of the next study, Study 6a.)

The results of Study 6 found no differences in learning outcomes across the three conditions (neither pre-to-posttest nor pre-to-delayed posttest), time spent on the game, or error rates while playing. In other words, our hypothesis of the benefits of mindfulness was not confirmed, i.e., embedding mindfulness prompts within the game did not enhance learning nor change students' gameplay behaviors. Thus, at least this particular study suggests that a mindfulness induction does not enhance learning within digital learning games. Alternatively, we may not have successfully induced a state of mindfulness in the students; we explored this topic in the next study, Study 6a.

9.4.6.1 Study 6a: Mindfulness Induction When Learning with an Online Game, with a Manipulation Check to Test for the Impact of the Mindfulness Induction

Because we were unsure whether our online approach to inducing mindfulness in Study 6 had the desired effect, we ran another study—Study 6a, conducted in the spring of 2022—in an attempt to replicate the findings of Study 6 but to examine whether we had, in fact, induced mindfulness in students (Bereczki, et al., 2024). Thus, besides the question we had already explored in Study 6, we also explored the following question:

Study 6a, RQ: Did we manage to induce mindfulness in students in the mindfulness condition?

We hypothesized that students in the *Mindfulness* condition would report higher state mindfulness immediately after the mindfulness manipulation than those in the *Story* and *Control* conditions. The materials and procedures of Study 6a were the same as in Study 6, except that at the beginning of each game session, after students in the *Mindfulness* and *Story* conditions engaged in the initial mindfulness manipulation and heard a story, respectively, the students completed a state mindfulness measure.

[8] Note that the final population of students reported in Bereczki, et al., 2024 – 227 – is larger than what is reported here and in Nguyen et al., 2022b – 166. This is because Bereczki, et al., 2024 applied a less stringent exclusion criteria: students were excluded from the analyses if they did not complete at least 80% of the intervention game (versus 100% completion of pretest, intervention, posttest, and delayed posttest, as reported in Nguyen et al., 2022b).

Students in the *Control* condition did not have any intro procedure, so they completed the state mindfulness measure at the beginning of each of their game sessions. The state mindfulness check was measured with a 5-item scale adapted from the MAAS-A (Brown et al., 2011), so that statements would reflect students' experience at the moment. Example items of the scale include: "Right now I find it difficult to stay focused on what's happening." or "Right now I'm doing things automatically, without being aware of what I'm doing." Items were answered on a seven-point scale (1 = Not at all; 7 = Very much so).

Study 6a was also conducted in 5th and 6th grade classrooms across 3 additional public chools. A total of 193 students originally participated in the study, but 16 were excluded from the analyses because they did not complete at least 80% of the games. Thus, the final sample included 177 students (86 females, 91 males), with 62 students randomly assigned to the *Mindfulness* condition, 61 to the *Story* condition and 54 to the *Control* condition.

Similar to our results in Study 6, we found no evidence that students in the *Mindfulness* condition learned more from pretest to posttest or from pretest to delayed posttest than those in the other two conditions. We also found no difference in problem-solving duration and errors made among the three conditions. We did find a marginally significant condition effect on correctness after reminder between the *Mindfulness* and *Story* conditions: Students in the *Mindfulness* condition made more correct steps after reminders than those in the *Story* condition. Finally, a univariate ANOVA showed no significant effect of condition on students' state mindfulness after inductions (*Mindfulness* or *Story*) or at the beginning of the game sessions in the *Control* condition, $F(2, 174) = 0.51$, $p = 0.60$, $\eta p^2 = 0.006$. Also, neither of the planned comparisons were significant: *Control* vs. rest ($p = 0.65$) and *Story* vs. *Mindfulness* treatment ($p = 0.37$). These results show that we did not manage to induce mindfulness.

In conclusion, the lack of a mindfulness effect in both Study 6 and Study 6a may be due to the classroom context. First, we conducted mindfulness as an online, self-guided activity, as opposed to the more common instructor-led group activity. It is also likely that the presence of classmates who were engaging with different versions of the game—recall that *Mindfulness*, *Story*, and *Control* students were mixed together in classrooms—introduced distractions that may have reduced mindfulness. It may also have been that students were self-conscious about closing their eyes and following the mindfulness instruction. Given these possibilities, we don't see our findings as conclusive with respect to whether mindfulness can enhance learning with a digital learning game; further research is needed, with changes made to the way mindfulness is induced.

9.4.7 The Gender Effect: A Replication Across Multiple Studies

As mentioned in the discussion of Study 1, we became interested in whether girls or boys benefited more from playing *Decimal Point*. This interest arose from our knowledge of the gap between girls and boys in math achievement (Breda et al., 2018; Wai et al., 2010) and a desire to lessen this gap, at least in a small way, with our learning game. The question we asked ourselves was:

> RQ across studies: "Do female students benefit more, less, or the same as their male counterparts playing the game?" (Nguyen et al, 2022a)

The math gender gap may be attributed to *stereotype threat,* in which reminders of social group stereotypes can impact the behavior and performance of members of that group (Spencer et al., 1999). Despite a reduction in gender-based differences in math achievement over recent decades (Lindberg et al., 2010; Reardon et al., 2019), early-emerging stereotypes, such as the perception that males excel in math, can persist from childhood through adulthood (Cvencek et al., 2011; Furnham et al., 2002; Nosek et al., 2002; Passolunghi et al., 2014). Consequently, these perceptions may impact the performance of female students in mathematics and influence their interests and, eventually, their career choices (Adams et al., 2019; Bian et al., 2017; Ochsenfeld, 2016). Ultimately, addressing stereotype threat involves the complex task of promoting self-efficacy, interest, and achievement among female students, while simultaneously mitigating math anxiety and stereotype threat.

Through data from six of our *Decimal Point* studies—in particular, Studies 1, 2, 2a, 3, 4, and 5—involving approximately 1,100 students, we identified a consistent gender effect that was first seen in Study 1 and then replicated across five other studies: male students tended to do better than female students at pretest, while female students tended to learn more from the game, catching up to their male counterparts by posttest. (The first 4 of these gender effect studies, involving more than 600 students, are reported in Nguyen et al, 2022a). In addition, female students were more careful in answering the self-explanation questions, which significantly mediated the relationship between gender and learning gains in two of the first four studies (Nguyen et al, 2022a). More specifically, we found that female students made less errors and "gamed" the self-explanation step of *Decimal Point* mini-games significantly less than male students, resulting in more learning for female students, less for male students (Baker et al., 2024). These findings show that digital learning games, in combination with prompted self-explanation, can be effective tools for bridging the gender gap in middle school math education, which in turn could lead to the design and development of more personalized and inclusive learning games. Given the complexity of gender and the need to conduct research that goes beyond a binary approach to gender (Hyde et al., 2019), we are currently conducting research that measures multiple dimensions of gender, including gender identity, gender typicality, and gender-typed interests, activities, and traits (Hyde et al., 2019), to understand which aspects of gender explain the differences we have observed in learning behaviors and outcomes (Liben & Bigler, 2002). Preliminary results suggest that this multidimensional approach of using gender-typed scales may better explain students' feelings toward and preferences about digital learning games than the binary gender (Nguyen et al., 2023c).

9.5 Key Take-Aways: Digital Learning Game Findings

The decade-long research program that the McLearn Lab has conducted with the *Decimal Point* learning game has led to some important, some not so important, but always intriguing learning science results. The wide variety of studies, all conducted with *Decimal Point* as the centerpiece, has afforded the opportunity to investigate many and varied issues. In this section I will highlight the most noteworthy findings of the McLearn Lab's research program with *Decimal Point*.

Our most *fundamental* research finding, from Study 1 and reported in McLaren et al. (2017a), uncovered that digital learning games can surpass conventional online methods in improving engagement and learning outcomes. Prior to our Study 1, educational technology research had presented mixed results regarding the comparative advantages of learning games for mathematics and more traditional learning technologies (Mayer, 2014). Thus, given the state of game-based learning science as of the publication of our seminal 2017 paper, this was an important and novel finding.

Our most *robust* finding has been that female students have exhibited greater learning gains from the *Decimal Point* game as compared to their male counterparts. This finding, again, first found in Study 1 and then replicated across five other studies—Studies 2, 2a, 3, 4, and 5—featuring diverse versions of the *Decimal Point* game, serves as the focal point of my student Huy Nguyen's PhD thesis and is extensively discussed in Nguyen et al., 2022a. That paper covers the first 4 of the gender effect studies. We continue to pursue this issue in our most recent studies, including two that have not yet been published. For one of those studies, we created a new game, *Ocean Adventure*, which has exactly the same decimal content and instruction as *Decimal Point*, but with an entirely new, masculine-oriented narrative (see Fig. 9.16), which we designed based on a survey conducted with 333 students, designed to probe the preferences of male and female students (Nguyen et al., 2023c). The goal was to see whether boys would be more engaged in the new game and thus learn as much, or more than girls. While there was some evidence that boys were more engaged in the new game, they did not learn more. Ultimately, we hope an important practical outcome that will emerge from this line of inquiry will be the identification of game-based learning guidelines for alleviating the stereotype threat in female students, thus resulting in better math learning outcomes—and eventually better career prospects—for female students.

Perhaps our most *surprising* finding in this line of research—although the oft-replicated gender effect would also be a good choice—emerged from Study 4 and McLaren et al. (2022b) where we reported that hints and error messages within

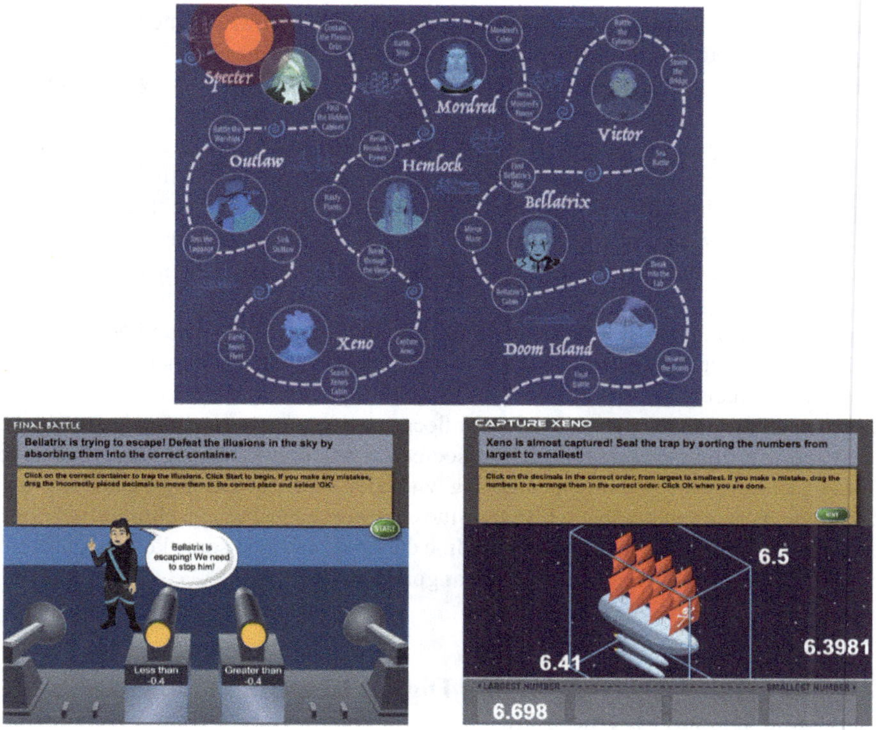

Fig. 9.16 Screenshot of the new *Ocean Adventure* learning game, along with screenshots of two of its mini-games. This game was specifically designed to be more male-oriented, according to a survey we conducted with over 300 middle school students

a digital learning game do *not* necessarily enhance learning outcomes. Given the fundamental use of hints and error feedback in other forms of educational technology, most especially intelligent tutoring systems (VanLehn, 2006, 2011; Xu et al., 2019), this was particularly unexpected. Our conjecture is that the interruption of game flow (Czikszentmihalyi, 1990) through hints and feedback may have led to unintended and negative learning consequences. This unexpected finding, on the other hand, provides evidence for the ICAP Framework (Chi & Wylie, 2014), suggesting that the struggle students may encounter—for instance, by not having hints and feedback to lean on—might contribute to deeper learning in the context of a digital learning game.

The most *central contribution to learning science* of this line of research comes from Study 5 and our findings reported in (McLaren et al., 2022a) regarding the advantages of prompted self-explanation within the context of game-based learning. Specifically, focused self-explanation — a form of prompted self-explanation in which students must generate their own explanations — emerged as the most effective form of self-explanation that we studied. This finding is important, as it presents a counter to the findings of Johnson and Mayer (2010) in which menu-based self-explanation led to better learning outcomes in the context of a digital learning game. They conjecture that minimizing impact to student game play and flow—as a menu-based approach surely does—led to their findings, whereas we conjecture that the constructive approach inherent in a focused, open-ended self-explanation led to productive student struggle (Chi & Wylie, 2014) and thus to our findings. Further exploring these different outcomes is an excellent direction for future studies. Perhaps most importantly, we discovered that prompted self-explanation likely holds the key to understanding the gender effect that we have found in many of our studies (Nguyen et al., 2022a).

Finally, our most *forward-looking* finding comes from Study 5a and Nguyen et al (2023b), in which we investigated the contribution that GPT could make in providing feedback to students who play and learn from *Decimal Point.* With AI, and especially large language models, providing an inflection point for how technology will be used and contribute to many aspects of society, it was important and timely for us to investigate how AI could impact learning with educational technology generally and our game more specifically. While the study we conducted was preliminary—done completely in post-hoc fashion with off-line data—it provided some key insights into how students might benefit from large language models.

9.6 Key Take-Aways: Use of a Digital Learning Game as a Research Platform

In essence, *Decimal Point* has functioned not only as a research tool but has become a more general *research platform*, pushing the boundaries of our understanding of how learning science can be effectively integrated into the design and implementation of learning games. We've discovered that a digital learning game can provide a rich environment for experimenting with many aspects of learning. The many and varied features of online games—both for learning and playing—furnish an excellent framework for systematic exploration, encompassing learning aspects such as the potential of student agency during game play, the tension between enjoyment and learning in game-based learning, and the benefits of hints and feedback in game-based learning, among other facets. We have leveraged the game as a platform for exploring all of these issues—and more.

A key to *Decimal Point* acting as a research platform has been its overall architecture and design. For instance, we've discovered that a learning game can be built with an underlying tutoring system engine and ITS principles (Aleven et al., 2016). The ITS model and approach has helped to structure instructional aspects of the game. Principles of ITSs, such as providing immediate feedback and on-demand hints, influenced the design of both the game and our studies with the game. While "gamification"—attempting to improve learners' engagement and experience with educational technology through, for instance, the inclusion of badges, points, leaderboards, and interactive playful agents (Landers & Landers, 2014; Landers et al., 2017)—is a popular approach to studying how game techniques can make learning more enjoyable and effective (Long & Aleven, 2014, 2018; Tahir et al., 2020), in this line of research we have shown what is possible when a game is built *from scratch* with underlying ITS principles, a more fundamental design approach than gamification. Essentially, we have shown that the degrees of freedom for experimentation are ultimately much wider (and arguably richer, as well) when a learning game is designed originally as a game, rather than as a gamified tutoring system.

Another important question that arose during the use of *Decimal Point* as a research platform is whether games are better suited for learning at home or in a classroom. Students are used to digital games being a fun, at-home activity. In contrast, they know that in school activities are more structured, less free and perhaps less fun. So, can we fully engage students in school with a perceived out of school activity? This is an excellent, open question. This issue came up in multiple instances over the years of experimenting with *Decimal Point.* For instance, in Studies 2 (Nguyen et al., 2018) and 2a (Harpstead et al., 2019)—in which we essentially studied the trade-offs between student autonomy and game system control, students in the classroom may not have really felt in control of their learning, due to the influence of the teacher and classroom context. Perhaps autonomy and agency would have been more greatly felt at home? We didn't have the opportunity to explicitly explore this, but it would be an interesting topic still to investigate. *Decimal Point*'s infrastructure and Internet implementation would allow for such a study. As another example, Study 4 (McLaren et al., 2022b)—the hints study that ended up being ½ conducted at school, ½ at home—was a step toward exploring this dichotomy that may lead to further contrasting studies.

A key aspect of intelligent tutoring systems, first articulated by Kurt van Lehn (2006), is the distinction between the "outer loop", in which problem ordering and selection is handled, and the "inner loop", in which student interactions within problems occur, is another way in which *Decimal Point* has acted as an excellent research platform. In our studies, student agency, indirect control, and mindfulness—all outer loop activities—did not yield significant differences between conditions. Conversely, the inner loop, which involves elements we tested such as hints and errors and self-explanation, emerged as a locus of noteworthy variations in learning outcomes. This is likely due to the learning aspect of *Decimal Point* being more prominent than the game aspect, which makes it harder for individual tweaks on the game mechanics to significantly change learning, but also makes the game "safer" and more robust to changes—we have never seen a condition that did not lead to significant pre-post learning gains.

Finally, a very interesting and important observation—since it has meaningful implications for game design and for how we should approach future game-based learning research—is that many of our interventions *did not actually show learning differences between conditions*. Our most significant learning difference was found in Study 1 when we compared the game to a conventional learning technology (McLaren et al., 2017a). There are surely different reasons for the lack of condition differences in each of the game versus game studies; this could be evidence that it is tricky to significantly alter learning outcomes by tweaking individual features of a game. This further suggests that perhaps students are more consistent in how they play learning games—or more resistant to our efforts to change their ways of playing—than we might think. This may have been due, at least in part, to the mostly-unchanging basic instructional approach of *Decimal Point* being more prominent than the game aspect. Throughout the decade of the game being used as a research platform, the basic precepts of *Decimal Point*'s instructional approach remained (a) a focus on decimal misconceptions and (b) an underlying ITS instructional approach. This surely made it difficult for individual tweaks to the game mechanics to significantly change learning. At the same time, it also likely made the game "safer" for and more robust to changes, as mentioned, we have so far not seen a condition that did not lead to significant pre-post learning gains. In short, a lesson for future game-based research platforms might be to create a more modifiable instructional component for experimentation.

9.7 Conclusions

In conclusion, I will propose a few possible future directions for the McLearn lab's continuing work with *Decimal Point* more specifically and for digital learning games research more generally. One direction that could be further explored in connection with digital learning games is the "Assistance Dilemma" (Koedinger & Aleven, 2007), reaching beyond the standard textual hints and feedback support we investigated in Study 4. The Assistance Dilemma raises the question of the trade-offs between giving and withholding help in the context of instructional technology. Giving help can move students forward who are stuck; it can also lead to shallow learning. Withholding help can push students to think and learn more deeply; it can also lead to frustration when they are truly stuck. The trade-offs in a game-based learning context may differ from other educational technology, however, given how games are intended to promote flow and engagement. Our Study 4 results, in which the students who received hints learned less, seemed to indicate that withholding help was the correct choice for learning with *Decimal Point,* perhaps because of how the particular help we provided might have disrupted student engagement. One aspect of the Assistance Dilemma that could be further investigated would be the value of using a different model of providing help than allowing students to simply request it and to receive standard textual hints. Perhaps, for instance, instead of providing on-demand hints, students could be prompted to ask for help when they have clearly demonstrated they need support. Such an approach might involve less disruption to a student's engagement with a game, yet still provide timely assistance. Another aspect of the Assistance Dilemma that could be explored—and which would fit the context of game-based learning well—would be the use of non-textual hints, such as animations (Berney & Bétrancourt, 2016; Nathan, 1998; Scheiter et al., 2010) or visual representations (Nagashima et al., 2021). Given the highly visual and engaging nature of learning games, not to mention evidence that visual models can support the learning of mathematics (Hegarty & Kozhevnikov, 1999; Luzón & Letón, 2015), animated or visual hints might provide better, easier to process, and more engaging help in digital learning games than standard textual hints.

Another intriguing avenue worthy of investigation involves the incorporation of learning from erroneous examples, which have been shown to be an effective learning technique in a variety of studies (Adams et al., 2012; Durkin & Rittle-Johnson, 2012; Grosse & Renkl, 2007; McLaren et al., 2012; Tsovaltzi et al., 2012), in the context of a learning game. Erroneous examples are worked examples of problem solving in which one or more of the steps has an error, typically a common error made by students. This is, in fact, how the McLearn Lab started this line of research with learning games (although we early on departed from this exploration). In particular, we originally set out to see if we could create a learning game around erroneous examples, which have a natural interactivity or playfulness associated with them in presenting students with the challenge of errors to fix. One could imagine a version of *Decimal Point* in which students don't (always) directly solve problems themselves but instead are challenged to find and fix errors made by the fantasy characters in a gameful way. Furthermore, providing badges and prizes to students as they manage to find and fix the errors could provide an even more gameful aspect to *Decimal Point*—and perhaps create a blueprint for a new type of learning game.

Of course, as discussed, the recent rise and huge steps forward in large language models and artificial intelligence raises some intriguing possibilities for AI applications in the context of digital learning games. In a recent book chapter (McLaren & Nguyen, 2023) we described the many ways that AI has already been used in digital learning games, including adapting game play and problems, AI-powered dashboards, educational data mining for game improvement and identifying cognitive, behavioral, and affective aspects of learning, and AI-powered non-player characters (NPCs). As discussed earlier, we have experimented with the first three of these AI approaches within the *Decimal Point* game, using older AI techniques than LLMs. LLMs present new and exciting opportunities to create and extend learning games with intelligent capabilities. Our Study 5a (Nguyen et al., 2023b) was a very promising first step toward incorporating the latest advancements in AI into digital learning games, but there are many other directions that could be pursued. For instance, a large language model could be called upon to not only provide cognitive feedback, as per our recent study, but also meta-cognitive (Hattie & Timperley, 2007) and affective (Howard, 2021) feedback, both of which are valuable to learners and for which there is extensive information on the Internet from which an LLM could generate feedback. An illuminating study would be one that compares a learning game that has manually-created feedback, the typical case, to feedback generated by an LLM. Another possibility for LLMs in the context of learning games would be replacing NPCs, which are currently implemented with earlier generation natural language processing (NLP) techniques, with LLMs. Given the superior language capabilities of LLMs, this could potentially be one of the most significant applications within learning games. A final suggestion for how LLMs could be employed in support of digital learning games is that they could be used as "helpers" in designing and developing new games. More specifically, LLMs could be used to rapidly generate new game ideas and narratives that game designers could build upon and to provide feedback on game ideas and early prototypes of game developers. Work in this direction has, in fact, already begun (Gatti Junior et al., 2023).

Finally, an important facet of *Decimal Point*—and digital games more generally—that warrants further investigation is the potential presence of unconscious bias embedded in the game's mechanics and artistic elements. The designers of learning games, who are typically and predominantly White—which is true for the designers of *Decimal Point* and cited as 78% the case for the game design industry more generally (Kumar et al., 2022)—are usually well meaning but often unaware of how their own biases frequently lead to design choices that subtly (or even overtly) create biased functionality, blatantly stereotypical game characters and environments, and player identities that turn away children of color, or more specifically, Black children (Peckham, 2020; Rankin & Henderson, 2021; Richard, 2017). In fact, *Decimal Point* has provided at least preliminary evidence of implicit bias. In a recent analysis of more than 700 students using the game, spanning three of the classroom studies reported in this chapter (i.e., Studies 5, 6, and 6a) and a new study, we found that well-represented students (White and Asian; $n = 578$) showed more engagement and less anxiety in using *Decimal Point* than under-represented students (Black, Hispanic or Latino, Indigenous, and multiracial; $n = 158$) (Ni et al., 2024). Unpacking potential biases is crucial for a nuanced understanding of how learning games can be designed and redesigned to support diverse learners. To address this, we recently proposed a project to the National Science Foundation in which we will engage 120 Black middle school students in co-design sessions with *Decimal Point* and in the analysis of 10 other STEM learning games, including *Math Blaster*, *Math Playground,* and *BrainPop*. By scrutinizing these games, and redesigning *Decimal Point* if and where necessary, we could contribute to the ongoing discourse on diversity, equity, and inclusion in digital learning games, paving the way for more informed and culturally sensitive game design practices.

The McLearn Lab's ten-year research program with *Decimal Point* has been thrilling, with some prominent successes, such as the gender effect and the self-explanation findings, but also some disappointing failures, such as the lack of impact of agency and mindfulness inductions in the context of the game. *Decimal Point* as a research platform has facilitated much of the work described in this chapter. The McLearn Lab looks forward to continuing this line of research not only with *Decimal Point*, but with two new games the lab has designed and developed: *Angle Jungle*, a game to help elementary and middle school students learn about angles that was reimplemented and extended for classroom use from a prior implementation (Khan et al., 2017) and *Ocean Adventure*, a game that is a "reskinning " of *Decimal Point* with precisely the same content and instructional approach but with a completely different narrative and art assets. The future possibilities of learning game design, development, and research are myriad, and we intend to pursue many of these possibilities with our various learning games.

Acknowledgements I have been fortunate to have had many incredible collaborators throughout the years. This work has been greatly influenced and guided by conversations and collaborations with many of these colleagues and students, including Ryan Baker, Rosta Farzan, Jodi Forlizzi, Jessica Hammer, Erik Harpstead, Xinying Hou, Richard E. Mayer, Michael Mogessie, Huy Nguyen, Jaclyn Ocumpaugh, Eleanor O'Rourke, Liz Richey, Jon Star, and Yeyu Wang. Thank you to Ilona Buchem, Imogene Casebourn, Huy Nguyen, Liz Richey, and Rupert Wegerif for reviewing this chapter. I would especially like to thank Huy Nguyen and Liz Richey who have been my closest and most significant collaborators throughout the many years of this research. ChatGPT 4.0 was used both to draft and to revise some paragraphs of the chapter, as well as to collect relevant information about learning games and past research. The work has been supported by three National Science Foundation grants (NSF Award #s 1238619, 1661121, and 2201796). The opinions expressed are those of the author and do not represent the views of NSF.

References

Adams, D., McLaren B.M., Durkin, K., Mayer, R.E., Rittle-Johnson, B., Isotani, S., & Van Velsen, M. (2012). Erroneous examples versus problem solving: Can we improve how middle school students learn decimals? In *Proceedings of the 34th Meeting of the Cognitive Science Society (CogSci 2012)*. Sapporo, Japan: Cognitive Science Society, pp. 1260–1265.

Adams, D., McLaren, B.M., Mayer, R.E., Goguadze, G., & Isotani, S. (2013). Erroneous examples as desirable difficulty. In Lane, H.C., Yacef, K., Mostow, J., & Pavlik, P. (Eds.). *Proceedings of the 16th International Conference on Artificial Intelligence in Education (AIED 2013)*. LNCS 7926. Springer, Berlin, pp. 803–806.

Adams, R.B., Barber, B.M., & Odean, T. (2019). The math gender gap and women's career outcomes. Available at SSRN 2933241.

Adams, D. M., & Clark, D. B. (2014). Integrating self-explanation functionality into a complex game environment: Keeping gaming in motion. *Computers and Education, 73*, 149–159.

Ahmadi, A., Noetel, M., Parker, P., Ryan, R. M., Ntoumanis, N., Reeve, J., Beauchamp, M., Dicke, T., Yeung, A., Ahmadi, M., Bartholomew, K., Chiu, T., Curran, T., Erturan, G., Flunger, B., Frederick, C., Froiland, J. M., González-Cutre, D., Haerens, L., ... Lonsdale, C. (2023). A classification system for teachers' motivational behaviors recommended in self-determination theory interventions. *Journal of Educational Psychology*. https://doi.org/10.1037/edu0000783

Ainsworth, S., & Burcham, S. (2007). The impact of text coherence on learning by self-explanation. *Learning and Instruction, 17*(3), 286–303.

Aleven, V.A.W.M.M., Koedinger, K. R., & Popescu, O. (2003). A tutorial dialog system to support self-explanation: Evaluation and open questions. In *Proceedings of the 11th International Conference on Artificial Intelligence in Education*, pp. 39–46.

Aleven, V. A. W. M. M., & Koedinger, K. R. (2002). An effective metacognitive strategy: Learning by doing and explaining with a computer-based Cognitive Tutor. *Cognitive Science, 26*(2), 147–179. https://doi.org/10.1016/S0364-0213(02)00061-7

Aleven, V., McLaren, B. M., Sewall, J., & Koedinger, K. R. (2009). A new paradigm for intelligent tutoring systems: Example-tracing tutors. *International Journal of Artificial Intelligence in Education, 19*(2), 105–154. https://doi.org/10.1109/DIGITEL.2010.55

Aleven, V., McLaren, B. M., Sewall, J., van Velsen, M., Popescu, O., Demi, S., Ringenberg, M., & Koedinger, K. R. (2016). Example-tracing tutors: Intelligent tutor development for non-programmers. *International Journal of Artificial Intelligence in Education, 26*(1), 224–269. https://doi.org/10.1007/s40593-015-0088-2

Aleven, V., McLaughlin, E. A., Glenn, R. A., & Koedinger, K. R. (2017). Instruction based on adaptive learning technologies. In R. E. Mayer & P. Alexander (Eds.), *Handbook of research on learning and instruction* (2nd ed., pp. 522–560). Routledge.

Aleven, V., Roll, I., McLaren, B. M., & Koedinger, K. R. (2010). Automated, unobtrusive, action-by-action, assessment of self-regulation during learning with an intelligent tutoring system. *Educational Psychologist, 45*(4), 224–233. https://doi.org/10.1080/00461520.2010.517740

Atkinson, R. K., Derry, S. J., Renkl, A., & Wortham, D. (2000). Learning from examples: Instructional principles from the worked examples research. *Review of Educational Research, 70*(2), 181–214.

Baker, R.S., Richey, J.E., Zhang, J., Karumbaiah, S., Andres-Bray, J.M., Nguyen, H.A., Andres, J.M.A.L., & McLaren, B.M. (2024). Gaming the system mediates the relationship between gender and learning outcomes in a digital learning game. *Instructional Science*. DOI: 10.1007/s11251-024-09679-3

Baker, R., Scruggs, R., Pavlik, P. I., McLaren, B. M., & Liu, Z. (2023). How well do contemporary knowledge tracing algorithms predict the knowledge carried out of a digital learning game? *Educational Technology Research & Development*. https://doi.org/10.1007/s11423-023-10218-z

Bard, G.V. (2007), Spelling-error tolerant, order-independent pass-phrases via the Damerau–Levenshtein string-edit distance metric, *Proceedings of the Fifth Australasian Symposium on ACSW Frontiers : 2007, Ballarat, Australia, January 30 - February 2, 2007*, Conferences in Research and Practice in Information Technology, vol. 68, Darlinghurst, Australia: Australian Computer Society, Inc., pp. 117–124, ISBN 978-1-920682-49-1.

Bauckhage, C. (2015). Numpy/scipy recipes for data science: k-medoids clustering. https://doi.org/10.13140/2.1.4453.2009.

Benton, L., Mavrikis, M., Vasalou, A., Joye, N., Sumner, E., Herbert, E., Revesz, A., Symvonis, A., & Raftopoulou, C. (2021). Designing for "challenge" in a large-scale adaptive literacy game for primary school children. *British Journal of Educational Technology, 52*, 1862–1880. https://doi.org/10.1111/bjet.13146

Bereczki, E., Takacs, Z. K., Richey, J. E., Nguyen, H., Mogessie, M., & McLaren, B. M. (2024). Mindfulness in a digital math learning game: Insights from two randomized controlled trials. *Journal of Computer Assisted Learning*. https://doi.org/10.1111/jcal.12971

Berney, S., & Bétrancourt, M. (2016). Does animation enhance learning? A meta-analysis. *Computers & Education, 101*, 150–167. https://doi.org/10.1016/j.compedu.2016.06.005

Bian, L., Leslie, S.-J., & Cimpian, A. (2017). Gender stereotypes about intellectual ability emerge early and influence children's interests. *Science, 355*(6323), 389–391.

Bodily, R., Kay, J., Aleven, V., Jivet, I., Davis, D., Xhakaj, F., & Verbert, K. (2018). Open learner models and learning analytics dashboards: a systematic review. In: *Proceedings of the 8th International Conference on Learning Analytics and Knowledge*, pp. 41–50.

Bouvier, P., Lavoué, E., Sehaba, K., & George, S. (2013). Identifying learner's engagement in learning games: A qualitative approach based on learner's traces of interaction. In *5th International Conference on Computer Supported Education (CSEDU 2013)*, May 2013, Aachen, Germany, pp. 339–350.

Bransford, J. D., Brown, A. L., & Cocking, R. R. (Eds.). (2000). *How people learn: Brain, mind, experience, and school*. National Academy Press.

Breda, T., Jouini, E., & Napp, C. (2018). Societal inequalities amplify gender gaps in math. *Science, 359*(6381), 1219–1220.

Brown, K. W., West, A. M., Loverich, T. M., & Biegel, G. M. (2011). Assessing adolescent mindfulness: Validation of an adapted mindful attention awareness scale in adolescent normative and psychiatric populations. *Psychological Assessment, 23*(4), 1023–1033. https://doi.org/10.1037/a0021338

Brueckner, L. J. (1928). Analysis of difficulties in decimals. *Elementary School Journal, 29*, 32–41.

Bull, S. (2020). There are open learner models about! *IEEE Transactions on Learning Technologies, 13*(2), 425–448. https://doi.org/10.1109/TLT.2020.2978473

Carvalho, P. F., & Goldstone, R. L. (2015). The benefits of interleaved and blocked study: Different tasks benefit from different schedules of study. *Psychonomic Bulletin & Review, 22*(1), 281–288. https://doi.org/10.3758/s13423-014-0676-4. PMID: 24984923.

Chang, K.-E., Wu, L.-J., Weng, S.-E., & Sung, Y.-T. (2012). Embedding game-based problem-solving phase into problem-posing system for mathematics learning. *Computers & Education, 58*, 775–786.

Cheng, M. T., Rosenheck, L., Lin, C. Y., & Klopfer, E. (2017). Analyzing gameplay data to inform feedback loops in The Radix Endeavor. *Computers & Education, 111*, 60–73.

Chi, M. T. H., Bassok, M., Lewis, M. W., Reimann, R., & Glaser, R. (1989). Self explanations: How students study and used examples in learning to solve problems. *Cognitive Science, 13*, 145–182.

Chi, M. T. H., DeLeeuw, N., Chiu, M.-H., & LaVancher, C. (1994). Eliciting self-explanations improves understanding. *Cognitive Science, 25*(4), 471–533.

Chi, M. T., & Wylie, R. (2014). The ICAP framework: Linking cognitive engagement to active learning outcomes. *Educational Psychologist, 49*(4), 219–243.

Clark, D. B., Tanner-Smith, E., & Killingsworth, S. (2016). Digital games, design, and learning: A systematic review and meta-analysis. *Review of Educational Research, 86*(1), 79–122. https://doi.org/10.3102/0034654315582065

Conati, C., & VanLehn, K. (2000). Toward computer-based support of meta-cognitive skills: A computational framework to coach self-explanation. *International Journal of Artificial Intelligence in Education, 11*, 398–415.

Corbett, A. T., & Anderson, J. R. (1994). Knowledge tracing: Modeling the acquisition of procedural knowledge. *User Modeling and User-Adapted Interaction, 4*(4), 253–278.

Cordova, D. I., & Lepper, M. R. (1996). Intrinsic motivation and the process of learning: Beneficial effects of contextualization, personalization, and choice. *Journal of Educational Psychology, 88*(4), 715.

Cragg, L., & Gilmore, C. (2014). Skills underlying mathematics: The role of executive function in the development of mathematics proficiency. *Trends in Neuroscience and Education, 3*(2), 63–68. https://doi.org/10.1016/j.tine.2013.12.001

Cvencek, D., Meltzoff, A. N., & Greenwald, A. G. (2011). Math–gender stereotypes in elementary school children. *Child Development, 82*(3), 766–779.

Czikszentmihalyi, M. (1975). *Beyond boredom and anxiety*. Jossey Bass.

Czikszentmihalyi, M. (1990). *Flow: The psychology of optimal experience*. Harper & Row.

Darling-Hammond, L., Flook, L., Cook-Harvey, C., Barron, B., & Osher, D. (2020). Implications for educational practice of the science of learning and development. *Applied Developmental Science, 24*(2), 97–140.

Deci, E. L. (1975). *Intrinsic motivation*. Plenum Press.

Deci, E. L., & Ryan, R. M. (1985). *Intrinsic Motivation and Self-Determination in Human Behavior*. Springer Science & Business Media. https://doi.org/10.1007/978-1-4899-2271-7

Dunning, D. L., Griffiths, K., Kuyken, W., Crane, C., Foulkes, L., Parker, J., & Dalgleish, T. (2019). Research review: The effects of mindfulness-based interventions on cognition and mental health in children and adolescents–a meta-analysis of randomized controlled trials. *Journal of Child Psychology and Psychiatry., 60*, 244–258.

Dunning, D., Tudor, K., Radley, L., Dalrymple, N., Funk, J., Vainre, M., Ford, T., Montero-Marin, J., Kuyken, W., & Dalgleish, T. (2022). Do mindfulness-based programmes improve the cognitive skills, behaviour and mental health of children and adolescents? An updated meta-analysis of randomised controlled trials. *Evidence Based Mental Health, 25*(3), 135–142. https://doi.org/10.1136/ebmental-2022-300464

Durkin, K., & Rittle-Johnson, B. (2012). The effectiveness of using incorrect examples to support learning about decimal magnitude. *Learning and Instruction, 22*, 206–214. https://doi.org/10.1016/j.learninstruc.2011.11.001

Dwyer, C., & Johnson, L. (1997). Grades, accomplishments, and correlates. In W. Willingham & N. Cole (Eds.), *Gender and Fair Assessment* (pp. 127–156). Erlbaum.

Else-Quest, N. M., Mineo, C. C., & Higgins, A. (2013). Math and science attitudes and achievement at the intersection of gender and ethnicity. *Psychology of Women Quarterly, 37*(3), 293–309.

Entwisle, D. R., Alexander, K. L., & Olson, L. S. (1997). *Children, Schools, and Inequality.* Westview Press.

Erhel, S., & Jamet, E. (2013). Digital game-based learning: Impact of instructions and feedback on motivation and learning effectiveness. *Computers & Education, 67*, 156–167. https://doi.org/10.1016/j.compedu.2013.02.019.

Forlizzi, J., McLaren, B. M., Ganoe, C. H., McLaren, P. B., Kihumba, G., & Lister, K. (2014). Decimal Point: Designing and developing an educational game to teach decimals to middle school students. In Busch, C. (Ed.) *Proceedings of the 8th European Conference on Games Based Learning (ECGBL-2014)*. Academic Conferences and Publishing International Limited, Reading, U.K, pp. 128–135.

Furnham, A., Reeves, E., & Budhani, S. (2002). Parents think their sons are brighter than their daughters: Sex differences in parental self-estimations and estimations of their children's multiple intelligences. *The Journal of Genetic Psychology, 163*(1), 24–39.

Gatti Junior, W., Marasco, E., Kim, B., Behjat, L., & Eggermont, M. (2023). How ChatGPT can inspire and improve serious board game design. *International Journal of Serious Games, 10*(4), 33–54. https://doi.org/10.17083/ijsg.v10i4.645

Gee, J. P. (2007). *Good video games and good learning Collected essays on video games, learning and literacy*, 2nd Edition. Peter Lang International Academic Publishers. https://doi.org/10.3726/978-1-4539-1162-4.

Gee, J. P. (2003). *What video games have to teach us about learning and literacy.* Palgrave/Macmillian.

Glasgow, R., Ragan, G., Fields, W. M., Reys, R., & Wasman, D. (2000). The decimal dilemma. *Teaching Children Mathematics, 7*(2), 89–93.

Graeber, A., & Tirosh, D. (1988). Multiplication and division involving decimals: Preservice elementary teachers' performance and beliefs. *Journal of Mathematics Behavior, 7*, 263–280.

Grosse, C. S., & Renkl, A. (2007). Finding and fixing errors in worked examples: Can this foster learning outcomes? *Learning and Instruction, 17*, 612–634. https://doi.org/10.1016/j.learninstruc.2007.09.008

Güldal, Ş., & Satan, A. (2020). The effect of mindfulness-based psychoeducation program on adolescents' character strengths, mindfulness and academic achievement. *Current Psychology*, pp. 1–12.

Habgood, M. P. J., & Ainsworth, S. E. (2011). Motivating children to learn effectively: Exploring the value of intrinsic integration in educational games. *Journal of the Learning Sciences, 20*(2), 169–206. https://doi.org/10.1080/10508406.2010.508029

Harpstead, E., Richey, J.E., Nguyen, H., & McLaren, B. M. (2019). Exploring the subtleties of agency and indirect control in digital learning games. In *Proceedings of the 9th International Conference on Learning Analytics & Knowledge (LAK'19)*, pp. 121–129). ACM. https://doi.org/10.1145/3303772.3303797.

Hattie, J., & Timperley, H. (2007). The power of feedback. *Review of Educational Research, 77*, 81–11.

Hegarty, M., & Kozhevnikov, M. (1999). Types of visual–spatial representations and mathematical problem solving. *Journal of Educational Psychology, 91*, 684–689. https://doi.org/10.1037/0022-0663.91.4.684

Hidi, S., & Renninger, K. A. (2006). The four-phase model of interest development. *Educational Psychologist, 41*(2), 111–127. https://doi.org/10.1207/s15326985ep4102_4

Hiebert, J., & Wearne, D. (1985). A model of students' decimal computation procedures. *Cognition and Instruction, 2*, 175–205.

Honey, M.A., & Hilton, M.L. (2011). Learning science through computer games and simulations. The National Academies Press. (http://www.nap.edu/openbook.php?record_id=13078&page=R1).

Hooshyar, D., Malva, L., Yang, Y., Pedaste, M., Wang, M., & Lim, H. (2021). An adaptive educational computer game: Effects on students' knowledge and learning attitude in computational thinking. *Computers in Human Behavior, 114.*

Hou, X., Nguyen, H.A., Richey, J.E., & McLaren, B.M. (2020). Exploring how gender and enjoyment impact learning in a digital learning game. In: Bittencourt, I., Cukurova, M., Muldner, K., Luckin, R., & Millán, E. (Eds.) *Proceedings of the 21st International Conference on Artificial Intelligence in Education. AIED 2020*. Lecture Notes in Computer Science (LNCS, Vol. 12163). Springer, Cham. https://doi.org/10.1007/978-3-030-52237-7_21.

Hou, X., Nguyen, H. A., Richey, J. E., Harpstead, E., Hammer, J., & McLaren, B. M. (2022). Assessing the effects of open models of learning and enjoyment in a digital learning game. *International Journal of Artificial Intelligence in Education., 32*, 120–150. https://doi.org/10.1007/s40593-021-00250-6

Howard, N. R. (2021). "How did i do?": Giving learners effective and affective feedback. *Educational Technology Research and Development, 69*(1), 123–126. https://doi.org/10.1007/s11423-020-09874-2

Hsu, C.-Y., & Tsai, C.-C. (2011). Investigating the impact of integrating self-explanation into an educational game: A pilot study. In Edutainment tech, pp. 250–254.

Huang, X., Zhang, J., & Hudson, L. (2019). Impact of math self-efficacy, math anxiety, and growth mindset on math and science career interest for middle school students: The gender moderating effect. *European Journal of Psychology of Education, 34*(3), 621–640.

Hussein, M. H., Ow, S. H., Elaish, M. M., & Jensen, E. O. (2022). Digital game-based learning in K-12 mathematics education: A systematic literature review. *Education and Information Technologies, 27*(2), 2859–2891. https://doi.org/10.1007/s10639-021-10721-x

Hyde, J. S., Bigler, R. S., Joel, D., Tate, C. C., & van Anders, S. M. (2019). The future of sex and gender in psychology: Five challenges to the gender binary. *American Psychologist, 74*(2), 171–193.

Irwin, K. C. (2001). Using everyday knowledge of decimals to enhance understanding. *Journal for Research in Mathematics Education, 32*(4), 399–420.

Isotani, S., McLaren, B. M., & Altman, M. (2010). Towards intelligent tutoring with erroneous examples: A taxonomy of decimal misconceptions. In *Proceedings of the 10th International Conference on Intelligent Tutoring Systems (ITS-10)*, Lecture Notes in Computer Science, 6094. Berlin: Springer, pp. 346–348.

Johnson, C. I., & Mayer, R. E. (2010). Applying the self-explanation principle to multimedia learning in a computer-based game-like environment. *Computers in Human Behavior, 26*(6), 1246–1252.

Johnston, K. (2021). Engagement and immersion in digital play: Supporting young children's digital wellbeing. *International Journal of Environmental Research and Public Health., 18*(19), 10179. https://doi.org/10.3390/ijerph181910179. PMID: 34639481 PMCID: PMC8507672.

Juraschka, R. (2019). How digital game-based learning improves student success. https://www.prodigygame.com/main-en/blog/digital-game-based-learning/

Kafai, Y. B. (1996). Learning design by making games: Children's development of strategies in the creation of a complex computational artifact. In Y. B. Kafai & M. Resnick (Eds.), *Constructionism in practice: Designing, thinking and learning in a digital world* (pp. 71–96). Erlbaum.

Khan, J., Wang, J., Wang, X., Zhang, Y., Hammer, J., Stevens, S., & Washington, R. (2017). Angle jungle: An educational game about angles. In *Extended Abstracts Publication of the Annual Symposium on Computer-Human Interaction in Play*, pp. 633–638.

Koedinger, K. R., & Aleven, V. (2007). Exploring the assistance dilemma in experiments with cognitive tutors. *Educational Psychology Review, 19*, 239–264.

Koedinger, K. R., Baker, R. S. J. D., Cunningham, K., Skogsholm, A., Leber, B., & Stamper, J. (2010). A Data Repository for the EDM community: The PSLC DataShop. In C. Romero, S. Ventura, M. Pechenizkiy, & R. S. J. D. Baker (Eds.), *Handbook of Educational Data Mining*. CRC Press.

Kumar, S., Kwan, E., Weststar, J., & Coppins, T. (2022). *International Game Developers Association (IGDA), Developer Satisfaction Survey 2021: Diversity in the Game*

Industry Report. https://igda-website.s3.us-east-2.amazonaws.com/wp-content/uploads/2022/11/15161607/IGDA-DSS-2021-Diversity-Report_Final.pdf.

Landers, R., Armstrong, M., & Collmus, A. (2017). How to use game elements to enhance learning: Applications of the theory of gamified learning. *Serious Games and Edutainment Applications.* https://doi.org/10.1177/1046878114563660

Landers, R., & Landers, A. (2014). An empirical test of the theory of gamified learning: The effect of leaderboards on time-on-task and academic performance. *Simulation & Gaming., 45*(6), 769–785. https://doi.org/10.1177/1046878114563662

Liben, L. S., & Bigler, R. S. (2002). The developmental course of gender differentiation: Conceptualizing, measuring, and evaluating constructs and pathways. *Monographs of the Society for Research in Child Development, 67*(2), vii–147. https://doi.org/10.1111/1540-5834.t01-1-00187

Lindberg, S. M., Hyde, J. S., Petersen, J. L., & Linn, M. C. (2010). New trends in gender and mathematics performance: A meta-analysis. *Psychological Bulletin, 136*(6), 1123.

Loderer, K., Pekrun, R., & Plass, J. L. (2019). Emotional foundations of game-based learning. In J. L. Plass, R. E. Mayer, & B. D. Homer (Eds.), *Handbook of Game-Based Learning* (pp. 111–151). MIT.

Lomas, D., Patel, K., Forlizzi, J.L., & Koedinger, K.R. (2013). Optimizing challenge in an educational game using large-scale design experiments. In *Proceedings of the SIGCHI Conference on Human Factors in Computing Systems*, pp. 89–98.

Long, Y., & Aleven, V. (2014). Gamification of joint student/system control over problem selection in a linear equation tutor. In: Trausan-Matu, S., Boyer, K.E., Crosby, M., Panourgia, K. (Eds.) *Intelligent Tutoring Systems. ITS 2014.* Lecture Notes in Computer Science, vol. 8474. Springer, Cham. https://doi.org/10.1007/978-3-319-07221-0_47.

Long, Y., & Aleven, V. (2018). Educational game and intelligent tutoring system: A classroom study and comparative design analysis. *Proceedings of CHI.* https://doi.org/10.1145/3057889

Luzón, J. M., & Letón, E. (2015). Use of animated text to improve the learning of basic mathematics. *Computers & Education, 88*, 119–128. https://doi.org/10.1016/j.compedu.2015.04.016

Malone, T. W. (1981). Toward a theory of intrinsically motivating instruction. *Cognitive Science, 5*, 333–369. https://doi.org/10.1207/s15516709cog0504_2

Malone, T. W., & Lepper, M. R. (1987). Making learning fun: A taxonomy of intrinsic motivations for learning. *Aptitude, Learning, and Instruction, 3*(1987), 223–253.

Mayer, R.E., & Johnson, C.I. (2010). Adding instructional features that promote learning in a game-like environment. *Journal of Educational Computing Research, 42*(3), 241–265. https://journals.sagepub.com/doi/https://doi.org/10.2190/EC.42.3.a.

Mayer, R. E. (2014). Computer games for learning: An evidence-based approach. Cambridge, MA: MIT Press. ISBN: 9780262027571

Mayer, R. E. (2019). Computer games in education. *Annual Review of Psychology, 70*, 531–549.

Maynard, B. R., Solis, M. R., Miller, V. L., & Brendel, K. E. (2017). Mindfulness-based interventions for improving cognition, academic achievement, behavior, and socioemotional functioning of primary and secondary school students. *Campbell Systematic Reviews., 13*, 1–144.

McLaren, B.M., Lim, S., & Koedinger, K.R. (2008). When and how often should worked examples be given to students? New results and a summary of the current state of research. In Love, B. C., McRae, K. & Sloutsky, V. M. (Eds.), *Proceedings of the 30th Annual Conference of the Cognitive Science Society.* Austin, TX: Cognitive Science Society. pp. 2176–2181.

McLaren, B.M., Adams, D., Durkin, K., Goguadze, G. Mayer, R.E., Rittle-Johnson, B., Sosnovsky, S., Isotani, S., & Van Velsen, M. (2012). To err is human, to explain and correct is divine: A study of interactive erroneous examples with middle school math students. In Ravenscroft, A., Lindstaedt, S., Delgado Kloos, C. & Hernández-Leo, D. (Eds.), *Proceedings of EC-TEL 2012: Seventh European Conference on Technology Enhanced Learning,* LNCS 7563. Springer, Berlin, pp. 222–235.

McLaren, B.M., Timms, T., Weihnacht, D., Brenner, D., Luttgen, K., Grillo-Hill, A., & Brown, D.H. (2014a). A web-based system to support inquiry learning: Towards determining how much assistance students need. In Zvacek, S., Restivo, M.T., Uhomoibhi, J. and Helfert, M. (Eds.) *Proceedings of the Sixth International Conference on Computer-Supported Education (CSEDU-2014).* SCITEPRESS – Science and Technology Publications. Vol. 1, pp. 43–52.

McLaren, B.M., van Gog, T., Ganoe, C., Yaron, D. & Karabinos, M. (2014b) Exploring the assistance dilemma: Comparing instructional support in examples and problems. In Trausan-Matu, S. et al. (Eds.) *Proceedings of the Twelfth International Conference on Intelligent Tutoring Systems* (ITS-2014). LNCS 8474. Springer International Publishing Switzerland, pp. 354–361.

McLaren, B. M., Farzan, R., Adams, D. M., Mayer, R. E., & Forlizzi, J. (2017b). Uncovering gender and problem difficulty effects in learning with an educational game. In André, E., Baker, R., Hu, X., Rodrigo, M.M.T., & du Boulay, B. (Eds.). In *Proceedings of the 18th International Conference on Artificial Intelligence in Education (AIED 2017).* LNAI 10331. Springer: Berlin, pp. 540–543.

McLaren, B.M., Richey, J.E., Nguyen, H.A., & Mogessie, M. (2022c). A digital learning game for mathematics that leads to better learning outcomes for female students: Further evidence. In: *Proceedings of the 16th European Conference on Game Based Learning (ECGBL 2022).* pp. 339–348.

McLaren, B.M., Nguyen, H.A., Richey, J.E., & Mogessie, M. (2022a). Focused self-explanations lead to the best learning outcomes in a digital learning game. In: *Proceedings of the 16th International Conference on Learning Science (ICLS 2022).* pp. 1229–1232.

McLaren, B. M., & Nguyen, H. A. (2023). Digital learning games in Artificial Intelligence in Education (AIED): A review. *Handbook of Artificial Intelligence in Education.* Chapter 20.

McLaren, B. M., Adams, D. M., Mayer, R. E., & Forlizzi, J. (2017a). A computer-based game that promotes mathematics learning more than a conventional approach. *International Journal of Game-Based Learning (IJGBL), 7*(1), 36–56. https://doi.org/10.4018/IJGBL.2017010103

McLaren, B. M., DeLeeuw, K. E., & Mayer, R. E. (2011a). Polite web-based intelligent tutors: Can they improve learning in classrooms? *Computers & Education, 56*(3), 574–584. https://doi.org/10.1016/j.compedu.2010.09.019

McLaren, B. M., DeLeeuw, K. E., & Mayer, R. E. (2011b). A politeness effect in learning with web-based intelligent tutors. *International Journal of Human Computer Studies, 69*(1–2), 70–79. https://doi.org/10.1016/j.ijhcs.2010.09.001

McLaren, B. M., Richey, J. E., Nguyen, H., & Hou, X. (2022b). How instructional context can impact learning with educational technology: Lessons from a study with a digital learning game. *Computers & Education.* https://doi.org/10.1016/j.compedu.2021.104366

McNamara, D. S., Jackson, G. T., & Graesser, A. C. (2010). Intelligent tutoring and games (ITaG). In Y. K. Baek (Ed.), *Gaming for classroom-based learning: Digital role-playing as a motivator of study* (pp. 44–65). IGI Global.

Mogessie M., Richey J. E., McLaren B. M., Andres-Bray J. M. L., & Baker R. S. (2020). Confrustion and gaming while learning with erroneous examples in a decimals game. In *Proceedings of the 21st International Conference on Artificial Intelligence in Education. AIED 2020.* Lecture Notes in Computer Science (LNCS, Vol. 12164). Springer, Cham. https://doi.org/10.1007/978-3-030-52240-7_38.

Moyer-Packenham, P. S., Lommatsch, C. W., Litster, K., Ashby, J., Bullock, E. K., Roxburgh, A. L., & Jordan, K. (2019). How design features in digital math games support learning and mathematics connections. *Computers in Human Behavior, 91*, 316–332.

Nagashima, T., Bartel, A. N., Yadav, G., Tseng, S., Vest, N. A., Silla, E. M., Alibali, M.W., & Aleven, V.A. (2021). Scaffolded self-explanation with visual representations promotes efficient learning in early algebra. *Annual Meeting of the International Society of the Learning Sciences (ISLS 2021).*

Namkung, J. M., Peng, P., & Lin, X. (2019). The relation between mathematics anxiety and mathematics performance among school-aged students: A meta-analysis. *Review of Educational Research, 89*(3), 459–496.

Nathan, M. J. (1998). Knowledge and situational feedback in a learning environment for algebra story problem solving. *Interactive Learning Environments, 5*, 135–159.

Nguyen, H., Harpstead, E., Wang, Y., & McLaren, B.M. (2018). Student agency and game-based learning: A study comparing low and high agency. In C. Rosé, R. Martínez-Maldonado, H.U. Hoppe, R. Luckin, M. Mavrikis, K. Porayska-Pomsta, B. McLaren and B. du Boulay (Eds.). *Proceedings of the 19th International Conference on Artificial Intelligence in Education (AIED 2018)*. LNAI 10947. Springer: Berlin, pp. 338–351.

Nguyen, H., Wang, Y., Stamper, J., & McLaren, B.M. (2019). Using knowledge component modeling to increase domain understanding in a digital learning game. In *Proceedings of the 12th International Conference on Educational Data Mining* (EDM 2019), pp. 139–148.

Nguyen, H. A., Takacs, Z.K. Bereczki, E., Richey, J. E., & Mogessie, M. & McLaren, B. M. (2022b). Investigating the effects of mindfulness meditation on a digital learning game for mathematics. In: *Proceedings of the 23rd International Conference on Artificial Intelligence in Education (AIED 2022)*. pp. 762–767. https://doi.org/10.1007/978-3-031-11644-5_80.

Nguyen, H., Hou, X., Stec, H., Di, S., Stamper, J., & McLaren, B.M. (2023a). Examining the benefits of prompted self-explanation for problem-solving in a decimal learning game. In *Proceedings of 24th International Conference on Artificial Intelligence in Education (AIED 2023)*.

Nguyen, H., Else-Quest, N., Richey, J.E., Hammer, J., Di, S., & McLaren, B.M. (2023c). Gender differences in learning game preferences: Results using a multi-dimensional gender framework. In *Proceedings of 24th International Conference on Artificial Intelligence in Education (AIED 2023)*. pp. 553–564.

Nguyen, H., Stec, H., Hou, X., Di, S., & McLaren, B.M. (2023b). Evaluating ChatGPT's decimal skills and feedback generation to students' self-explanations in a digital learning game. *Proceedings of Eighteenth European Conference on Technology Enhanced Learning (ECTEL 2023)*.

Nguyen, H., Hou, X., Richey, J. E., & McLaren, B. M. (2022a). The impact of gender in learning with games: A consistent effect in a math learning game. *International Journal of Game-Based Learning (IJGBL).*, *12*(1), 1–29. https://doi.org/10.4018/IJGBL.309128

Ni, X., Nguyen, H.A., Else-Quest, N., Pagano, A., & McLaren, B.M. (2024). Investigating racial and ethnic differences in learning with a digital game and tutor for decimal numbers. *The Nineteenth European Conference on Technology Enhanced Learning (ECTEL 2024)*. Krems, Austria, September 16-20, 2024

Noël, M. P., Grégoire, J., Meert, G., & Seron, X. (2008). The innate schema of natural numbers does not explain historical, cultural, and developmental differences. *Behavioral and Brain Sciences, 31*(6), 664–665.

Nokes, T. J., Hausmann, R. G., VanLehn, K., & Gershman, S. (2011). Testing the instructional fit hypothesis: The case of self-explanation prompts. *Instructional Science, 39*(5), 645–666.

Nosek, B. A., Banaji, M. R., & Greenwald, A. G. (2002). Math= male, me= female, therefore math≠ me. *Journal of Personality and Social Psychology, 83*(1), 44.

O'Neil, H. F., & Perez, R. S. (2008). *Computer games and team and individual learning*. Elsevier.

O'Rourke, E., Ballweber, C., & Popovic, Z. (2014). Hint systems may negatively impact performance in educational games. In *Proceedings of the First Annual ACM Conference on Learning @ Scale (L@S '14)*, pp. 51–60. https://doi.org/10.1145/2556325.2566248.

Ochsenfeld, F. (2016). Preferences, constraints, and the process of sex segregation in college majors: A choice analysis. *Social Science Research, 56*, 117–132.

Passolunghi, M. C., Ferreira, T. I. R., & Tomasetto, C. (2014). Math–gender stereotypes and math-related beliefs in childhood and early adolescence. *Learning and Individual Differences, 34*, 70–76.

Peckham, E. (2020). Confronting racial bias in video games. *Tech Crunch*. Downloaded from https://techcrunch.com/2020/06/21/confronting-racial-bias-in-video-games/.

Piaget, J. (1962). *Play, dreams, and imitation in childhood*. Norton.

PlayToday (2023). Gamer demographics: 2023 Game-changing statistics worth checking https://playtoday.co/blog/stats/gamer-demographics/.

Putt, I. J. (1995). Preservice teachers ordering of decimal numbers: When more is smaller and less is larger! *Focus on Learning Problems in Mathematics, 17*(3), 1–15.

Rankin, Y. A., & Henderson, K. K. (2021). Resisting racism in tech design: Centering the experiences of Black youth. *Proceedings of the ACM on Human-Computer Interaction, 5*(CSCW1), 1–32.

Read, J. C., & MacFarlane, S. (2006). Using the fun toolkit and other survey methods to gather opinions in child computer interaction. In *Proceedings of the 2006 Conference on Interaction Design and Children*, pp. 81–88.

Reardon, S. F., Fahle, E. M., Kalogrides, D., Podolsky, A., & Zárate, R. C. (2019). Gender achievement gaps in US school districts. *American Educational Research Journal, 56*(6), 2474–2508.

Renkl, A. (2014). Learning from worked examples: How to prepare students for meaningful problem solving. In V. A. Benassi, C. E. Overson, & C. M. Hakala (Eds.), *Applying science of learning in education: Infusing psychological science into the curriculum* (pp. 118–130). Society for the Teaching of Psychology.

Resnick, L. B., Nesher, P., Leonard, F., Magone, M., Omanson, S., & Peled, I. (1989). Conceptual bases of arithmetic errors: The case of decimal fractions. *Journal for Research in Mathematics Education, 20*(1), 8–27.

Ribner, A. D., Willoughby, M. T., Blair, C. B., & Family Life Project Key Investigators (2017). Executive function buffers the association between early math and later academic skills. Frontiers in Psychology 869 https://doi.org/10.3389/fpsyg.2017.00869

Rice, J. W. (2007). New media resistance: Barriers to implementation of computer video games in the classroom. *Journal of Educational Multimedia and Hypermedia* 16 (3), July 2007 ISSN 1055–8896 Publisher: Association for the Advancement of Computing in Education (AACE).

Richard, G. T. (2017). Video games, gender, diversity, and learning as cultural practice: Implications for equitable learning and computing participation through games. *Educational Technology*, pp. 36–43.

Richey J. E., Zhang, J., Das, R., Andres-Bray, J. M. Scruggs, R., Mogessie, M., Baker R. S., & McLaren, B. M. (2021). Gaming and confrustion explain learning advantages for a math digital learning game. In: *Proceedings of the 22nd International Conference on Artificial Intelligence in Education (AIED 2021)*.

Rittle-Johnson, B., & Loehr, A. M. (2017). Eliciting explanations: Constraints on when self-explanation aids learning. *Psychonomic Bulletin Review, 24*, 1501–1510. https://doi.org/10.3758/s13423-016-1079-5

Roll, I., Aleven, V., McLaren, B. M., & Koedinger, K. R. (2011). Metacognitive practice makes perfect: Improving students' self-assessment skills with an intelligent tutoring system. In Biswas, G., Bull, S., Kay, J. & Mitrovic, A. (Eds.), *Proceedings of the 15th International Conference on Artificial Intelligence in Education (AIED-2011)*. Lecture Notes in Computer Science, 6738. Berlin: Springer, pp. 288–295.

Ryan, R. M., Rigby, C. S., & Przybylski, A. (2006). The motivational pull of video games: A self-determination theory approach. *Motivation and Emotion, 30*(4), 347–363. https://doi.org/10.1007/s11031-006-9051-8

Sackur-Grisvard, C., & Léonard, F. (1985). Intermediate cognitive organizations in the process of learning a mathematical concept: The order of positive decimal numbers. *Cognition and Instruction, 2*, 157–174.

Samuel, T. S., & Warner, J. (2021). "I can math!": Reducing math anxiety and increasing math self-efficacy using a mindfulness and growth mindset-based intervention in first-year students. *Community College Journal of Research and Practice, 45*, 205–222. https://doi.org/10.1080/10668926.2022.2050843

Sawyer, R., Smith, A., Rowe, J., Azevedo, R., Lester, J. (2017). Is more agency better? The impact of student agency on game-based learning. In: André, E., Baker, R., Hu, X., Rodrigo, M.M.T., du Boulay, B. (eds.) AIED 2017. LNCS, vol. 10331, pp. 335–346. Springer, Cham. https://doi.org/10.1007/978-3-319-61425-0 28.

Scheiter, K., Gerjets, P., & Schuh, J. (2010). The acquisition of problem-solving skills in mathematics: How animations can aid understanding of structural problem features and solution procedures. *Instructional Science, 38*, 487–502. https://doi.org/10.1007/s11251-009-9114-9

Schell, J. (2008). Story and game structures can be artfully merged with indirect control. *The Art of Game Design: A Book of Lenses*. Taylor & Francis, pp. 317–334.

Schell, J. (2005). Understanding entertainment. *Computers in Entertainment., 3*, 6. https://doi.org/10.1145/1057270.1057284

Schunk, D. H., & Zimmerman, B. J. (Eds.). (1998). *Self-regulated learning: From teaching to self-reflective practice*. Guilford Press.

Shute, V. J., Rahimi, S., & Smith, G. (2019). Chapter 4: Game-based learning analytics in physics playground. In Tlili, A. & Chang, M. (Eds.), *Data Analytics Approaches in Educational Games and Gamification Systems, Smart Computing and Intelligence*, https://doi.org/10.1007/978-981-32-9335-9_4.

Singh, N. N., Lancioni, G. E., Nabors, L., Myers, R. E., Felver, J. C., & Manikam, R. (2018). Samatha meditation training for students with attention deficit/hyperactivity disorder: Effects on active academic engagement and math performance. *Mindfulness, 9*, 1867–1876.

Sitzmann, T. (2011). A meta-analytic examination of the instructional effectiveness of computer-based simulation games. *Personnel Psychology, 64*, 489–528. https://doi.org/10.1111/j.1744-6570.2011.01190.x

Snow, E. L., Allen, L. K., Jacovina, M. E., & McNamara, D. S. (2015). Does agency matter?: Exploring the impact of controlled behaviors within a game-based environment. *Computers & Education, 82*, 378–392.

Spencer, S. J., Steele, C. M., & Quinn, D. M. (1999). Stereotype threat and women's math performance. *Journal of Experimental Social Psychology, 35*(1), 4–28.

Squire, K. (2005). Changing the game: What happens when video games enter the classroom? *Innovate: Journal of Online Education, 1*(6).

Stacey, K., Helme, S., & Steinle, V. (2001). Confusions between decimals, fractions and negative numbers: A consequence of the mirror as a conceptual metaphor in three different ways. In Heuvel-Panhuizen, M. V. D. (Ed.), *Proceedings of the 25th Conference of the International Group for the Psychology of Mathematics Education*. Utrecht: PME., vol. 4, pp. 217–224.

Tahir, F., Mitrovic, A., & Sotardi, V. (2020). Investigating the effects of gamifying SQL-Tutor. In: So, H. J. et al. (Eds.) *Proceedings of the 28th International Conference on Computers in Education*. Asia-Pacific Society for Computers in Education. pp. 416–425, ISBN978-986-97214-5-5.

Takacs, Z. K., & Kassai, R. (2019). The efficacy of different interventions to foster children's executive function skills: A series of meta-analyses. *Psychological Bulletin, 145*, 653.

The NPD Group (2019). Retail tracking service, 2019 Entertainment Survey. https://www.npd.com/news/press-releases/2019/according-to-the-npd-group-73-percent-of-u-s-consumers-play-video-games/

Tobias, S., & Fletcher, J. D. (2011). *Computer games and instruction*. Charlotte NC: Information Age. https://eric.ed.gov/?id=ED529495.

TrueList (2023). 33 Evolutionary Gaming Statistics of 2023. https://truelist.co/blog/gaming-statistics/.

Tsovaltzi, D., Melis, E., & McLaren, B. M. (2012). Erroneous examples: Effects on learning fractions in a web-based setting. *International Journal of Technology Enhanced Learning (IJTEL)*. V4 N3/4 2012 pp. 191–230.

Van Eck, R., & Dempsey, J. (2002). The effect of competition and contextualized advisement on the transfer of mathematics skills in a computer-based instructional simulation game. *Educational Technology Research and Development, 50*, 23–41.

VanLehn, K. (2006). The behavior of tutoring systems. *International Journal of Artificial Intelligence in Education, 16*(3), 227–265.

VanLehn, K. (2011). The relative effectiveness of human tutoring, intelligent tutoring systems, and other tutoring systems. *Educational Psychologist, 46*(4), 197–221. https://doi.org/10.1080/00461520.2011.611369

Vekety, B., Kassai, R., & Takacs, Z. K. (2022). Mindfulness with children: A content analysis of evidence-based interventions from a developmental perspective. *The Educational and Developmental Psychologist, 39*(2), 231–244. https://doi.org/10.1080/20590776.2022.2081072

Vygotsky, L. S. (1978). In: Cole, M., John-Steiner, V., Scribner, S. & Souberman, E. (Eds.), *Mind in society: The development of higher psychological processes*. Cambridge: Harvard University Press. ISBN 9780674576292.

Wai, J., Cacchio, M., Putallaz, M., & Makel, M. C. (2010). Sex differences in the right tail of cognitive abilities: A 30-year examination. *Intelligence, 38*(4), 412–423.

Walker, E., McLaren, B. M., Rummel, N., & Koedinger, K. R. (2007). Who says three's a crowd? Using a cognitive tutor to support peer tutoring. In Luckin, R. Koedinger, K. R. & Greer, J. (Eds.), In *Proceedings of the 13th International Conference on Artificial Intelligence in Education* (AIED-07), Artificial Intelligence in Education: Building Technology Rich Learning Contexts That Work. Amsterdam: IOS Press, pp. 399–406.

Walsh, G. (2009). Wii can do it: Using co-design for creating an instructional game. In *CHI'09 Extended Abstracts on Human Factors in Computing Systems*, pp. 4693–4698.

Wang, Y., Nguyen, H. A., Harpstead, E., Stamper, J. & McLaren, B. M. (2019). How does order of gameplay impact learning and enjoyment in a digital learning game? In: Isotani S., Millán E., Ogan A., Hastings P., McLaren B., Luckin R. (Eds). *Proceedings of the 20th International Conference on Artificial Intelligence in Education (AIED 2019)*. LNAI 11625. Springer, pp. 518–531.

Wang, L. H., Chen, B., Hwang, G. J., Guan, J. Q., & Wang, Y. Q. (2022). Effects of digital game-based STEM education on students' learning achievement: A meta-analysis. *International Journal of STEM Education, 9*, 26. https://doi.org/10.1186/s40594-022-00344-0

Wechselberger, U. (2013). Learning and enjoyment in serious gaming-contradiction or complement? In: *DiGRA Conference*, pp. 26–29.

Wittwer, J., & Renkl, A. (2010). How effective are instructional explanations in example-based learning? A meta-analytic review. *Educational Psychology Review, 22*(4), 393–409.

Wolfram, S. (2023). What is ChatGPT doing... and why does it work? Wolfram Media, Inc. ISBN-13: 978-1-57955-081-3 (paperback).

Woolf, B. P. (2008). *Building intelligent interactive tutors: Student-centered strategies for revolutionizing e-learning*. Morgan Kaufmann.

Wouters, P., & van Oostendorp, H. (Eds.). (2017). *Instructional techniques to facilitate learning and motivation of serious games*. Springer.

Wylie, R., & Chi, M. T. H. (2014). The self-explanation principle in multimedia learning. In R. E. Mayer (Ed.), *The Cambridge Handbook of Multimedia Learning* (pp. 413–432). Cambridge University Press.

Xu, Z., Wijekumar, K., Ramirez, G., Hu, X., & Irey, R. (2019). The effectiveness of intelligent tutoring systems on K-12 students' reading comprehension: A meta-analysis. *British Journal of Educational Technology, 50*(6), 3119–3137.

Yáñez-Gómez, R., Cascado-Caballero, D., & Sevillano, J. L. (2017). Academic methods for usability evaluation of serious games: A systematic review. *Multimedia Tools and Applications., 76*(4), 5755–5784. https://doi.org/10.1007/s11042-016-3845-9.hdl:11441/74400.ISSN1380-7501.S2CID254833872

Ye, J., Chen, X., Xu, N., Zu, C., Shao, Z., Liu, S., Cui, Y., Zhou, Z., Gong, C., Shen, Y., Zhou, J., Chen, S., Gui, T., Zhang, Q., & Huang, X. (2023). A comprehensive capability analysis of GPT-3 and GPT-3.5 series models. arXiv preprint arXiv:2303.10420.

Zimmerman, B. J. (2008). Investigating self-regulation and motivation: Historical background, methodological developments, and future prospects. *American Educational Research Journal, 45*(1), 166–183.

Chapter 10
Leveraging AI to Advance Science and Computing Education Across Africa: Challenges, Progress and Opportunities

George Boateng

Abstract Across the African continent, students grapple with various educational challenges, including limited access to essential resources such as computers, internet connectivity, reliable electricity, and a shortage of qualified teachers. Despite these challenges, recent advances in AI such as BERT, and GPT-4 have demonstrated their potential for advancing education. Yet, these AI tools tend to be deployed and evaluated predominantly within the context of Western educational settings, with limited attention directed towards the unique needs and challenges faced by students in Africa. In this chapter, we discuss challenges with using AI to advance education across Africa. Then, we describe our work developing and deploying AI in Education tools in Africa for science and computing education: (1) SuaCode, an AI-powered app that enables Africans to learn to code using their smartphones, (2) AutoGrad, an automated grading, and feedback tool for graphical and interactive coding assignments, (3) a tool for code plagiarism detection that shows visual evidence of plagiarism, (4) Kwame, a bilingual AI teaching assistant for coding courses, (5) Kwame for Science, a web-based AI teaching assistant that provides instant answers to students' science questions and (6) Brilla AI, an AI contestant for the National Science and Maths Quiz competition. Finally, we discuss potential opportunities to leverage AI to advance education across Africa.

Keywords AI · Generative AI · Tutoring · Question answering · Science education · Computing education · NLP · BERT · GPT-4

10.1 Introduction

In Africa, a significant portion of students grapples with formidable educational barriers arising from a multitude of challenges, including limited access to essential

G. Boateng (✉)
Kwame AI Inc., Claymont, DE, USA

ETH Zurich, Zurich, Switzerland
e-mail: jojo@kwame.ai

P. Ilic et al. (eds.), *Artificial Intelligence in Education: The Intersection of Technology and Pedagogy*, Intelligent Systems Reference Library 261,
https://doi.org/10.1007/978-3-031-71232-6_10

resources such as computers (UNESCO, 2020), internet connectivity (IEA, 2023; UNESCO, 2020), reliable electricity (Ehl and Grün, 2020), and a shortage of qualified teachers (World Bank Group, 2020). For instance, as of 2018, the average student-teacher ratio in Sub-Saharan Africa stood at 35:1, starkly contrasting with the more favorable ratio of 14:1 observed in Europe (World Bank Group, 2020). The absence of these resources impedes the quality and accessibility of education across the continent.

Despite these challenges, recent strides in Artificial Intelligence (AI) technology, exemplified by sophisticated systems like BERT (Devlin et al., 2019) and GPT-4 (Achiam et al., 2023), have showcased their potential to revolutionize education globally. Major players in the global EdTech landscape, such as Duolingo (2023), Quizlet (2023), Chegg (2023), and KhanAcademy (2023), have actively integrated these AI advancements into their platforms to enhance learning experiences. However, the deployment and evaluation of these AI systems predominantly occur within the context of Western educational settings, with limited attention directed towards the unique needs and challenges faced by students in Africa. For example, the release of GPT-4 in March 2023 featured various academic exams as benchmarks with none from Africa (Achiam et al., 2023). This oversight underscores the tendency for the African continent to be marginalized in the application of cutting-edge AI advancements, with little consideration given to the diverse educational landscapes and requirements of African students. As a consequence, the potential of AI to address the educational inequities and barriers prevalent in Africa remains largely untapped, highlighting the urgent need for greater inclusivity and tailored solutions that account for the specific needs and challenges faced by students on the continent.

In this chapter, we discuss challenges with using AI to advance education across Africa. Then, we describe our work developing and deploying AI in Education (AIED) tools in Africa for science and computing education: (1) SuaCode, an AI-powered app that enables Africans to learn to code using their smartphones, (2) AutoGrad, an automated grading, and feedback tool for graphical and interactive coding assignments, (3) a tool for code plagiarism detection that shows visual evidence of plagiarism, (4) Kwame, a bilingual AI teaching assistant for coding courses, (5) Kwame for Science, a web-based AI teaching assistant that provides instant answers to students' science questions and (6) Brilla AI, an AI contestant for the National Science and Maths Quiz competition (NSMQ). Finally, we discuss potential opportunities to leverage AI to advance education across Africa and then conclude.

10.2 Challenges

One of the key challenges with leveraging AI to improve education in Africa is access to resources (like computers) and infrastructure (Internet, electricity) to make use of AIED tools. Across sub-Saharan Africa, Internet data is expensive (Ehl and Grün, 2020) and electricity is either not available or erratic in some countries (IEA, 2023). According to UNESCO, 11% of students in sub-Saharan Africa have access to household computers, and only 18% have access to the Internet (UNESCO, 2020). So, how can students in a rural part of Ghana, for example, access and use an AI tutor if they do not have computers or smartphones or constant electricity? If these tutors are deployed to students in urban areas who have the necessary resources and infrastructure, it could increase the divide between the haves and the have-nots, further exacerbating the existing inequities in the education sector in various African countries.

A lack of regulatory support for innovation could hamper efforts to deploy AIED tools. For example, several Senior High Schools in Ghana are boarding schools that do not allow their students to have access to their devices (smartphones and computers) (Boateng et al., 2024). Since students spend about 9 months of the 12 months in school, they mostly do not have access to devices. Only a few private schools have computers in their libraries for students to use. This issue is a significant barrier to the potential impact of AIED tools on students at the senior high school level. How do you give access to an AI tutor to students if the Ghana Education Service, for example, restricts access to smart devices in boarding schools at the secondary school level?

There is heterogeneity in educational systems across different parts of Africa making it difficult to deploy AIED tools scalably. For example, Anglophone West Africa has the West African Secondary School Certificate Exam (WASSCE) and Kenya has the Kenya Certificate of Secondary Education (KCSE) exam. Some countries use English whereas others use French as their official language of instruction. Yet, several local languages abound which may need to be supported to reach students who have low literacy in official languages. Hence, AIED tools developed in one country or region of the continent would need to be adapted to the local educational system of other countries. Considering there are 54 countries in Africa, this process would also entail jumping through different regulatory hoops making it time-consuming and costly.

Educational materials in various parts of Africa needed as input for building AIED tools tend to be available in hardcopy format which poses challenges for digitization. One challenge we encountered revolved around the formatting of scientific and mathematical symbols and equations. Regrettably, the open-source optical character recognition (OCR) technologies we experimented with proved inadequate in accurately extracting these symbols (Boateng et al., 2024). Consequently, manual intervention became necessary to ensure an additional layer of quality control. There is a pressing need for further advancements in developing systems capable of seamlessly converting scanned scientific documents into outputs that maintain correct representations, facilitating compatibility with standard formats such as markdown. Until such solutions are realized, substantial time and financial investments will continue to be essential for generating high-quality usable data.

One of the critical challenges facing the development and deployment of AIED tools is the pervasive lack of representative data, particularly from regions in the Global South in pretrained AI models. There are challenges accessing copyrighted, local textbooks to build AIED tools (Boateng et al., 2024). The absence of diverse and inclusive datasets undermines the effectiveness and fairness of AI technologies, leading to biased outcomes that fail to adequately address the needs and realities of communities in these regions. Some AI tools are built, evaluated, and deployed generally and also for educational use cases without incorporating data from Africa such as GPT-4's evaluation which missed education data from Africa (Achiam et al., 2023). These result in biased systems that do not well in the African context. Examples include speech-to-text systems that do not work well for African accents, or text-to-speech systems that do not speak with an African accent (Boateng et al., 2023b). If such AI systems are integrated into an AI tutor for African students, for example, they will not work well, defeating the goal of the AI tutor.

Students could end up using these AIED tools as a crutch either for cheating or just being overreliant on them such as to solve homework assignments, write their essays for them and even cheat in national exams (Dzodzegbe, 2023). If students use these tools in this way, it defeats the pedagogical goals of the assigned homework. Relying excessively on AIED tools for completing homework assignments poses a significant risk of undermining the educational objectives set by instructors. When students resort to these tools as a crutch, they miss out on the opportunity to engage critically with the material and develop essential problem-solving skills. Instead of grappling with the complexities of the subject matter, students may opt for the path of least resistance, simply inputting their assignment prompts into AI platforms and passively accepting the generated solutions. This not only diminishes the intellectual rigor of the learning process but also fosters a culture of academic dishonesty.

Generative AI models that power AIED tools have an issue of sometimes confidently generating responses that are factually incorrect, referred to as "Hallucination" which could affect their utility in improving education. In instances where students rely heavily on AI-generated content for learning and study purposes, the propagation of misinformation can undermine the integrity of their educational experiences. Instead of fostering a deep understanding of the subject matter, the presence of inaccuracies may cultivate a false sense of comprehension, ultimately hindering students' ability to engage critically with the material.

10.3 Progress: AIED Solutions

In this section, we discuss AIED tools that we developed and deployed for science and computing education, and how they address some of the challenges previously discussed.

10.3.1 SuaCode

SuaCode [1] is an AI-powered smartphone-based application that enables students across Africa to learn to code using smartphones (Figs. 10.1 and 10.2). The app has monthly online coding courses with lesson notes, exercises, quizzes, and fun

[1] http://suacode.ai/

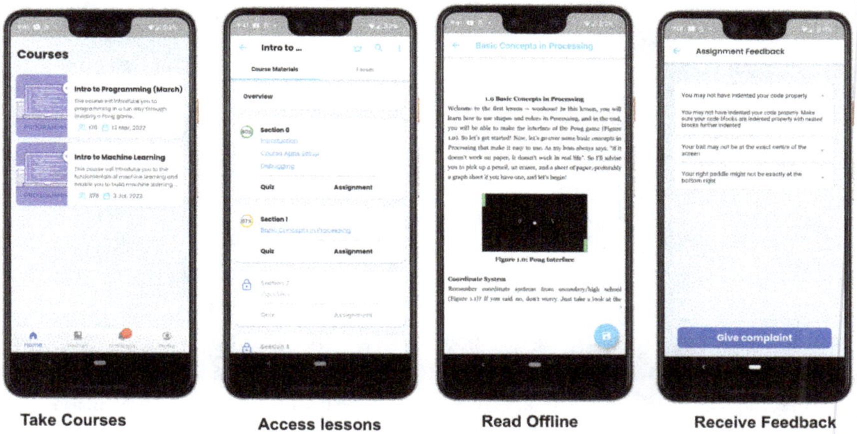

Fig. 10.1 Screenshots of the SuaCode App with course materials and assignment feedback

coding assignments in English and French which cover official languages across Africa. We use simplified lesson notes designed for offline reading to reduce Internet data usage since they are expensive across Africa (Ehl and Grün, 2020). Our courses adopt a project-based learning approach where learners build and interact with a game on their phones as assignments. Each section of the course ends with multiple-choice quizzes and a coding assignment. The assignments are graded by our automated grading software, AutoGrad (Annor et al., 2021) which also checks for plagiarism (John and Boateng, 2021) and provides detailed individualized explanations for wrong answers with an option for learners to submit a complaint in cases where they disagree with the provided answers and feedback. Our course-specific, AI-powered, human-in-the-loop forum is designed to allow learners to ask questions (even anonymously) and get quick and accurate answers from their peers, facilitators, and our AI teaching assistant, Kwame (Boateng, 2021). Kwame enables learners to get individualized learning support so they do not quit when it gets tough. Each person's name shows by the side an earned badge to encourage helpful engagement. Each of the 3 badges (bronze, silver, and gold) is received based on a "helpfulness score" that we calculate using various metrics such as upvotes on questions and answers contributed by each learner. Our leaderboard celebrates helpful learners and facilitators thereby encouraging others to be helpful. Students who complete receive certificates and mentoring by African tech professionals in companies such as Google and Amazon enabling our learners to receive advice from experienced individuals who have similar backgrounds as them. SuaCode has over 2.5K learners across 43 African countries and 117 globally.

We created SuaCode to address the problem that less than 1% of African children leave school with basic coding skills with several jobs struggling to fill IT-related roles despite Africa being home to the youngest workforce in the world (SAP, 2016). In 2017, while running the 4th edition of our annual innovation bootcamp, Project

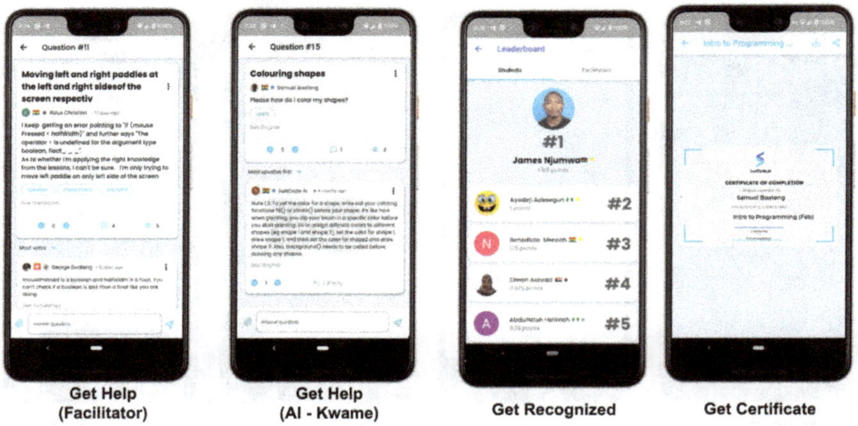

Fig. 10.2 Screenshots of the SuaCode App with forum, leaderboard, and certificate

iSWEST, in Ghana, we noticed from our pre-survey that out of our 27 students, 25% had laptops, yet 100% had smartphones. This situation of limited access to computers led us to modify our coding course and deliver it using smartphones, the first of its kind in Ghana. Our students built pong games on their phones with several students working on coding assignments while in traffic (Boateng and Kumbol, 2018). Realizing the potential of our smartphone-based course, in 2018, we created SuaCode, a smartphone-based online coding program aiming to teach millions across Africa how to code by exploiting the proliferation and untapped capabilities of smartphones (Boateng et al., 2019).

Between 2018 and 2020, we ran 4 pilots of SuaCode while growing exponentially (Fig. 10.3) that reached 3K learners across 69 countries (42 in Africa) (Boateng et al., 2019; Boateng, 2020; Boateng et al., 2021). The 2018 version focused on students in Ghana and had over 30 students enrolled with only 23% completing (Boateng et al., 2019). It was delivered using Google Classroom as the course management platform (forum and assignment submission), Google Docs for lesson notes and APDE for the coding app. We had 4 lessons and assignments, and an optional project at the end. The first version in 2019 also focused on Ghanaians. The second version in 2019 expanded beyond Ghana to all Africans and was dubbed SuaCode Africa (Boateng et al., 2021). It had 709 applicants across 37 African countries. We introduced an acceptance criterion (submit the first 2 assignments) to filter for highly motivated students. Out of that, 210 were accepted and 151 completed resulting in a 72% completion rate. The 2020 version was dubbed SuaCode Africa 2.0 with the course being offered in both English and French (Boateng, 2020; SuaCode Africa 2.0, 2021). We used Piazza for that cohort which had a better forum experience than Google Classroom. Over 2300 students across 69 countries, 42 of which were in Africa applied. We accepted and trained 740 students and 62% completed. We then used the learnings to build our AI-powered smartphone app, SuaCode and then launched it in 2022 to scale the impact of SuaCode.

These courses were structured in cohorts that created a sense of community and fulfilled learners' need for affiliation, support, and interaction and were also optimized to encourage completion resulting in high completion rates (62%, n=1000) vs industry standards (less than 20%). Our 600+ past learners significantly improved their understanding of fundamental coding concepts despite using only smartphones to learn. We collected qualitative and quantitative feedback from our learners in various cohorts. Our quantitative analysis showed that students had an average of 17 out of 20 across all 4 assignments indicating mastery of the content (Boateng et al., 2021). Also, there was no statistically significant difference in assignment scores between males and females, and educational level (high school, university, and high school graduates) suggesting that our course might be adequate for different demographics (Boateng et al., 2021). Furthermore, there was a statistically significant improvement in students' self-reported proficiency in the programming concepts (Boateng et al., 2021). In user surveys about the smartphone coding experience, 85% of our learners (n = 457) rated the coding experience 4+ on a 5-point Likert scale (SuaCode Africa 2.0, 2021) with feedback such as *"It was really convenient, honestly. I didn't have to necessarily sit behind a desk to do it so I could do it when I was on my bed, eating, even using the bathroom. So it was fun and convenient coding on my phone"*. Overall, there were several positive feedbacks about the experience such as *"Suacode has been a very great experience for me. I got to learn processing and actually code on my phone. I also had help from the tutors and my fellow course mates which made it easier. I learnt a lot and I'm glad I had the opportunity to be part of the first batch of suacode initiative"* and *"SuaCode helped improve my algorithmic thought process. I had lots of practice with thinking in a step by step process and working through challenges"*.

An innovation like SuaCode is an example of ways to get around the previously mentioned challenge of limited access to resources such as computers and infrastructure like affordable Internet which could hamper efforts to deploy AIED tools. By leveraging tools that are much more accessible to Africans like smartphones, AIED tools could be made available to these students and eventually improve their learning outcomes like we have done with SuaCode. Furthermore, our use of lesson notes rather than videos for lesson delivery addresses the challenge of expensive Internet data for African students. Intentional design of educational experiences could help to address the effect of some of the infrastructural challenges in Africa.

10.3.2 AutoGrad

AutoGrad is a novel cross-platform software for automatically grading and evaluating graphical and interactive programs written in the Processing programming language in SuaCode courses. (Annor et al., 2021). It uses APIs to retrieve assignments from the course platform, conducts both static and dynamic analyses on these assignments

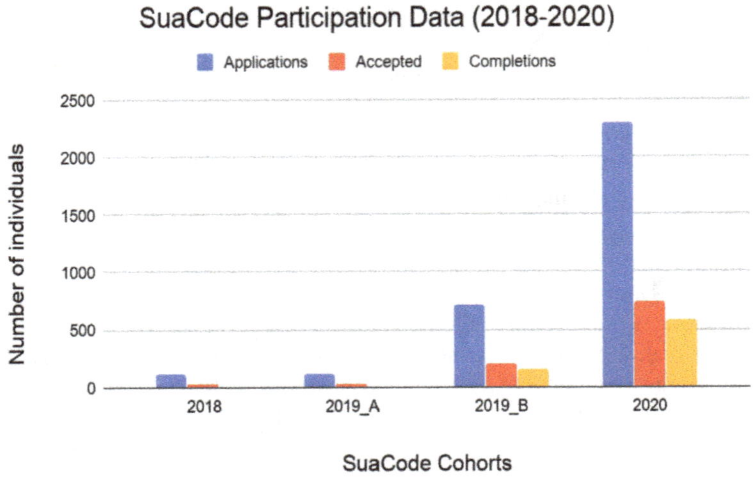

Fig. 10.3 Growth of SuaCode between 2018 and 2020

to assess their graphical and interactive outputs, and subsequently furnishes students with grades and feedback (Fig. 10.4). The AutoGrad system itself was written using Python and Processing. The Python component manages assignment retrieval and the dissemination of grades and feedback to students, while Processing oversees the grading module, ensuring the effective execution of checks on Processing code. AutoGrad has been successfully deployed in multiple iterations of SuaCode cohorts, servicing over 1000 students across Africa and evaluating more than 3000 code files. These deployments involved running AutoGrad as software on a computer. Notably, in the latest cohort, students were allowed to submit complaints in cases where they believed their assignments were inaccurately graded. This approach not only ensured fairness to all students but also provided valuable insights for addressing instances where AutoGrad's performance fell short and facilitated ongoing enhancements to the software.

We assessed AutoGrad's grading accuracy using both test assignment scripts and actual student scripts from previous cohorts of the SuaCode course. Using 10 student scripts for Assignment 1, AutoGrad yielded identical grades to those assigned by manually grading instructors in 8 scripts, resulting in a mean absolute error of 0.7. However, subsequent assignments exhibited a higher frequency of discrepancies than Assignment 1 (Annor et al., 2021). We also collected feedback from students regarding their experiences with AutoGrad. Quantitative feedback was gathered by asking students to rate their agreement with the statement "I liked the feedback from AutoGrad" on a 5-point Likert scale, ranging from strongly disagree to strongly agree. Among the 457 students who completed the course and responded in the 2020 cohort, 75.9% agreed or strongly agreed with the statement, indicating that a majority of students found AutoGrad's feedback beneficial, albeit with some room

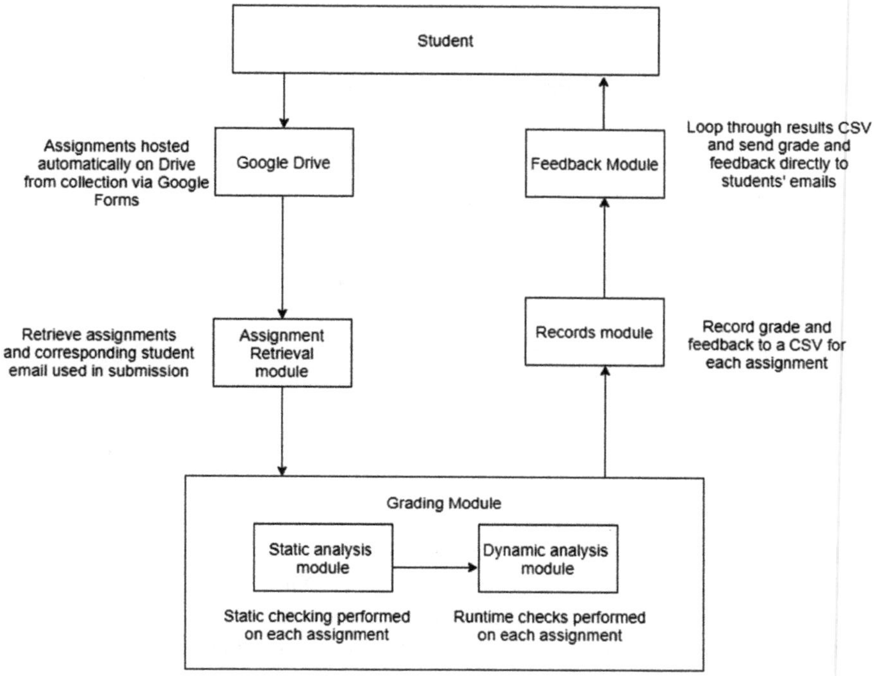

Fig. 10.4 System design of AutoGrad *source* Annor et al., (2021)

for improvement, as indicated by a mean rating of 4 out of 5. Qualitative feedback was also collected, focusing on suggestions for enhancing AutoGrad's feedback mechanism. A common theme among responses was the request for more detailed explanations accompanying the feedback provided by AutoGrad, as well as the identification of specific areas within their code where improvements could be made (Annor et al., 2021).

10.3.3 Code Plagiarism Detector

We developed a tool that performs code plagiarism detection in students' assignment submissions in SuaCode courses and also shows visual evidence of the detected plagiarism (John and Boateng, 2021). We built this tool to address the issues of code plagiarism in the 2020 SuaCode cohort where we previously identified that 27% of the 431 students had cases of plagiarism via manual inspection. We trained machine learning models on three (3) cosine similarity-based scores extracted from the TF-IDF (with n-grams) feature vector of the code files to detect pairs of code files that had cases of plagiarism. We used as features the cosine similarity of (1)

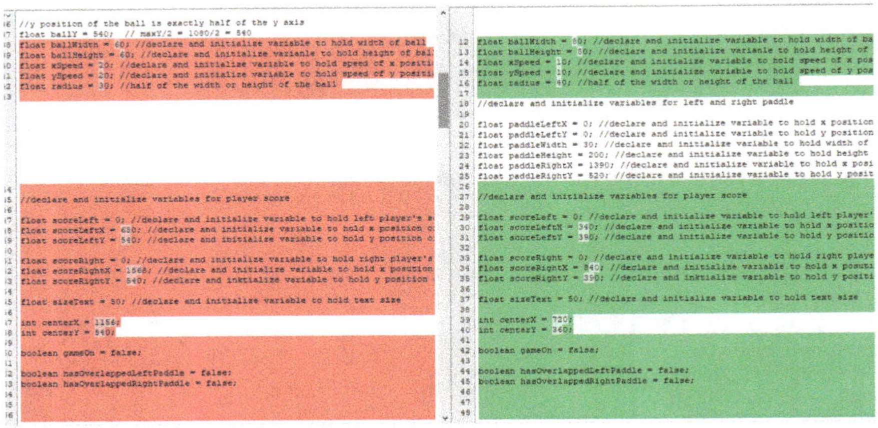

Fig. 10.5 GUI tool highlighting plagiarized code sections in two files *source* Annor et al., (2021)

student 1 code and student 2 code, (2) student 1 code and example code template, and (3) student 2 code and example code template. Our evaluation using 431 code files showed a balanced accuracy of 84% using random forest with less than 1% false positive. Also, the system provides proof of plagiarism via a GUI tool that displays side-by-side pairs of code files while highlighting sections with overlapping code. This work is the first that builds a complete end-to-end system that uses (1) machine learning algorithms to detect plagiarized source code containing English and French texts while taking the code example provided by the instructors into consideration and (2) provides visual evidence of the plagiarism (Fig. 10.5) (John and Boateng, 2021).

10.3.4 Kwame

Kwame is a bilingual AI teaching assistant that provides answers to students' coding questions in English and French for SuaCode courses (Boateng, 2021). Kwame is a deep learning-based question-answering system that was trained using the SuaCode course materials (lesson notes and past questions and answers) and evaluated using accuracy and time to provide answers. It finds the paragraph most semantically similar to a question via cosine similarity using Sentence-BERT (SBERT) (Reimers and Gurevych, 2019), a large language model that uses siamese and triplet network architecture with BERT to train models such that semantically similar sentences are closer in vector space (Fig. 10.6). We compared Kwame with other approaches and performed a real-time implementation which showed fast response time and superior accuracy of top 1, 3, and 5 accuracies of 58.3% (58.3%), 83.3% (80%) and 100% (91.7%) for English (French) respectively. Kwame has been integrated into

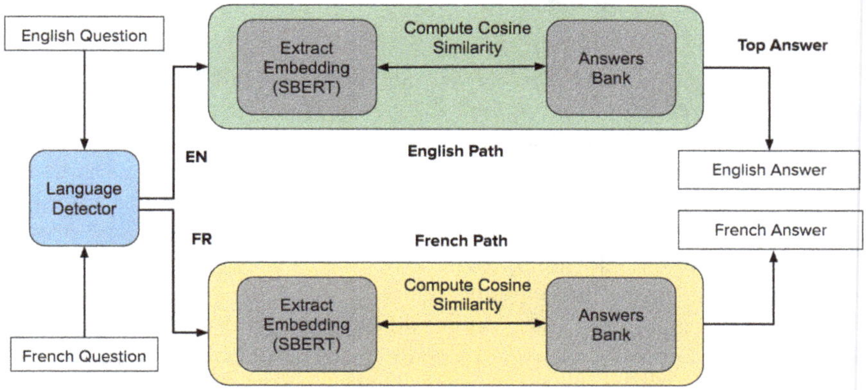

Fig. 10.6 System architecture of Kwame *source* Boateng, (2021)

the SuaCode app and answers students' questions by posting an answer on the course forum in reply to students' questions. Kwame's pipeline is a retrieval system that consists of an ElasticSearch vector store of the embeddings of passages from the lesson notes. When a question is asked, our system computes an embedding of the question using SBERT, computes cosine similarity scores with all saved embeddings, and retrieves the top 3 passages which are then posted on the forum as answers. Kwame is named after Dr. Kwame Nkrumah the first President of Ghana and a Pan-Africanist whose vision for a developed Africa motivates this work.

Kwame was developed to address the issue where our learners needed a lot of assistance given it was the first coding course for most learners. We relied on human facilitators to provide support and answer students' questions. For example, in SuaCode Africa 2.0, facilitators contributed over 1,000 hours of assistance time for 8 weeks and helped to achieve an average response time of 6 minutes through the course (SuaCode Africa 2.0, 2021). This approach was however not scalable as the number of students applying to SuaCode increased exponentially year on year. Hence, in 2020, we built Kwame, an AI teaching assistant to provide accurate and quick answers to students' questions which would reduce the burden on human teaching assistants, and provide an opportunity to scale learning support. Kwame addresses the diversity of languages used for learning in Africa by offering answers in both English and French. This caters to both Anglophone and Francophone regions, covering the official languages spoken across different parts of the continent.

10.3.5 Kwame for Science

Kwame for Science[2] is an AI-powered web application that offers two primary functionalities: (1) question answering and (2) viewing past national exam questions (Boateng et al., 2022, 2024). These features are enabled by our curated knowledge base of content from textbooks and past national exams over the past 28 years (for Integrated Science at the Senior High School level) with answers from certified teachers. We built Kwame for Science to address the impact of the shortage of qualified teachers (World Bank Group, 2020) across Africa which makes it difficult for students to have adequate learning support.

The question-answering (QA) feature allows students to pose science-related queries and receive three passages as responses, each accompanied by a confidence score (Fig. 10.7). Additionally, this feature provides the top five related past national exam questions from the Integrated Science subject of the West African Secondary School Certificate Exam (WASSCE), along with their corresponding expert answers. Users also can review their question history. When a student asks a question, our system extracts an embedding from the text using SBERT, computes cosine similarity between the embedding and the embeddings of passages from textbooks stored in ElasticSearch on Google Cloud Platform, and then returns the top answers based on the cosine similarity scores. Students could rate the helpfulness of answers and related questions.

Moreover, the View Past Questions feature permits students to explore past national exam questions and answers from the Integrated Science subject. This feature includes filters for refining the displayed questions based on parameters such as examination year, specific exam, question type, and automatically categorized topics generated by a custom topic detection model (Fig. 10.8). All these criteria could be inferred easily from the metadata of the original exam files except the topic for which we developed a model to automatically categorize each question according to one of the syllabus topics (Boateng et al., 2024). We trained a machine-learning model that used a support vector machine and SBERT embeddings of passages from the Integrated Science subject syllabus to classify each of the past exam questions into one of the 48 syllabus topics. We then used the model to automatically categorize questions into topics in the syllabus for all 28 years of exams.

We launched the web app in beta from 10th June 2022 to 19th February. During the 8-month deployment, we had 750 users across 32 countries (15 in Africa) asking 1.5K questions with Kwame's helpfulness scores being top 1 and top 3 of 72.6% and 87.2% respectively. For the viewing past exam questions features, users most frequently used the filtering by year feature (237 times) (Boateng et al., 2024). Future work will assess how Kwame for Science can improve learning outcomes.

In building Kwame for Science, we faced challenges (previously highlighted) related to accessing and using educational content from Ghana. We were unable to get official partnerships with local textbook publishers to use their copyrighted Science textbooks due to trust issues in the ecosystem. We addressed this issue

[2] http://ed.kwame.ai/

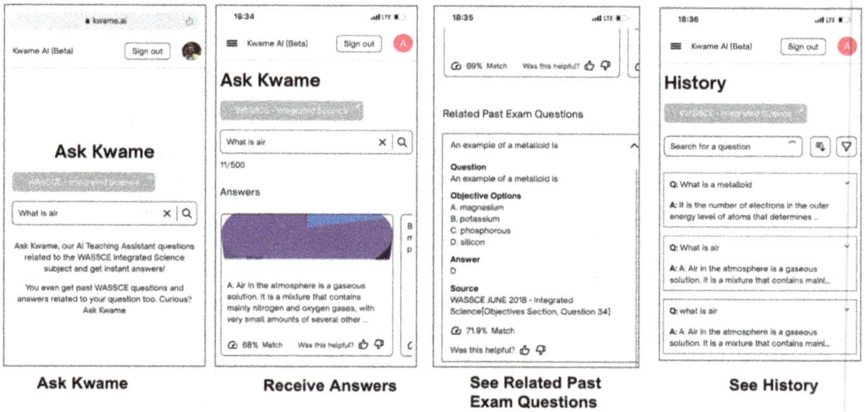

Fig. 10.7 Screenshots of QA feature of Kwame for Science *Source* Boateng et al., (2024)

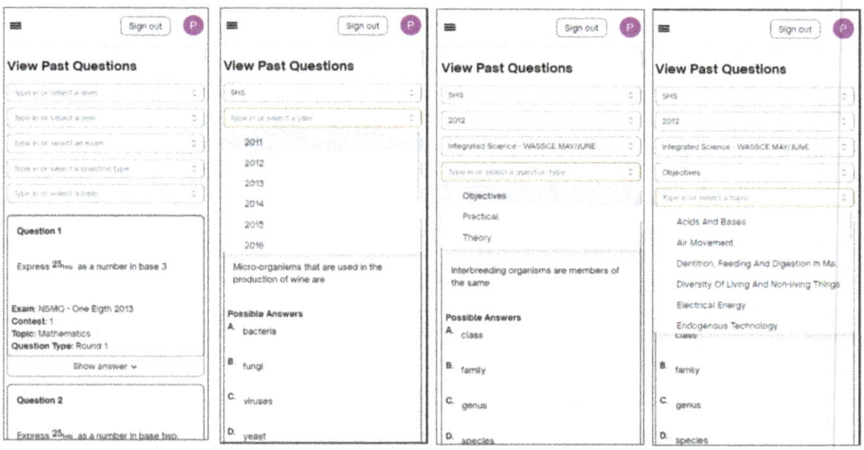

Fig. 10.8 Screenshots of View Past Questions feature of Kwame for Science *source* Boatengetal., (2024)

by using global open-source Science textbooks and hiring local experts to provide answers to past exam questions. Furthermore, past national exam questions were only available in hardcopy formats and open-source OCR technologies we experimented with proved inadequate in accurately extracting scientific and mathematical symbols and equations from scanned copies of the documents. We addressed this issue by hiring individuals to annotate the content and exploring commercial solutions with advanced AI technologies. Future work will explore the use of Generative AI to generate contextual, local content based on local syllabi and content from our experts, and also extract well-formatted content from scanned documents.

10.3.6 Billa AI

Brilla AI is an AI contestant that we developed and deployed to unofficially compete remotely and live in the Riddles round of the 2023 NSMQ Grand Finale, the first of its kind in the 30-year history of the competition (Boateng et al., 2023b, 2024). This work is motivated by the lack of enough qualified teachers in Africa (World Bank Group, 2020) which hampers the provision of adequate learning support. An AI could potentially augment the efforts of the limited number of teachers, leading to better learning outcomes. Yet, there exists no robust, real-world benchmark to evaluate such an AI. Towards that end, we built Brilla AI as the first key output for the **NSMQ AI Grand Challenge**, which proposed a robust, real-world challenge in education for such an AI: *"Build an AI to compete live in Ghana's National Science and Maths Quiz (NSMQ) competition and win - performing better than the best contestants in all rounds and stages of the competition"* (Boateng et al., 2023). The NSMQ is an annual live science and mathematics competition for senior secondary school students in Ghana in which 3 teams of 2 students compete by answering questions across biology, chemistry, physics, and math in 5 rounds over 5 progressive stages until a winning team is crowned for that year (National Science and Maths Quiz., 2024).

Brilla AI is currently available as an open-source web app (built with Streamlit (Streamlit. A faster way to build and share data apps., 2024)) that live streams the Riddles round of the contest, and runs 4 machine learning systems: (1) speech-to-text (using Whisper (Radford et al., 2023)) (2) question extraction (using BERT (Devlin et al., 2019)) (3) question answering (using Mistral (Jiang et al., 2023)) and (4) text-to-speech (using VITS (Kim et al., 2021)) that work together in real-time to transcribe Ghanaian accented English speech, extract the question, provide an answer, and then say it with a Ghanaian accent (Fig. 10.9) (Boateng et al., 2024). In its debut in October 2023, our AI answered one of the 4 riddles ahead of the 3 human contesting teams, unofficially placing second (tied) (Quaicoe, 2023; Boateng et al., 2024). Improvements and extensions of this AI could potentially be deployed to offer science tutoring to students and eventually enable millions across Africa to have one-on-one learning interactions, democratizing science education.

Brilla AI addresses the challenge of biased AI systems that do not work well in the African context. Brilla AI contains models such as speech-to-text systems that work for African accents, text-to-speech systems that speak with African accents, and question-answering systems that provide answers using African education materials ensuring that AIED tools address the African context.

10.4 Opportunities

There is a proliferation of mobile devices, particularly, smartphones in Africa with a projection of over 600 million smartphones in Africa by the year 2025 (GSMA, 2020). Hence, smartphones provide a unique opportunity to deliver AIED tools just

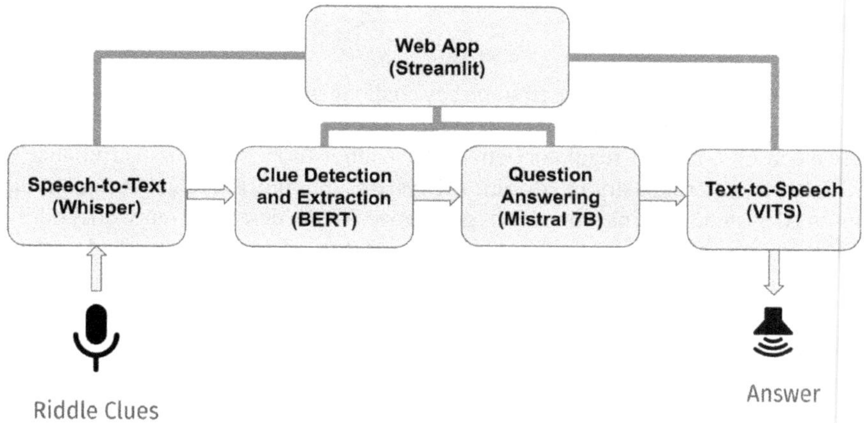

Fig. 10.9 Brilla AI System *source* Boateng et al., (2024)

as we have done with our SuaCode, a smartphone-based app for learning to code (Boateng et al., 2019). For such a context, it is important that these tools are built to have a great mobile user experience that fits different screen sizes.

Large language Models (LLMs) and generative LLMs in particular could be leveraged to develop AI teaching assistants (Boateng et al., 2021, 2024) to augment the efforts of teachers, and AI tutors to offer personalized learning interactions with students. Leveraging approaches such as retrieval augmented generation (RAG) (Lewis et al., 2020) can help to address the hallucination problem of generative AI, enabling their use to generate lesson plans, lesson content, exercises, and exam questions, all grounded in local syllabi. Furthermore, RAG can be used to generate open-source textbooks using the local syllabus as the grounding, which could then be edited by local teachers to make various sections contextually relevant. These models could also be used to better convert scanned scientific and mathematical documents into well-formatted outputs.

Various initiatives could be run to crowdsource ideas to build and develop AIED tools in Africa. We started such an initiative called AfricAIED [3], a workshop on AI in Education in Africa. We ran the first version in 2023 which had 35 people in attendance at Google Research Ghana in Accra, Ghana, and 40 people online (Boateng and Kumbol, 2024). Initiatives like this will enable the sharing of best practices and potential pitfalls toward developing AIED tools that benefit all students across Africa. Such efforts like Brilla AI are already resulting in the building of localized models (Boateng et al., 2023b) that could be used to develop and deploy a conversational AI tutor available via mobile devices, even non-smartphones through calling which transcribes local accented speech, provides answers to science questions using local examples, and says them out with a local accent.

[3] https://www.africaied.org

10.5 Conclusion

In this book chapter, we discussed some key challenges with leveraging AI to improve education in Africa such as limited access to computers, affordable Internet, and reliable electricity, lack of regulatory support for innovation, students overreliance on AIED tools, heterogeneity in educational systems, undigitized education materials, biased AI systems and inaccuracies of generative AI. We described our work building and deploying AIED tools in the African context to advance science and computing education. In particular, we highlighted (1) SuaCode, an AI-powered app that enables Africans to learn to code using their smartphones, (2) AutoGrad, an automated grading, and feedback tool for graphical and interactive coding assignments, (3) a tool for code plagiarism detection that shows visual evidence of plagiarism, (4) Kwame, a bilingual AI teaching assistant for coding courses, (5) Kwame for Science, a web-based AI teaching assistant that provides instant answers to students' science questions and (6) Brilla AI, an AI contestant for the NSMQ. We described opportunities to use AI to improve education in Africa such as the proliferation of smartphones, generative AI to build AI teaching assistants and tutors, and AfricAIED, a workshop on AI in Education in Africa. We are bullish on the potential of AI to transform education across Africa and enable equitable, accessible, high-quality education for millions of students. It is the reason we have been leveraging resources that are more accessible like smartphones, training machine learning models with local content, and building various AI-powered education apps for the African context since 2020, long before tools like ChatGPT made AI assistants popular.

Acknowledgements These works have been supported over the years with grants from the Processing Foundation, the African Union, the Africa Prize for Engineering Innovation program of the U.K.'s Royal Academy of Engineering, Dartmouth College, and ETH Zurich. The following generative AI tools were used in preparing this manuscript: ChatGPT, Perplexity, and Consensus.

References

Achiam, J., Adler, S., Agarwal, S., Ahmad, L., Akkaya, I., Aleman, F.L., Almeida, D., Altenschmidt, J., Altman, S., Anadkat, S., et al. (2023). Gpt-4 technical report. arXiv preprint arXiv:2303.08774

Annor, P.S., Kayang, E., Boateng, S., Boateng, G. (2021). Autograd: Automated grading software for mobile game assignments in suacode courses. In: Proceedings of the 10th Computer Science Education Research Conference, pp. **79**–85.

Boateng, G. (2020). Kwame: A bilingual AI teaching assistant for online Suacode courses. arXiv preprint arXiv:2010.11387

Boateng, G. (2021). Kwame: A bilingual ai teaching assistant for online Suacode courses. In: International Conference on Artificial Intelligence in Education, Springer, pp. **93**–97.

Boateng, G., Annor, P.S., Kumbol, V.W.-A. (2021). Suacode Africa: Teaching coding online to Africans using smartphones. In: Proceedings of the 10th Computer Science Education Research Conference, pp. **14**–20

Boateng, G., John, S., Boateng, S., Badu, P., Agyeman-Budu, P., Kumbol, V. (2024). Real-World Deployment and Evaluation of Kwame for Science, an AI Teaching Assistant for Science Education in West Africa. In: Olney, A.M., Chounta, IA., Liu, Z., Santos, O.C., Bittencourt, I.I. (eds) Artificial Intelligence in Education. AIED 2024. Lecture Notes in Computer Science, vol 14830. Springer, Cham. https://doi.org/10.1007/978-3-031-64299-9_9

Boateng, G., John, S., Glago, A., Boateng, S., Kumbol, V. (2022). Kwame for science: An AI teaching assistant based on sentence-BERT for science education in west Africa. iTextbooks@ AIED

Boateng, G., Kumbol, V. (2018). Project isWEST: Promoting a culture of innovation in Africa through stem. In: 2018 IEEE Integrated STEM Education Conference (ISEC), IEEE, pp. **104**–111.

Boateng, G., Kumbol, V. (2024). AfricAIED 2024: 2nd Workshop on Artificial Intelligence in Education in Africa. In: Olney, A.M., Chounta, IA., Liu, Z., Santos, O.C., Bittencourt, I.I. (eds) Artificial Intelligence in Education. Posters and Late Breaking Results, Workshops and Tutorials, Industry and Innovation Tracks, Practitioners, Doctoral Consortium and Blue Sky. AIED 2024. Communications in Computer and Information Science, vol. 2151. Springer, Cham. https://doi.org/10.1007/978-3-031-64312-5_5

Boateng, G., Kumbol, V.W.-A., Annor, P.S. (2019). Keep calm and code on your phone: A pilot of Suacode, an online smartphone-based coding course. In: Proceedings of the 8th Computer Science Education Research Conference, pp. **9**–14.

Boateng, G., Kumbol, V., Kaufmann, E.E. (2023). Can an AI win Ghana's national science and maths quiz? an AI grand challenge for education. arXiv preprint arXiv:2301.13089

Boateng, G., Mensah, J.A., Yeboah, K.T., Edor, W., Mensah-Onumah, A.K., Ibrahim, N.D., Yeboah, N.S. (2023). Towards an AI to win Ghana's national science and maths quiz. In: Deep Learning Indaba 2023

Boateng, G. et al. (2024). Brilla AI: AI Contestant for the National Science and Maths Quiz. In: Olney, A.M., Chounta, IA., Liu, Z., Santos, O.C., Bittencourt, I.I. (eds.) Artificial Intelligence in Education. Posters and Late Breaking Results, Workshops and Tutorials, Industry and Innovation Tracks, Practitioners, Doctoral Consortium and Blue Sky. AIED 2024. Communications in Computer and Information Science, vol 2150. Springer, Cham. https://doi.org/10.1007/978-3-031-64315-6_17

Chegg (2023). Chegg announces CheggMate, the new AI companion, built with GPT-4. https://www.businesswire.com/news/home/20230417005324/en/Chegg-announces-CheggMate-the-new-AI-companion-built-with-GPT-4

Devlin, J., Chang, M.-W., Lee, K., Toutanova, K. (2019). BERT: Pre-training of deep bidirectional transformers for language understanding. In: Burstein, J., Doran, C., Solorio, T. (eds.) Proceedings of the 2019 Conference of the North American Chapter of the Association for Computational Linguistics: Human Language Technologies, vol. 1 (Long and Short Papers), pp. **4171**–4186. Association for Computational Linguistics, Minneapolis, Minnesota. 10.18653/v1/N19-1423 . https://aclanthology.org/N19-1423

Duolingo (2023). Introducing Duolingo Max, a learning experience powered by GPT-4. https://blog.duolingo.com/duolingo-max/

Dzodzegbe, C. (2023). Some subject results of candidates from 235 schools withheld for using AI-generated answers. https://www.myjoyonline.com/some-subject-results-of-candidates-from-235-schools-withheld-for-using-ai-generated-answers/

Ehl, D., Grün G. (2020). Why mobile internet is so expensive in Africa. https://www.dw.com/en/why-mobile-internet-is-so-expensive-in-some-african-nations/a-55483976

GSMA (2020). Mobile Economy Sub-Saharan Africa. https://www.gsma.com/solutions-and-impact/connectivity-for-good/mobile-economy/wp-content/uploads/2024/11/GSMA_ME_SSA_2024_Infographic_Spread.pdf

IEA (2023). SDG7: data and projections, IEA, Paris. https://www.iea.org/reports/sdg7-data-and-projections

Jiang, A.Q., Sablayrolles, A., Mensch, A., Bamford, C., Chaplot, D.S., Casas, D.d.l., Bressand, F., Lengyel, G., Lample, G., Saulnier, L., et al. (2023). Mistral 7b. arXiv preprint arXiv:2310.06825

John, S., Boateng, G. (2021). "i didn't copy his code": Code plagiarism detection with visual proof. In: International Conference on Artificial Intelligence in Education, Springer, pp. **208**–212.

Khan Academy (2023). Harnessing GPT-4 so that all students benefit. A nonprofit approach for equal access. https://blog.khanacademy.org/harnessing-ai-so-that-all-students-benefit-a-nonprofit-approach-for-equal-access/

Kim, J., Kong, J., Son, J. (2021). Conditional variational autoencoder with adversarial learning for end-to-end text-to-speech. In: International Conference on Machine Learning, PMLR, pp. **5530**–5540.

Lewis, P., Perez, E., Piktus, A., Petroni, F., Karpukhin, V., Goyal, N., Küttler, H., Lewis, M., Yih, W.-T., & Rocktäschel, T. (2020). Retrieval-augmented generation for knowledge-intensive NLP tasks. *Advances in Neural Information Processing Systems,33*, 9459–9474.

National Science and Maths Quiz. https://nsmq.com.gh/

Quaicoe, E. (2023). NSMQ 2023: AI answered one riddle correctly ahead of contestants in grand-finale. https://www.myjoyonline.com/nsmq-2023-ai-answered-one-riddle-correctly-ahead-of-contestants-in-grand-finale/

Quizlet (2023). Introducing Q-Chat, the world's first AI tutor built with OpenAI's ChatGPT. https://quizlet.com/blog/meet-q-chat

Radford, A., Kim, J.W., Xu, T., Brockman, G., McLeavey, C., Sutskever, I. (2023). Robust speech recognition via large-scale weak supervision. In: International Conference on Machine Learning, PMLR, pp. **28492**–28518.

Reimers, N., Gurevych, I. (2019). Sentence-BERT: Sentence embeddings using Siamese BERT-networks. arXiv preprint arXiv:1908.10084

SAP: Africa Code Week—Bridging the Digital Skills Gap in Africa (2016). https://africacodeweek.org/fr/blog/https-www.linkedin.com-pulse-africa-code-week-bridging-digital-skills-gap

Streamlit. A faster way to build and share data apps. https://streamlit.io/

SuaCode Africa 2.0: Teaching coding online to Africans using smartphones during COVID-19. https://www.c4dhi.org/news/lecture-by-boateng-suacode-africa-20210122/ (2021)

UNESCO. (2020). Startling digital divides in distance learning emerge. https://www.unesco.org/en/articles/startling-digital-divides-distance-learning-emerge

World Bank Group (2020). Pupil-teacher ratio Sub-Saharan Africa. https://data.worldbank.org/indicator/SE.PRM.ENRL.TC.ZS?locations=ZG

Chapter 11
Educating Manufacturing Operators by Extending Reality with AI

Paul-David Zuercher, Michel Schimpf, Slawomir Tadeja, and Thomas Bohné

Abstract Emerging user interface technologies such as Augmented Reality (AR) and and Virtual Reality (VR), falling under the umbrella term Extended Reality (XR), may leverage Artificial Intelligence (AI) to allow the creation of virtual learning spaces enabling unique pedagogical tools and educational experiences. In this Chapter, we present three approaches utilising AR and VR environments to illustrate the potential of integrating these emerging technologies with AI for manufacturing education. We first demonstrate how an AR interface facilitated with a see-through head-mounted display enhanced with computer vision capabilities enables assisted contextual learning using in-situ guidance by projecting immediate support for novice operators performing common maintenance and repair tasks on 3D printers. We then demonstrate how VR and Head-Mounted Display Virtual Reality (HMDVR)—with different levels of immersion—enable ex-situ learning spaces to enable remote learning of assembly tasks. Together, the AR, Desktop-based Virtual Reality (DVR), and HMDVR examples demonstrate the potential of rethinking the design of pedagogic learning spaces using XR technology. We complement our results with a discussion on the future trajectory of AI-based education.

Keywords Virtual reality · Augmented reality · Extended reality · Artificial intelligence · Computer vision · Manufacturing education

P.-D. Zuercher · M. Schimpf · S. Tadeja · T. Bohné (✉)
Cyber-Human Lab, Department of Engineering, University of Cambridge, Trumpington Street, Cambridge, Cambridgeshire CB2 1PZ, UK
e-mail: tmb35@eng.cam.ac.uk
URL: https://www.ifm.eng.cam.ac.uk/people/tmb35/

P.-D. Zuercher
e-mail: pdz20@eng.cam.ac.uk
URL: https://pauldavidzuercher.com

M. Schimpf
e-mail: ms2957@eng.cam.ac.uk

S. Tadeja
e-mail: skt40@eng.cam.ac.uk

P. Ilic et al. (eds.), *Artificial Intelligence in Education: The Intersection of Technology and Pedagogy*, Intelligent Systems Reference Library 261,
https://doi.org/10.1007/978-3-031-71232-6_11

11.1 Introduction

The fast pace of modern manufacturing process innovation leads to rapid skills demand changes, resulting in gaps between the skills workers have and should have (Wellener et al., 2021). To close such skills gaps, education is vital. Concurrently, the increased digitalisation of the manufacturing sector offers new opportunities for pedagogues to combine digitalised assets, operational knowledge, and information to create digital learning content that provides users with the necessary scalable, actionable, and specialised training (Moencks et al., 2020; Roth et al., 2022).

Modern digital training solutions allow combining interactive technologies such as Augmented Reality (AR) and Virtual Reality (VR) enhanced with AI (Artificial Intelligence) to allow the designing of novel immersive learning spaces for self-guided but technologically assisted learning. AR and VR applications enable advancing educational experiences with pedagogical methods in distinct ways. For AR, digital objects are rendered according to the user's environment. For VR applications, the user is exposed to completely virtual environments, allowing them to train remotely. Thus, AR educational benefits depend on the in-situ value of the presented information while VR training allows educators to immerse learners into realistic and educational learning scenarios. The arising fundamental pedagogic difference stems from the influence of the user environment on the teaching content and warrants different AI approaches to maximise the training effectiveness. Hence, the two virtual learning spaces require distinct pedagogical approaches for personalising feedback using intelligent learning systems that we address in this chapter.

First, in the remainder of this Introduction section we motivate the need for scalable manufacturing education. We will introduce principles for alleviating the underlying pedagogic challenges with Extended Reality (XR) learning environments. Additionally, the section contains the technical background for the presented AI-enhanced AR and VR applications. In the second section, we describe an AR system that assists and guides novices in repairing one of the most common causes of a 3D printer malfunction. Subsequently, in the third section, we present experiment results showing how virtual education can leverage pedagogic design elements. Lastly, in the Discussion and Conclusion section, we provide an overview of selected future research opportunities.

11.1.1 Innovating Manufacturing Needs Continuous Learning

Companies stay competitive by adopting innovative manufacturing techniques. These process innovations result in specialised manufacturing techniques increase the demand for specialised education across sectors (Benešová et al., 2018). Particularly in manufacturing, innovations such as 3D printing have revolutionised traditional production methods but require specialised skills and knowledge (Bozzi et al., 2023). Introducing new technologies requires retraining and upskilling the current workforce (Mital et al.,1999) as manufacturing professionals must understand new mechanics, relevant parts of the underlying software, materials science, and design requirements. Consequently, competitive companies adapting innovative manufacturing techniques, experience a growing need for scalable and adaptive education.

11.1.2 Shortage of Skilled Labour Force

The shortage of skilled workers is estimated at 2.1 million unfilled positions by the year 2030 in the US manufacturing sector alone (Wellener et al., 2021). The shortage is compounded by a generational shift. As older experts retire, the younger generation often lacks familiarity with the existing production methods and processes (Wellener et al., 2021). They also represent diverse educational backgrounds that require thorough onboarding and training (Wellener et al., 2021). To address the challenge of educating the next generation of workers, scalable learning environments facilitated through new technology-based tools have a chance to play a substantial role. Emerging XR interfaces introduced in the following sections are promising tools that can help pedagogies to innovate education with novel education spaces.

11.1.3 Education Augmented with Pedagogic XR

Educating with intelligent user interfaces enables new pedagogical principles for adapting content to the (1) user's abilities, skills, and roles, (2) manufacturing processes or systems, and (3) business objectives. We will specifically focus on unique opportunities for XR interfaces. For instance, AR and VR allow their users to engage in digital content as they would with physical objects through hand-tracking and gesture recognition. A selection of other opportunities is listed below:

- **Interactive, Immersive, In-Context, Multi-Modal Learning Experience:** XR provides an immersive learning context, eliminating the need to switch between traditional resources like paper manuals or screens and the core learning objectives. Integrating learning context and application can promote the learner's focus and engagement (Beck, 2019). Interactive, immersive, in-context training that virtualises decision-making, sensations, and context-dependent judgments integral to the curriculum improves the learning transfer and real-task performance compared to other, cognitively unrealistic training (Jones, 2021; Krathwohl, 2002). Hence, immersive technology can bridge the transfer gap by integrating 3D objects, videos, and other media, enabling multi-modal content presentation for comprehensive conceptual understanding (Bozzi et al., 2023; Tadeja et al., 2023).

- **Learning with Immediate Feedback:** XR applications can support active learning. They interact with users by analysing and verifying different tasks' execution and subsequently provide feedback based on the user's actions (Nkulu-Ily, 2023). This interaction is similar to a training with a personal human instructor. Immersive technology can simulate and mimic real-life experiences in high fidelity. This allows safe learning where collating feedback in the real world might be dangerous for novices to explore (Zhao & Lucas, 2014). This extends to ergonomic benefits where XR grant users the flexibility to place information as per their comfort, freeing them from the limitations posed by physical constraints (Lim et al., 2021).
- **Customization:** XR-based educational applications can be tailored to individual users' needs and wants based on age or experience level. There's even potential for automatic detection and adaptation to the user's needs (Fu & Fischer, 2023). Additionally, XR can be inclusively designed to provide customised learning experiences for those who are immobile or cannot afford to travel to specific learning locations, ensuring that quality education is accessible to everyone, everywhere (Nesenbergs et al., 2020).
- **Scalability:** Once a XR application is developed, it can be easily scaled to multiple devices. This can reduce the amount of human intervention needed and, therefore, can maximise the effectiveness of a human educators' work (Siltanen & Heinonen, 2020). By recording and analysing learning statistics, immersive systems can offer valuable insights for addressing individual difficulties in personalised curriculums (Bozzi et al., 2023; Tadeja et al., 2023).

11.1.4 Pedagogic Learning Spaces in XR Leveraging AI

Utilising AI in XR has created additional advancements. Below are listed some opportunities that have arisen from the synergy of XR and AI:

- **Adaptive Education:** AI-powered XR applications can dynamically change and adjust their content to the user's skills and abilities using learner models, and business objectives to optimize the learning process (Fu & Fischer, 2023).
- **Environmentally Aware Applications:** By integrating AI, XR applications can gain a detailed understanding of the surrounding environment. This allows for a more immersive and context-aware experience, easing learning transfer as the XR content can adapt and respond to real-world stimuli and scenarios as demonstrated in our AR application.
- **Natural Language Interface:** Incorporating Natural Language Processing enables XR devices with voice interactivity (Fu & Fischer, 2023). Users can, for example, input commands in the XR applications while having hands available for task execution.

11.1.5 *Technical Background*

In what follows, we introduce the concepts and technologies discussed in this chapter, including *3D printing* and AR.

11.1.5.1 3D Printing

3D printing is an additive process that produces physical objects from digital designs (Bozzi et al., 2023; Tadeja et al., 2023). Unlike traditional subtractive methods, which remove material from a solid block, 3D printing builds objects additively layer-by-layer. The process can utilise various materials, from plastics to metals, and is typically guided by computer-aided design (CAD) files (Shahrubudin et al., 2019). The flexibility of 3D printing enables the relatively easy and prompt production of complex items and, therefore, finds its use in rapid prototyping. However, working with 3D printers differs from many other techniques and may requires extra training, especially when deployed on a large industrial scale. The 3D printer used for the AR application is a Creality CR-20 Pro (shown in Fig. 11.1).

Fig. 11.1 3D printer for our AR-based repair system

11.1.5.2 Reality—Virtuality Continuum

Immersive user interface technologies are classified on the reality-virtuality continuum into AR and VR (Milgram et al., 1995) (see Fig. 11.2). On the left side of the continuum is reality as experienced with human senses and the pure virtual world is on the other end. The reality-virtuality continuum illustrates the location of the applications our presented AR and VR training environments. We implemented the AR application to work with the Microsoft HoloLens 2 headset (see Fig. 11.3) and the VR applications with Meta's Quest 2 (see Fig. 11.4) as the Head-mounted Display VR (HMDVR) and a laptop-based Desktop-based VR (DVR).

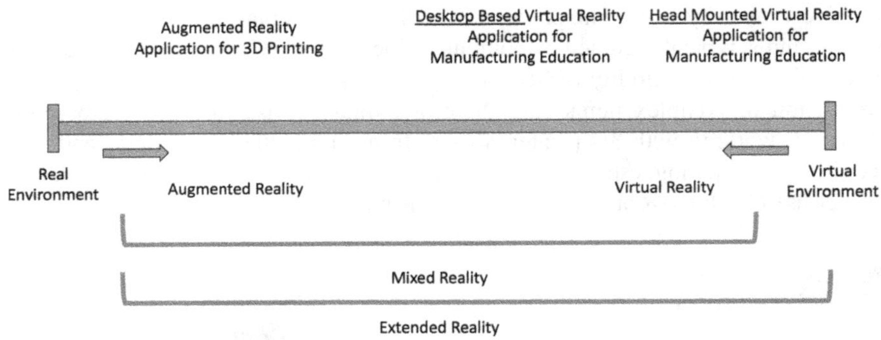

Fig. 11.2 The Reality—Virtuality continuum with examples (compare to Milgram et al. 1995)

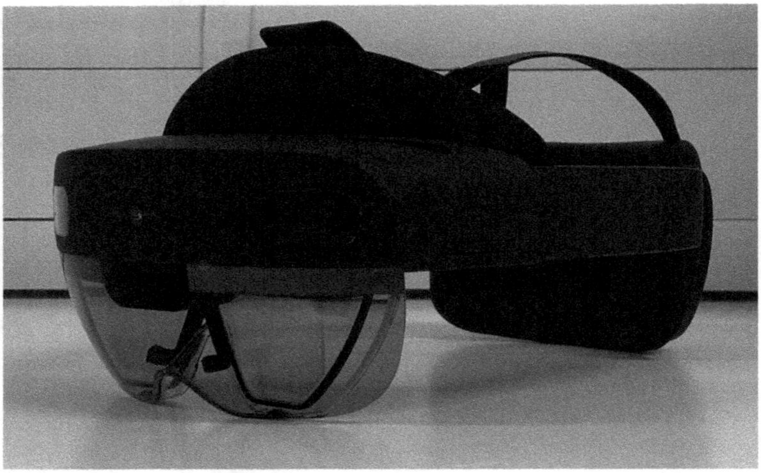

Fig. 11.3 The Microsoft HoloLens2 used for the 3D printing assistance application

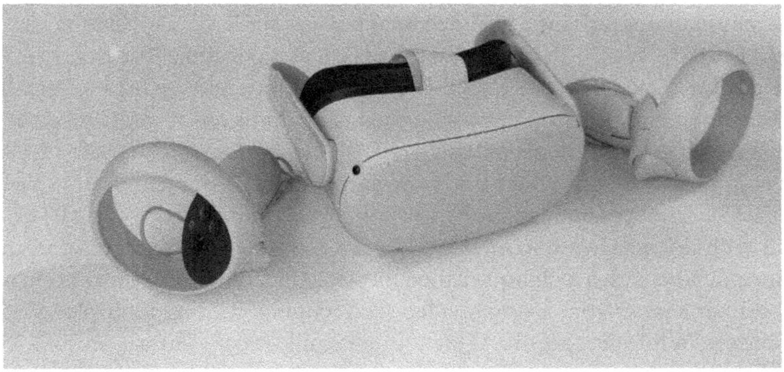

Fig. 11.4 The Meta Quest 2 used for the virtual training

Augmented Reality

AR allows overlaying digital artefacts into the user's field of view (Bozzi et al., 2023; Tadeja et al., 2023). Examples of digital content are 3D objects, images, and animations or videos. Together with various user input possibilities, AR enables operators to interact with digital instructions and guidance embedded in physical working environments. Moreover, AR can be realised with different computing architectures, including mobile devices (such as modern smartphones and tablets) or specifically designed AR headsets that ensure free, unconstrained hand movement when carrying out a given task (Arena et al., 2022).

Virtual Reality

VR allows the creation of virtual worlds through artificial sensory stimulation using displays (LaValle, 2023). For example, audio displays are used to artificially stimulate the hearing sense. The higher the quantity and quality of the artificially stimulated senses, the higher the immersion level of the technology (Miller & Bugnariu, 2016), allowing users to immerse in realistic virtual worlds. Modern VR systems are portable and typically stimulate users' visual, audio, and tactile senses. In the VR example, the Meta Quest 2 was used (compare to Fig. 11.4).

11.1.5.3 Computer Vision

Computer Vision (CV) is a subfield of AI that extracts and processes information from videos, images and other visual inputs like lidar sensors. It is foundational to AR and VR technologies. CV can be divided into several problem categories, including object detection and tracking, image segmentation, and image classification (Cyganek & Siebert, 2011). Algorithms solving these tasks can be applied to many applications, such as autonomous driving, medical imaging, or surveillance (Gao et al., 2018; Janai et al., 2020; Wang, 2013). Although classical algorithms, for example, Canny Edge Detector (Canny, 1986) for edge detection were successful historically, the recent development of neural network-based algorithms led to rapid advancements in the field (Voulodimos et al., 2018). *3D Pose estimation* is a subfield of CV. It is the process of finding out where and how an object is positioned in 3D space (Fan et al., 2022). In our demonstration, we utilise the Vuforia software (PTC, 2023) coupled with AR hardware to enable 3D pose estimation of parts of a 3D printer. More specifically, with a technique called monocular estimation, our solution obtains the 3D printer's pose from a single video feed without needing multiple cameras or depth sensors. While the specific AI techniques employed by Vuforia are undisclosed, many alternative open monocular 3D pose estimation approaches based on neural networks are publicly available (Fan et al., 2022).

11.2 Manufacturing Education with AR and AI

In this section, we present a project showing the potential and some limitations of coupling AR with AI for education and training purposes. To create an immersive learning application, we incorporate AR with 3D pose estimation for locating objects in the user's environment and displaying relevant information relative to these objects (compare to Fig. 11.5).

The developed application aims to assist novices in navigating the complex 3D printer nozzle replacement process. Next, we will (1) explain the task, (2) elicit the system development process, and (3) report the results of a domain expert's assessment.

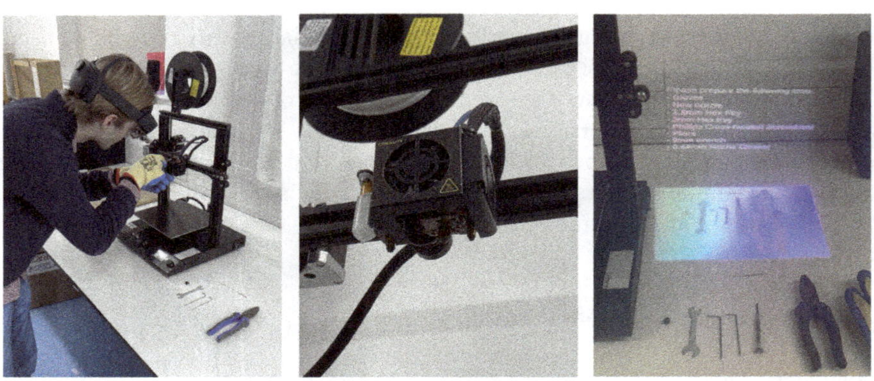

Fig. 11.5 From left to right: the operator with HoloLens 2, 3D printing nozzle, and virtual instruction step as seen by the operator

11.2.1 3D Printer Repair Task

Nozzle replacement is a frequent task when dealing with 3D printers and requires training with detailed instructions to allow 3D printing novices to successfully complete it. Nozzle replacement is necessary when the printer stops layering the filament because the nozzle is blocked and other measures to clean the nozzle from the outside are ineffective.

The nozzle replacement process involves thirty-two subsequent steps, for example, covering, turning on the printer and removing screws, wearing protective gloves to remove preheated parts, replacing the nozzle, and the steps to reassemble the printer. This task was selected in collaboration with an additive manufacturing expert as the nozzle replacement is one of the most common problems in 3D printing maintenance that is complex, varies between specific printers, and is complex but with proper instructions solvable by a novice.

11.2.2 Computer Vision Enabled AR System

To foster an engaging, visual, and spatial learning environment conducive to constructivist principles, the application provides instructions for each repair step directly on or near the 3D printer. This method, referred to as *in-situ guidance*, encourages learners to actively construct their understanding and skills by directly interacting with the 3D environment, thereby facilitating a hands-on, experiential learning process that aligns with constructivist theories of knowledge acquisition.

In-situ guidance is enabled by determining location and orientation through a computer vision (Łysakowski et al., 2023) 3D pose estimation algorithm and then displaying virtual elements relative to this position. Thanks to this, when the user carries out a repair task, there is no need to consult other instructions, such as paper or tablet-based manuals. Instead, the user can look at the printer to see the instructions (shown in Fig. 11.5).

While preparing the instructions, we consulted a domain expert on optimally extracting the filament from the printer. The expert revealed that the procedure requires a specific hand placement, demonstrated through virtual hands (depicted in Fig. 11.6). Illustrating such movements may be challenging in non-immersive interfaces.

Since safety is crucial in any industrial operation, we incorporated safety warnings within our system. Furthermore, through computer vision-enabled in-situ guidance, it is possible to project these warnings directly onto the object being operated on, making them more visually prominent than, for example, a paper manual that is not consistently in the user's field of vision. (shown in Fig. 11.6). Additionally, to improve safety, we overlaid a red virtual heating element during the steps where the heating element is hot, aiming to augment the user's perception by visualising the heat.

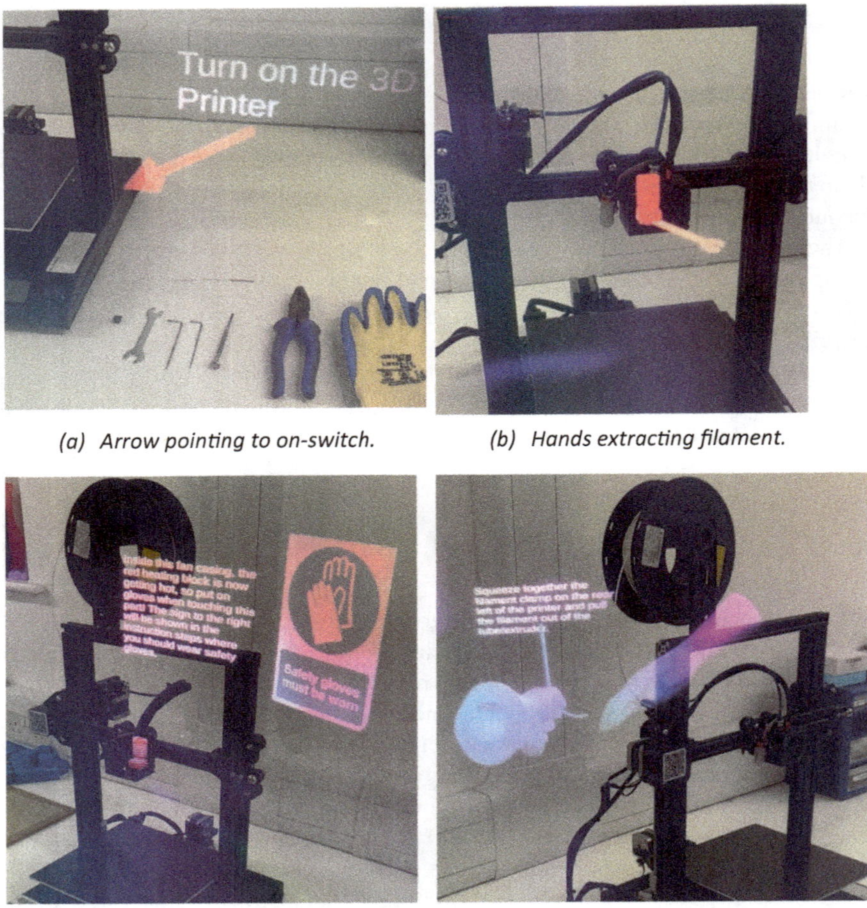

(a) *Arrow pointing to on-switch.* (b) *Hands extracting filament.*

(c) *Safety warnings on the printer.* (d) *Wrench turning left*

Fig. 11.6 Examples of various instruction steps

Moreover, the application displays the right tools and the orientation to turn them to simplify the identification and use of the appropriate tools in each instruction step. Therefore, we also illustrated repair steps like turning the nozzle with 3D models of the tools and added animations for further clarity (see Fig. 11.6).

11.2.3 Computer Vision Enabled AR System Development

During the application development, we created multiple iterations that were evaluated and improved with the feedback of a postdoctoral researcher and an expert

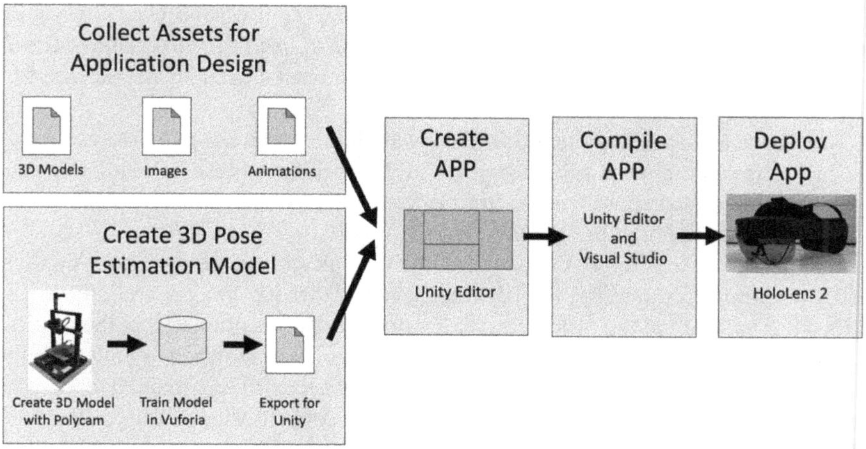

Fig. 11.7 Overview of the development process

in additive manufacturing. The application was built using the *Unity* (Juliani, et al., 2020) game engine and the *Microsoft Mixed Reality Toolkit* (MRTK) (Microsoft, 2022). The former is a video game engine that facilitates the creation of 3D scenes for XR software. MRTK, on the other hand, is a software framework tailored for developing VR and AR applications.

For the computer vision pose detection, particularly to identify parts of the 3D printer, we relied on *Vuforia* engine (PTC & "Vuforia"., 2023). It is a suite of tools for AR application development. The used *Model Target Module* operates by taking a computer-aided design (CAD) file and training a CV model to recognise the position and orientation of the object depicted in the CAD file. In our case, we generated the CAD file of the 3D printer with photogrammetry (Mikhail et al., 2001) using *PolyCam* (Polycam, 2023), an iOS application. This file can be seen in the bottom left of Fig. 11.7. To construct the 3D printer's 3D model with PolyCam, we rotated a phone camera of an iPhone 11 Pro around an object and captured 250 images of the printer that were subsequently used.

11.2.4 Computer Vision Enabled AR System Assessment

We assessed the AR system for 3D printer repair with a second domain expert participant, who is a postdoctoral researcher in the academic 3D printing lab. The expert found the application particularly beneficial for novices, although she also reflected that there is no need for such assistance for experienced professionals like herself.

Regarding the CV integration into the application, she appreciated the *in-situ guidance*. She liked that the virtual instructions were rendered close to or on top of the currently used physical parts and tools. Thus, there was no need for context switching between a virtual screen and the task at hand. Additionally, she mentioned that the *in-situ guidance* helps to locate where subtasks need to be executed by displaying the information at the specific location.

Moreover, a noteworthy point she raised was about safety. Despite her experience, she had grown complacent about using gloves for specific subtasks. By utilising pose estimation, the system could place reminders to wear gloves right on the object where needed, making them, in her opinion, hard to ignore.

In conclusion, the expert assessed that the CV integration supported the application's purpose. Furthermore, the participant noted that the images and animations helped her understand the task better. However, she also mentioned that the detailed guidance slowed the task execution. Her other concern was the application's inability to automatically recognise the completion of a step, necessitating manual input. This could be a potential area for improvement where a sophisticated CV algorithm may enhance the user's experience.

11.3 VR for Hands-On Manufacturing Education

In the prior section, we provided an example of how manufacturing education could be facilitated by AR enhanced with AI. In this section, we present and discuss two *virtual* education examples. We demonstrate how design elements can support the transfer of educational content between DVR and HMDVR. We then discuss the virtuality-reality continuum and the potential of enhancing education by applying AI to XR.

11.3.1 Research Design

The virtual training example in our study conveys the assembly of a low-frequency converter. We mapped the training procedure to DVR and VR aligned with pedagogic design elements (Bohné et al., 2022; Farr et al., 2023; Radianti et al., 2020) (compare to Table 11.1). Subsequently, we collected the time to complete and the number of mistakes in the training from over 90 participants.

11.3.2 Research Strategy

We report the training's performance determined in the experiment depending on the user interface technology (DVR or HMDVR). The used technology is the independent variable and the training's major effectiveness metrics (i.e., number of mistakes and time for completing the training) are the dependent variables (Fisher, 1992;

Table 11.1 Overview of the DVR and HMDVR groups' pedagogic design elements

Design element	Platform		Implementation equivalence
	Head-Mounted Display VR	Desktop VR	
Instructions			
Cognitive load effects			
Modality effect	Text and audio	Text and audio	Exactly same
Split-attention	Attention cue: colour and arrow	Attention cue: colour and arrow	Exactly same
Persona effect	Yes	Yes	Exactly same
Object Placement	Place with Rotation Gimbal	Auxiliary placement	Based on best-practices
Immediate feedback			
Audio-based feedback	Yes	Yes	Exactly same
Text-based feedback	Supportive	Supportive	Exactly same

Sitzmann & Yeo, 2013). Comparing these groups indicates the effects of different user interfaces on the training effectiveness. We recruited the participants from the targeted manufacturing apprentices of a training school and randomly assigned each participant to either DVR or VR training.

11.3.2.1 Experimental Design

The experiment consists of six scenes, structuring the user's tasks and environments (see Fig. 11.8).

Ethics and Legal Surveys

Provide the participant with information about the experiment and collate their consent to participate in the experiment.

Familiarisation of Environment and Controls

Is implemented as the DVR and HMDVR the controls differ (compare to Table 11.2). To ensure participants understand the controls and the basic task operations, we implemented a familiarisation scene teaching all necessary interactions and operations with guided instructions and an example of a simple assembly task.

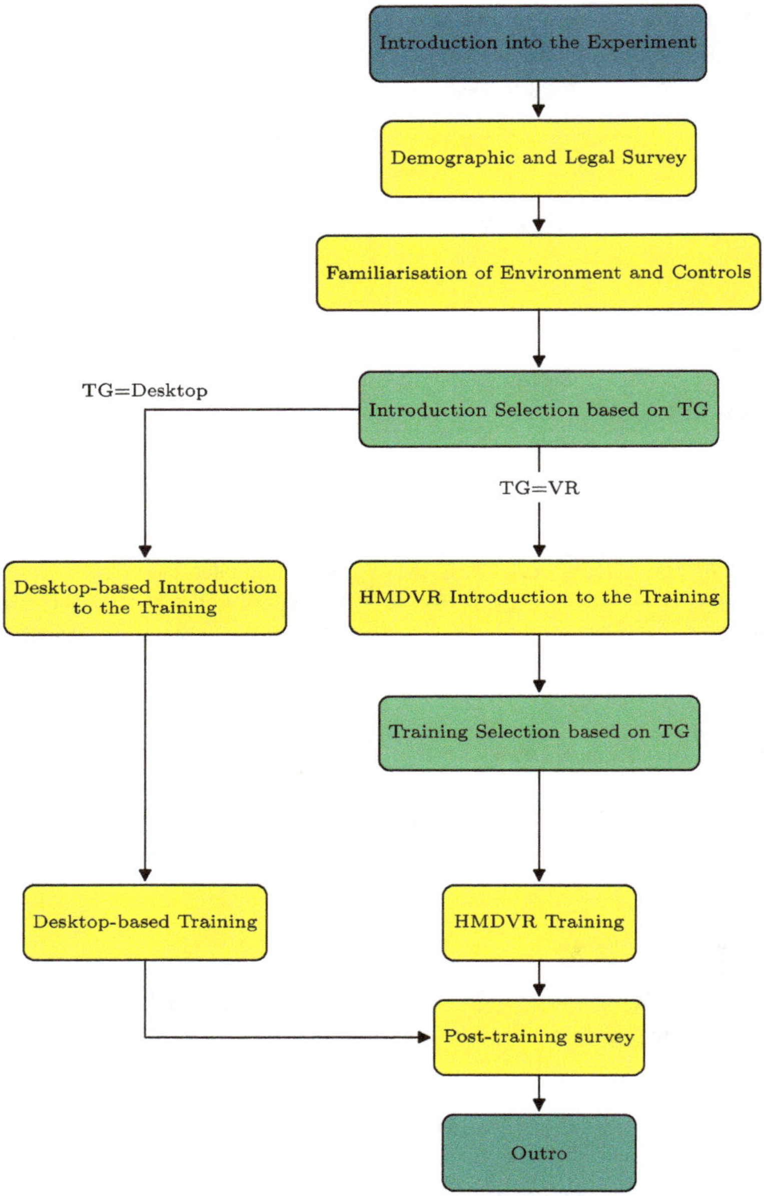

Fig. 11.8 Scenes of the training

Table 11.2 Interactions possible in the virtual education environment

Interaction	DVR control	HMDVR control
Moving around		
Movement-forward, left, right, back	Press W, A, S, D or arrow keys	Move head
Zoom in/Zoom out	Scroll up or down	Move head
Lean forward	Press Q	Lean head forward
Basic interaction with objects		
Picking something up/Putting something down	Left mouse-click	Primary trigger
Assembling objects		
Rotate an object in place	Rotation axis Gizmo; dragging to rotate	Rotate controller
Activate the function of (or place) an object	Right mouse-click	Secondary trigger

Assembly Training

The scene consists of multiple instruction steps designed to help train participants to assemble a frequency converter. We collect the time to complete, and the number of mistakes made in the training.

11.3.2.2 Learning Environment Design

Movement

Is enabled in the DVR environment by pressing W to move forward, A to move to the left, S to move backwards, or D to move to the right. Alternatively, users can use ↑ to move forward, ← to move left, ↓ to move backwards, or → to move right. In VR, users can change their physical position to update their virtual position via trackers in the headset.

Zooming in and Out

In the desktop version refers to the respective decreasing and increasing of the field of view while keeping a constant projection size. In the desktop version, this is controlled via the mouse wheel or multi-touch gesture (Technologies & "Unity 2020). For HMDVR, users can increase the size of objects by moving closer to them so that they naturally appear bigger due to perspective projection.

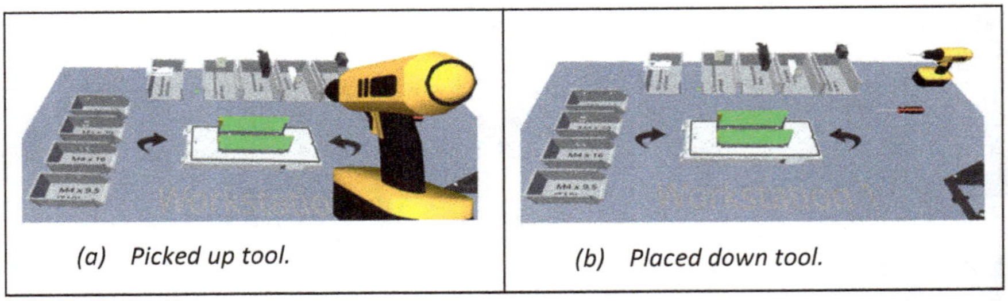

Fig. 11.9 Picking virtual objects up **a** and putting them down **b**

Lean Forward

In the desktop version, forward-leaning can be enabled by holding Q. Leaning forward in HMDVR, users can change their view by just moving their head to the desired position. All tasks are completable without leaning forward, but it can provide a better perspective on the object.

Picking up or Putting Down Objects

In the desktop version is possible with left mouse clicks. An object that is picked up appears on the right side of the user's view. If an object is already picked up and the left mouse is clicked again, the currently held object is translocated back to the original location (compare to Fig. 11.9). Objects can be put down at the correct location by right-clicking with the cursor in the centre of the view on the aimed location. If the location is wrong, a text popup and sound are feedback to the user that the location is invalid. In the HMDVR version, users can pick up objects by putting the controller close to the virtual object and holding the primary trigger. They can then move their hand to the appropriate location and release the primary trigger to place the object. If the position is valid, the part is placed. Otherwise, the part is translocated back to the original position.

11.3.2.3 Object Assembly

Object assembly refers to the combination of constituents into an object. In the training, the parts are collated in trays surrounding the object. At the start of the assembly procedure, the first part, on which all other parts need to be placed, is already at the centre of the table. The assembly process consists of assembly steps in which new parts are added to the iteratively assembled object. Each assembly step consists of three to four steps:

1. Picking up the valid assembly part;
2. Placing the assembly part in the valid location;

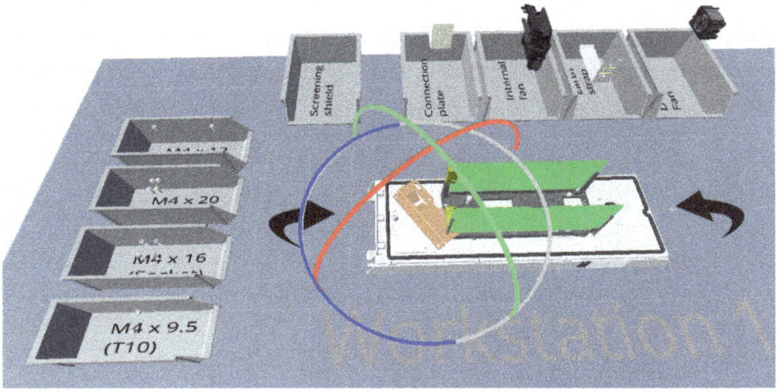

Fig. 11.10 Rotating parts post placement. Users can rotate placed objects by dragging on the gimbal manipulator's axis

3. Rotating the part into the correct orientation;
4. If screwing is required, screw the object into the screw holes using the screwdriver.

Placing an Object or Fixing an Object Using Screws

Refers to a similar process as picking something up or putting something down. Users can activate an object entelechy in DVR via a right mouse click or in HMDVR via a controller's secondary trigger. For example, while holding a screwdriver, the participant can screw in a screw using a screwdriver.

Rotation of Objects

To finish the assembly process, an object must be in the correct orientation. After an object has been placed in the desktop version, users can rotate it by dragging the mouse over a visible gimbal (compare to Fig. 11.10). Additionally, users can rotate the assembled components by left-clicking the black arrows on its side (compare to Fig. 11.10). In the HMDVR version, users can rotate their hand holding the object until the part is in the correct position.

11.4 Results

Educational design elements supported the transfer between media (compare to Table 11.1). From the 91 participants, outliers with more than three standard deviations from the mean were removed. The sample was obtained from *engineering and electrical apprenticeship* students and instructors, with a median age of 20 years. In the presented case, we transferred an optimised desktop-based to a VR training, yielding marginal performance differences. The mean time to completion is not significantly different under an α-level of 0.05 ($p = 0.826$, $t_{Welch}(88.65) = -0.22$) with a mean time to completion in the DVR group of 16.03 minutes and for the HMDVR group of 16.23 minutes (compare to Fig. 11.11a) leading to an average treatment effect of 12.22 s. Similarly, the numbers of mistakes made are not significantly different under an α-level of 0.05 ($p = 0.3570$, $t_{Welch}(86.06) = -0.93$) with the average number of mistakes of 10.51 with DVR and 11.27 with HMDVR (compare to Fig. 11.11b),yielding an average treatment effect of 0.76 mistakes.

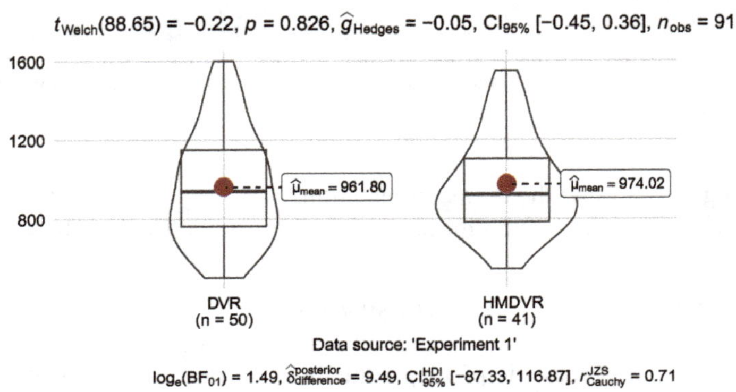

(a) Time to Completion.

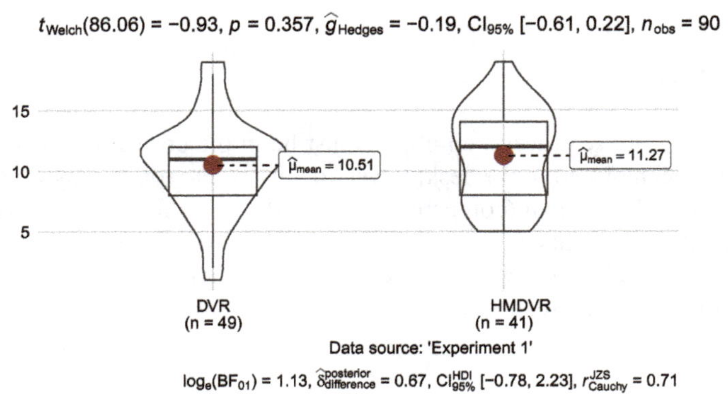

(b) Numbers of Mistakes.

Fig. 11.11 Time to **a** and Mistakes in completing VR versus Desktop-based lesson

11.5 Discussion and Conclusion

Through our 3D printing repair training we provide a proof-of-concept solution, demonstrating that AR and AI can provide immersive and scalable, in-context learning environments. With our DVR and HMDVR assembly training, we demonstrate the effectiveness of remote training using DVR and HMDVR. These two examples show that XR-based approaches allow pedagogues to provide scalable learning environments. However, the studies presented also highlighted limitations to consider when implementing future XR and AI applications in educational contexts.

11.5.1 Domain Specialised Pedagogy

While XR and AI frameworks offer tools to create virtual learning environments, domain-specialised pedagogues are needed to leverage these tools potential. For example, through the 3D printing use case scenario, we could observe the exact movements required to complete the repair task that are not found in regular user manuals. In the assembly training, methods of feedback, attention guidance methods, and subject-specific interactions like object placement and rotation are crucial for pedagogical success. Both these examples show the need for interdisciplinary research focusing on pedagogically beneficial design principles for subject-specific XR training.

11.5.2 Learning Environment Creation

Developing XR training requires in-depth technological knowledge and skills. For example, developing in-situ guidance for educational AR requires the skill of selecting, training, and deploying adequate AI models for the special case of 3D printer maintenance. Thus, lowering the skill barriers for deploying AI models to learning applications is critical for enabling widespread research on pedagogic in-contextual guidance. But even in VR training where AI models are embedded into hardware, creating VR training bridging pedagogic design to viable applications requires in-depth technical skills. Thus, interdisciplinary research projects remain crucial for creating guidance to building pedagogically valuable XR training.

On the XR spectrum, AR and VR are suited for different use cases. The 3D printing learning application shows that creating immersive learning environments enhanced with AI could augment educational applications by providing in-situ information to guide the learning process. VR environments are conversely most suited for training in which physical training is unsafe, inaccessible, unscalable, or inviable where VR training can offer an alternative safe, accessible, and scalable learning space.

11.6 Future Research Opportunities

XR-based education in manufacturing may be enhanced with AI to provide learned with more personalised and adaptive content. Pedagogic approaches are key to guiding AI integration towards effective and affective educational applications (Sitzmann & Weinhardt, 2019). For AR applications, scene understanding and state estimation as part of CV are key capabilities for providing in-context feedback to the learner. For VR, AI could be used to generate virtual environments based on user models, operation models, and objectives to maximise the relevance and effectiveness of trainings (Hooshyar et al., 2020; Jin et al., 2021; Tran O'Leary et al., 2021). Specific ideas for future research are discussed below:

11.6.1 Personalised Learning with Learner Models

The content display can benefit personalisation by modelling the learner's capabilities, skills, and knowledge (Bloom, 1984). Learner models have a long history, and past research could demonstrate that the modelling of abilities, skills, misconceptions, and traits allows the creation of more effective, individualised learning experiences (Abyaa et al., 2019).

11.6.2 Curriculum Generation

For effective education, manufacturing organisations must educate their workforce with relevant curricula. Simple but effective approaches target spaced repetition to maximise the retention rate of relevant content (Farr et al., 2023; Tabibian et al., 2019). AI methods can infer which knowledge gaps to construct the necessary curriculum (Krathwohl, 2002) for the workforce to reach and maintain the desired manufacturing goals (Jin et al., 2021). Furthermore, curriculum learning for AR can provide learners with relevant knowledge based on their environment, while VR learning experiences maximise value if they train in relevant scenarios and skills.

11.6.3 Machine Learning to Optimise Educational Tasks

The task-specific design improvements are possible by learning optimal design features for specific tasks (Shahriari et al., 2015; Snoek, 2015). For complex educational environments, the design space can be large, requiring scalable data acquisition methods to optimise designs (Dudley et al., 2019).

11.6.4 Dialogue Learning with Large Language Models

With the recent advancements in Large Language Models (LLMs), there has been a notable improvement in the capabilities of natural language generation, enhancing the interaction between humans and computers. Specifically, LLMs can facilitate discussions on diverse topics by analysing provided documents, such as user manuals, thereby offering context-aware responses and enriching the user experience (Pereira et al., 2022). Moreover, the integration of image input capabilities in LLMs, exemplified by models such as GPT-4 (OpenAI, 2023), allows applications to interpret and engage with the user's environment. This may enable scalable personalised dialogue by leveraging environmental context.

Acknowledgements This work was supported by the Engineering and Physical Sciences Research Council [EP/S023917/1].

References

Abyaa, A., Khalidi Idrissi, M., & Bennani, S. (2019). Learner modelling: Systematic review of the literature from the last 5 years. *Educational Technology Research and Development, 67*, 1105–1143.

Arena, F., Collotta, M., Pau, G., & Termine, F. (2022). An overview of augmented reality. *Computers, 11*(2), 28. https://doi.org/10.3390/computers11020028

Beck, D. (2019). Special issue: Augmented and virtual reality in education: Immersive learning research. *Journal of Educational Computing Research, 57*(7), 1619–1625. https://doi.org/10.1177/0735633119854035

Benešová, A., Hirman, M., Steiner, F., & Tupa, J. (2018). Analysis of education requirements for electronics manufacturing within concept industry 4.0. In *2018 41st International Spring Seminar on Electronics Technology (ISSE)*, pp. 1–5. https://doi.org/10.1109/ISSE.2018.8443681

Bloom, B. S. (1984). The 2 sigma problem: The search for methods of group instruction as effective as one-to-one tutoring. *Educational Researcher, 13*(6), 4–16.

Bohné, T., Heine, I., Mueller, F., Zuercher, P.-D.J., & Eger, V. M. (2022). Gamification intensity in web-based virtual training environments and its effect on learning. *IEEE Transactions on Learning Technologies, 16*, 1–19. https://doi.org/10.1109/TLT.2022.3208936

Bozzi, L. O. S., Samson, K. D., Tadeja, S., Pattinson, S., & Bohné, T. (2023). Towards augmented reality guiding systems: an engineering design of an immersive system for complex 3D printing repair process. In *2023 IEEE Conference on Virtual Reality and 3D User Interfaces Abstracts and Workshops (VRW)*, pp. 384–389. https://doi.org/10.1109/VRW58643.2023.00084

Canny, J. (1986). A computational approach to edge detection. *IEEE Transactions on Pattern Analysis and Machine Intelligence, 6*, 679–698.

Cyganek, B., & Siebert, J. P. (2011). *An introduction to 3D computer vision techniques and algorithms*. John Wiley & Sons.

Dudley, J. J., Jacques, J. T., & Kristensson, P. O. (2019). Crowdsourcing interface feature design with bayesian optimization. In *Proceedings of the 2019 CHI Conference on Human Factors in Computing Systems*, in CHI '19. New York, NY, USA: Association for Computing Machinery, pp. 1–12. https://doi.org/10.1145/3290605.3300482

Fan, Z., Zhu, Y., He, Y., Sun, Q., Liu, H., & He, J. (2022). Deep learning on monocular object pose detection and tracking: A comprehensive overview. Available: https://arxiv.org/abs/2105.14291

Farr, A., Pietschmann, L., Zürcher, P., Bohné, T., & Yapici, G. G. (2023). Skill retention after desktop and head-mounted-display virtual reality training. *Experimental Results, 4*, e2.

Fisher, R. A. (1992). The arrangement of field experiments. *Breakthroughs in statistics* (pp. 82–91). Springer.

Fu, L. C., & Fischer, R. (2023). Collaboration assistance through object based user intent detection using gaze data. In *Proceedings of the 2023 Symposium on Eye Tracking Research and Applications*, in ETRA '23. New York, NY, USA: Association for Computing Machinery. https://doi.org/10.1145/3588015.3590130

Gao, J., Yang, Y., Lin, P., & Park, D. S. et al., (2018).Computer vision in healthcare applications. *Journal of Healthcare Engineering*, vol. 2018.

Hooshyar, D., Pedaste, M., Saks, K., Leijen, Ä., Bardone, E., & Wang, M. (2020). Open learner models in supporting self-regulated learning in higher education: A systematic literature review. *Computers & Education, 154*, 103878.

Janai, J., Güney, F., Behl, A., Geiger, A., et al. (2020). Computer vision for autonomous vehicles: Problems, datasets and state of the art. *Foundations and Trends in Computer Graphics and Vision, 12*(1–3), 1–308.

Jin, Y., Lu, J., Wang, G., Wang, R., & Dimitris, K. (2021). Semantic modeling supports the integration of concept-decision-knowledge. In *Advances in Production Management Systems. Artificial Intelligence for Sustainable and Resilient Production Systems: IFIP WG 5.7 International Conference, APMS 2021, nantes, france, Proceedings, Part IV*, Springer, 2021, pp. 208–217.

Jones, M. (2021). Effects of simulation fidelity on learning transfer. *Journal of Educational Informatics, 2*(1), 24–34.

Juliani A. et al., (2020). Unity: A general platform for intelligent agents. Available: https://arxiv.org/abs/1809.02627

Krathwohl, D. R. (2002). A revision of bloom's taxonomy: An overview. *Theory into Practice, 41*(4), 212–218.

LaValle, S. M. (2023). *Virtual reality*. Cambridge University Press.

Lim, A. K., Ryu, J., Yoon, H. M., Yang, H. C., & Kim, S. (2021). Ergonomic effects of medical augmented reality glasses in video-assisted surgery. *Surgical Endoscopy, 36*(2), 988–998. https://doi.org/10.1007/s00464-021-08363-8

Łysakowski, M., Żywanowski, K., Banaszczyk, A., Nowicki, M. R., Skrzypczyński, P., & Tadeja, S. K. (2023). Real-time onboard object detection for augmented reality: enhancing head-mounted display with YOLOv8. In *2023 IEEE International Conference on Edge Computing and Communications (EDGE)*, pp. 364–371. https://doi.org/10.1109/EDGE60047.2023.00059

Microsoft, (2022). Mixed reality toolkit. Available: https://learn.microsoft.com/en-us/windows/mixed-reality/mrtk-unity/mrtk2/?view=mrtkunity-2022-05

Mikhail, E. M., Bethel, J. S., & McGlone, J. C. (2001). *Introduction to modern photogrammetry*. John Wiley & Sons.

Milgram, P., Takemura, H., Utsumi, A., & Kishino, F. (1995). Augmented reality: A class of displays on the reality-virtuality continuum. in *Telemanipulator and Telepresence Technologies*, Spie, pp. 282–292.

Miller, H., & Bugnariu, N. (2016). Level of immersion in virtual environments impacts the ability to assess and teach social skills in autism spectrum disorder. *Cyberpsychology, Behavior and Social Networking, 19*, 246–256. https://doi.org/10.1089/cyber.2014.0682

Mital, A., et al. (1999). The need for worker training in advanced manufacturing technology (AMT) environments: A white paper. *International Journal of Industrial Ergonomics, 24*(2), 173–184. https://doi.org/10.1016/s0169-8141(98)00024-9

Moencks, M., Roth, E., & Bohné, T. (2020). Cyber-physical operator assistance systems in industry: Cross-hierarchical perspectives on augmenting human abilities. In *2020 IEEE International Conference on Industrial Engineering and Engineering Management (IEEM)*, pp. 419–423. https://doi.org/10.1109/IEEM45057.2020.9309734

Nesenbergs, K., Abolins, V., Ormanis, J., & Mednis, A. (2020). Use of augmented and virtual reality in remote higher education: A systematic umbrella review. *Education Sciences, 11*(1), 8. https://doi.org/10.3390/educsci11010008

Nkulu-Ily, Y. S. (2023). Combining XR and AI for integrating the best pedagogical approach to providing feedback in surgical medical distance education. In *Lecture notes in Computer Science*, Cham, Switzerland: Springer Nature Switzerland, pp. 452–466. https://doi.org/10.1007/978-3-031-32883-1_41

OpenAI. (2023). GPT-4V(ision) system card. Available: https://cdn.openai.com/papers/GPTV_System_Card.pdf

Pereira, J., Fidalgo, R., Lotufo, R., & Nogueira, R. (2023). Visconde: Multi-document QA with GPT-3 and neural reranking. Available: https://arxiv.org/abs/2212.09656

Polycam, (2023). Polycam. Available: https://poly.cam

PTC, (2023). Vuforia. Available: https://developer.vuforia.com

Radianti, J., Majchrzak, T. A., Fromm, J., & Wohlgenannt, I. (2020). A systematic review of immersive virtual reality applications for higher education: Design elements, lessons learned, and research agenda. *Computers & Education, 147*, 103778.

Roth, E., Moencks, M., Beitinger, G., Freigang, A., & Bohné, T. (2022). Microlearning in human-centric production systems. In *2022 IEEE International Conference on Industrial Engineering and Engineering Management (IEEM)*, pp. 0037–0041. https://doi.org/10.1109/IEEM55944.2022.9989589

Shahriari, B., Swersky, K., Wang, Z., Adams, R. P., & De Freitas, N. (2015). Taking the human out of the loop: A review of bayesian optimization. *Proceedings of the IEEE, 104*(1), 148–175.

Shahrubudin, N., Lee, T. C., & Ramlan, R. (2019). An overview on 3D printing technology: Technological, materials, and applications. *Procedia Manufacturing, 35*, 1286–1296.

Siltanen, S., & Heinonen, H. (2020). Scalable and responsive information for industrial maintenance work: Developing XR support on smart glasses for maintenance technicians. In *Proceedings of the 23rd International Conference on Academic Mindtrek*, in AcademicMindtrek '20. New York, NY, USA: Association for Computing Machinery, pp. 100–109. https://doi.org/10.1145/3377290.3377296.

Sitzmann, T., & Weinhardt, J. M. (2019). Approaching evaluation from a multilevel perspective: A comprehensive analysis of the indicators of training effectiveness. *Human Resource Management Review, 29*(2), 253–269. https://doi.org/10.1016/j.hrmr.2017.04.001

Sitzmann, T., & Yeo, G. (2013). A meta-analytic investigation of the within-person self-efficacy domain: Is self-efficacy a product of past performance or a driver of future performance? *Personnel Psychology, 66*(3), 531–568.

Snoek J. et al., (2015). Scalable bayesian optimization using deep neural networks. In *International Conference on Machine Learning*, PMLR, pp. 2171–2180.

Tabibian, B., Upadhyay, U., De, A., Zarezade, A., Schölkopf, B., & Gomez-Rodriguez, M. (2019). Enhancing human learning via spaced repetition optimization. *Proceedings of the National Academy of Sciences, 116*(10), 3988–3993.

Tadeja, S. K., Bozzi, L. O. S., Samson, K. D., Pattinson, S. W., & Bohné, T. (2023). Exploring the repair process of a 3D printer using augmented reality-based guidance. *Computers & Graphics, 117*, 134–144. https://doi.org/10.1016/j.cag.2023.10.017

Tran O'Leary, J., Nandi, C., Lee, K., & Peek, N. (2021). Taxon: A language for formal reasoning with digital fabrication machines. In *The 34th Annual ACM Symposium on User Interface Software and Technology*, pp. 691–709.

Unity Technologies (2020). Unity documentation version: 2020.3 - input.GetAxis. Accessed Jan 2021. [Online]. Available: https://docs.unity3d.com/ScriptReference/Input.GetAxis.html

Voulodimos, A., Doulamis, N., Doulamis, A., & Protopapadakis, E. (2018). Deep learning for computer vision: A brief review. *Computational Intelligence and Neuroscience, 2018*, 1–13. https://doi.org/10.1155/2018/7068349

Wang, X. (2013). Intelligent multi-camera video surveillance: A review. *Pattern Recognition Letters, 34*(1), 3–19.

Wellener, P., Reyes, V., Ashton, H., & Moutray, C. (2021) Creating pathways for tomorrow's workforce today, *Deloitte Insights*.

Zhao, D., & Lucas, J. (2014). Virtual reality simulation for construction safety promotion. *International Journal of Injury Control and Safety Promotion, 22*(1), 57–67. https://doi.org/10.1080/17457300.2013.861853

Chapter 12
Pedagogical Restructuring of Business Communication Courses: AI-Enhanced Prompt Engineering in an EFL Teaching Context

Debopriyo Roy

Abstract ChatGPT represents a significant advancement in the field of EFL teaching pedagogy, offering both challenges and opportunities for redefining student interactions with AI chatbots and reshaping expectations for various assignments. This chapter explores the potential of utilizing prompt engineering as a methodology, in conjunction with Google search, to establish innovative approaches to foreign language education, similar to the strategies employed in language instruction. The aim is to foster critical thinking and assessment while using ChatGPT, with a focus on content authoring, the development of critical reasoning skills, and document production. In this chapter, we delve into a case study centered on project-based language learning within a business communication course tailored for EFL pedagogy. The primary objective is to investigate how prompt engineering can cultivate an entrepreneurial mindset in students, resulting in document production as a natural byproduct of critical business thinking and problem-solving. The emphasis shifts away from simply teaching the mechanics of language for technical and business reports and documents. Instead, it centers on guiding students to pose pertinent questions and construct logical layers of argumentation leading to the intended document as the final outcome. While our case study focuses on business communication, the principles of engagement discussed here can be extrapolated to various domains and diverse logical reasoning contexts.

Keywords Business communication · Generative AI · Pedagogy · Prompt engineering

---

D. Roy (✉)
Center for Language Research, School of Computer Science & Engineering, The University of Aizu, Aizuwakamatsu, Japan
e-mail: droy@u-aizu.ac.jp

P. Ilic et al. (eds.), *Artificial Intelligence in Education: The Intersection of Technology and Pedagogy*, Intelligent Systems Reference Library 261,
https://doi.org/10.1007/978-3-031-71232-6_12

12.1 Introduction

With the advent of ChatGPT, there is supposedly an increasing fear in English language teaching circles that hands-on critical thinking, writing, and documentation skills may be largely compromised. Such fears are equally prevalent in many other educational domains. But this wave of using AI chatbots is practically unstoppable, and it will only further develop over time. Now, more than ever, it's vital for higher education institutions to understand and harness the power of AI. Research has provided empirical evidence that supports the potential of ChatGPT as a powerful language-learning tool that EFL learners should utilize to participate in ecological CALL creatively and productively (Liu & Ma, 2023). So, we will need to explore how AI chatbots could be successfully integrated into the EFL classroom in a way where the focus is on cognitive comprehension of the content and content processing rather than focusing exclusively on the students' ability to self-generate text content.

However, research is upbeat about the potential for ChatGPT to aid and promote autodidactic learning and best utilize the power of chatbots and AI for technology-assisted learning (Biswas, 2023). The question that we need to answer is how the language teaching pedagogy should adapt to ChatGPT's ability to produce conversational and natural responses to user input. How should students best utilize the conversational experience that feels so natural and intuitive? How should we take advantage of the personalized and interactive assistance provided and tailored recommendations made in response to specific queries? To what extent is such information trustworthy, specific, and up to date? What are some domains and query types where the response from ChatGPT seems inadequate? Research until now has accepted that such tailored and interactive responses can enhance the learning experience and increase student engagement in online courses while developing independence and self-study skills. However, the question remains as to how such independence and self-study skills would work in contexts that demand the production of language and documents and developing abilities for language learning. The prevalence of ChatGPT raises concern for language teaching professionals because this large language model represented by ChatGPT exhibits the ability to think and answer questions like a human being, manifesting a creative capacity that was previously unavailable to artificial intelligence, marking a qualitative leap from quantitative change (Kosinski, 2023).

You in an article aptly summarized the following:

> *A recent survey revealed that nearly 89% of American college students use Chat GPT to complete homework tasks, with 53% using the tool for writing papers. Additionally, 48% of students use Chat GPT during exams and 22% use Chat GPT to generate paper outlines. However, it is worth noting that some students are not only able to successfully complete assignments using Chat GPT but also achieve high scores. Nevertheless, it is difficult for teachers to determine whether students are using Chat GPT, which has a negative impact on students' over-reliance on this tool, gradually causing them to lose their ability to think critically, explore, verify, and summarize actively. If this trend continues, it will greatly affect students' learning outcomes and development (*Kasneci et al., 2023*).*

With the widespread use of Chat GPT, many schools and institutions worldwide have implemented measures to limit or prohibit its use (Lund & Wang, 2023; Szefer and Deshpande, 2023; Yadava, 2023). Chan and Hu (2023) went on to mention that the implementation of these measures aims to ensure that students use Chat GPT tools correctly, avoid over-reliance and abuse, and safeguard academic integrity and education quality. However, in this article, we question the practicality of such an approach where the institution or the educational system decides to prohibit the use of a tool that will or has already become an integral part of human conversation, documentation, and language production in different domains. Further, even when such restrictions are imposed, it's well accepted that it's difficult for teachers to determine whether students are using Chat GPT, and to what extent. However, the blatant simplistic, and direct use of ChatGPT indeed creates a negative impact on students' over-reliance on this tool, gradually causing them to lose their ability to think critically, explore, verify, and summarize actively (Kasneci et al., 2023).

A fundamental question that we need to ask ourselves as language teachers in a non-native EFL context is how do we handle such instructional and pedagogical situations where the inclination for non-native speakers to use ChatGPT would probably be higher, indiscriminate, and more so for direct language production scenarios?

For example, until now, we could not successfully discourage our students from using translation software to the extent we would have liked. Research (Tsai, 2022) concluded that students using Google Translate as a revision tool displayed better L2 Performance in written language and content than in their self-writing, and especially non-English major students showed significantly more positive attitudes towards the use of Google Translate than English major students. On the contrary, a South African professional translation context in a higher education context concluded that the quality based on Google Translate was still below average, and the texts would require extensive post-editing for their function to be met (Rensburg et al., 2012). So, the point here is that Google Translate is extensively used in non-native contexts, and for translation assignments, but the drafts need extensive editing to bring it up to the professional quality desired. For classroom assignments that could be a major headache for students who depend on such translation software but have to invest efforts in post-editing. Now, this is where ChatGPT brings in huge relief for such students who are not willing to invest in traditional learning through constructivist language production and documentation.

An article by Timothy (2023) explained how ChatGPT could be particularly impactful—although currently getting little attention—in the domain of machine translation. Until now, Google Translate was the elephant in the room, and almost everyone else was playing catch-up. The author has pointed out that unfortunately, a good translation is hard, even from big translation services like Google Translate. Machine translation is tough because languages aren't the same in their approach to constructing sentences. Machine translation tools find it hard to get the context of a statement right. Rather than simply prompting ChatGPT to translate, one can ask ChatGPT to provide meaning in the target language. ChatGPT will then provide a literal translation and interpretation of the idiom. Timothy (2023) went on to say that this could be very useful when translating a large body of text that contains an idiom. In such cases, translating an idiom literally could be a source of confusion when read together with the surrounding text. Timothy (2023) further said that one of ChatGPT's greatest advantages is its ability to adjust its translation based on the context or extra provided by the user. Google Translate cannot currently do this.

These situations may at times (often) make it easier for EFL teachers to identify and differentiate a sincere attempt made by the student to produce language without translation versus with the aid of translation software. But when ChatGPT is used, it may not be easily identifiable unless the teacher has a general understanding of how well the students can perform on their own (without using either machine translation or ChatGPT).

So, this brings us to the question of how best to use ChatGPT in a classroom context, without having to put restrictions that may not be realistic and practically a wave that is unstoppable. Moreover, it's quite premature to create a holistic negative perception of ChatGPT without comprehensively comprehending how the tool could be successfully used in situations and for processes that necessitate constructivism and a design thinking approach.

In language learning and documentation contexts, we often situate our pedagogy on cognitive constructivism, pedagogic constructivism, and psychological constructivism. In such contexts, we need to explore how ChatGPT could be used for learning that is focused on the following fundamentals:

1. Iterative (as new learning draws on existing understanding)—create prompts that are based on existing understanding of the assignment and the context of the application.
2. Interpretative (as the individual draws upon internal interpretive resources (such as their existing concepts) for sense-making—interpret the responses based on the prompts created and ask subsequent questions. The correct interpretation is the key here.
3. Incremental (as the human cognitive system can only construct new knowledge at a limited rate, and it takes time for new learning to be consolidated sufficient for it to act as robust foundations for further learning)—create incremental prompts based on existing and acquired knowledge now. This could be a work in progress until the desired outcome is achieved, not only in terms of the content but how best to structure it in the context optimally.

So, fundamentally the idea is for EFL language instructors to worry less about the response generated by AI in itself but use the responses to structure the most optimal set of questions and tasks (prompts) in the correct sequence for ChatGPT, and use Google search in the process to add credibility to the AI-generated responses. This article is focused on demonstrating why and how the focus should shift from content production to prompt engineering, with the knowledge that prompt generation is a function of all intermediate and final content generated.

12.2 Fundamentals of Prompt Engineering

This brings us to the core discussion of prompt engineering and how it could be used successfully in different contexts of use in an EFL teaching classroom. Prompt engineering is the process of structuring text that can be interpreted and understood by a generative AI model (Diab et al., 2022).

A prompt is natural language text describing the task that an AI should perform. Prompt engineering may consist of a single prompt that includes a few examples for a model to learn from and may involve phrasing a query, specifying a style, providing relevant context, or assigning a role to the AI such as "act as a doctor" (Robinson, 2023; Greenburg & Laura, 2023).

Research has mentioned with clarity that depending on the user's goal, a prompt could be a question, a statement, a short story, a task, or a topic. They can be as easy as one word or as hard as an entire essay. A good prompt should be specific, clear, and give enough background information for GPT-4 to understand the topic and come up with a good answer (Kapoor, 2023).

The advantages of prompt engineering are as follows (Gouws-Stewart, 2023):

1. Enhanced control and relevance
2. Improved efficiency
3. Targeted Information
4. Adaptability and Iteration for Optimal Results
5. Mitigation of limitations and customization
6. Consistency and Fine-Grained Control

The types of prompts are as follows (Kapoor, 2023):

1. Question Prompts—Questions meant to get a specific answer or piece of information.
2. Completion Prompts—Statements that act as a starting point or a small piece of text and tell them how to finish it.
3. Story Prompts—Ask GPT to come up with a story or narrative about a certain topic or theme.
4. Dialogue Prompts—Ask GPT to come up with conversations between characters in a certain situation or setting.
5. Creative Prompts—Open-ended questions that ask GPT to make something creative, like a poem, song, or script.

Research (Yalalov & Gaszcz, 2023) has also suggested that we may also look at basic prompting in the following categories such as -

Summarization, Information Extraction from a Given Text, Question Answering, Information Classification (e.g., positive, negative sentiment, etc.), Conversation Analysis (role-playing), Code Generation, Reasoning, Reviewing Text.

Recent literature has created a ChatGPT prompt guide (Robinson, 2023) as follows:

1. Provide a prompt that offers a context.

2. Provide helpful information upfront to assist ChatGPT in creating a better and more customized response.
3. Provide examples in the prompt.
4. Guide ChatGPT about the length of the expected response.
5. Define the expected format if there are any.
6. Ask it to help you come up with an effective prompt.
7. Use some of the following expressions such as follows:

   - Let's think step by step.
   - Thinking backward (in case ChatGPT is providing inaccurate responses)
   - Ask for responses in the style of …..
   - Ask them to write for/as a specific professional.
   - Write for a specific age group.
   - For a specific type of company/institution
   - Ask ChatGPT to assume a specific role.

Research (Gouws-Stewart, 2023) suggested that a well-crafted prompt could be generated based on the following:

1. Context—A brief introduction or background information to set the context for the conversation. It's also important to give the bot an identity.
2. Instructions—Explain exactly what is required from the AI model to do the role to play, or the type of response expected. The sentences should be framed positively as part of the instructions.
3. Input Data: Include specific examples that you want the Generative AI model to consider or build on. Ask the user open or closed-ended questions with or without examples as input.
4. Output Indicator: Specify the format you want the response in, such as a bullet-point list, paragraph, code snippet, or any other preferred format. It's important to ensure that the instructions are concise, unambiguous, and actionable.

Off-course it's important to understand that in an EFL context, instructors could teach such prompting as part of different assignments geared towards certain expected outputs.

Our students and faculty in the EFL context should also consider a few advanced prompting techniques that will allow students to learn how specific assignments could be completed in stages and a gradual process.

Research (Yalalov & Gaszcz, 2023) has mentioned the following:

1. Zero-shot prompting—LLMs today can do tasks in a single attempt because they have been trained on a lot of data and are adjusted to following directions.
2. Few-shot prompting is a technique that involves providing the model with a small number of examples or demonstrations to improve its performance in tasks where zero-shot capabilities are not enough. This approach can be particularly useful in scenarios where the model needs to learn new concepts quickly.

3. Chain-of-thought (CoT) prompting, which was first described by Wei et al. (2022), permits complicated reasoning abilities through intermediary reasoning processes. On more difficult jobs that demand deliberation before replying, you can combine it with a few-shot prompting to achieve greater outcomes.

There are other layers of complexity in prompt engineering, which are beyond the scope of this chapter. However, these fundamentals will help us explore and analyze our discussed assignments shortly.

The idea of using prompt engineering to shape assignments for EFL pedagogy comes from the fundamental premise that if the assignments are incremental based on certain kinds of complexity that necessitate asking queries in stages, and with diverse and in-depth analysis, and in a constructivist platform of learning, then students will likely develop the ability to structure their queries with increasing awareness of the context and situations. So, we would like to ensure a process of meaningful dialogue between ChatGPT and the student with incremental complexity and knowledge acquirement. Moreover, we would like to create a situation, where students will still need to identify, select, and put together different kinds of information, some obtained from ChatGPT contributing to the fundamental structure of the response, and some kinds of information obtained from search engines such as Google that are very specific and obtained from institutional websites etc. We need students to be able to put these pieces together towards a comprehensive response. Further, language teachers should take advantage of the fact that content produced by ChatGPT could still be questioned in terms of accuracy and authenticity for specific instances, and depending on the complexity, may not be specific enough to generate responses that are actionable by businesses and institutions, beyond a point. It's important to note that ChatGPT comes with confirmation bias. Confirmation bias is the tendency for people to search for, interpret, and remember information in a way that confirms their preexisting beliefs or expectations. This bias can lead people to selectively attend to information that supports their views while ignoring or discounting information that contradicts them. EFL teachers should not shy away from raising these issues where such confirmation bias indicates the use of ChatGPT in a way where the authenticity of the information has not been questioned, or where the prompts are biased. Further, ChatGPT can provide wrong answers, and when confronted, it acts as if it understands that a wrong response was provided but continues to provide wrong responses. This suggests that when it didn't understand the initial prompt, it also struggled to understand any corrections. So, as EFL teachers, some fundamental understanding of prompt engineering is necessary so that it enables the language teachers to frame assignments in a way that necessitates prompting in a range and with certain expertise. As EFL teachers, we should not lose sight of the fact that the ability to ask pertinent questions in the target language and in stages is also a skill that needs to be harnessed. Thus, pedagogy must readjust and reinvent itself for such teaching situations and objectives.

12.3 Using Google Search Together with ChatGPT

So, if we propose teaching prompt engineering in an EFL classroom to ensure more complex assignments are being handled where the output is generated by ChatGPT, but the input is tailored and customized for more complex responses, where does it leave us in terms of using Google search vis-a-vis ChatGPT? A valid question that's being considered is whether ChatGPT will change how people use the internet forever. Is it a threat to Google? (Gohil, 2023). An interesting difference between Google and ChatGPT that we should consider as language teachers in the EFL context is the fact that although it has limited knowledge, ChatGPT is insanely good at understanding the user's prompts and generating responses and solutions through NLP, while Google relies on the available data in its index to answer the user's queries and searches (Gohil, 2023). Further, the reliability of the provided information is questionable in ChatGPT, while Google offers multiple options to choose from for the search results and provides a complete source of information on the result pages. Finally, it should be considered that ChatGPT responses are still in text format, while Google includes text, video, images, etc. as responses.

Interestingly, this controversy related to the use of AI chatbots did not arise when we were exclusively using Google search because the difference lies in the ability to present single-processed and definitive content with ChatGPT, vs. multi-directional and open-ended content processing with Google search. Google search brings in screen clutter and diverse navigation, allowing the user to decide the optimal choice of information, and helping judge its authenticity and credibility. The chapter is an exploratory analysis discussing examples of how with a broad question in mind, ChatGPT could be the starting point for Google search, and Google search along with contextual design thinking, could help create pertinent questions to be asked of ChatGPT, moving towards a comprehensive response.

12.4 ChatGPT in Business Communication Contexts: The Method

As part of this chapter, we indulged in complex assignments in the domain of business and entrepreneurship as part of an advanced graduate course in project management in an EFL context (taught as an independent study course with extensive one-to-one interaction between the teacher and the student). The idea behind the class was to explore the ability of the student to customize content for very specific scenarios that lead to the development of transfer skills for the industry, industrial content production, and generation of documents for employment purposes.

There are plenty of literature resources that allow us to understand and structure the use of ChatGPT for completing business plans (Parsons, 2023); ways entrepreneurs are using ChatGPT during their workday such as for inspiration and ideas, to improve or repurpose content, to ramp up content output, and to improve their processes; how ChatGPT can revolutionize the future of entrepreneurship with market research and business analysis, customer engagement and support, content creation and marketing, product development and innovation, and efficient decision-making and time management (Alqaheri, 2023). ChatGPT could provide incredible assistance when it comes to generating and refining business ideas and market research. However, research also mentioned that such models can provide a wide range of ideas and suggestions, but it is ultimately up to the entrepreneur to choose the best ones and ensure that they are viable and feasible. Additionally, Chat GPT is not yet able to fully understand and analyze complex market conditions or consumer behavior. It is best suited to generating and refining general business ideas, rather than highly specialized or technical ones. It is also important to note that Chat GPT, like any AI technology, is only as good as the data it is trained on (Mia, 2023).

As part of this chapter, we explored how graduate students in technical disciplines used Google search and ChatGPT for complex assignments where responses could not be obtained (in most cases) with zero-shot prompting, but rather involved few-shot prompting and chain-of-thought prompting to successfully complete the assignments. However, it's important to note that as part of this exercise with the specific graduate student, the student was not tutored on the prompting techniques. Rather, the assignments were chosen in a way that would discourage any instant output that would be efficient. Any productive output would have to be obtained based on a gradual process of prompting, combined with the use of Google search in cases where applicable. So, the idea was to encourage the use of both ChatGPT and Google search most optimally, depending on the assignment necessity. The assignments attempted to explore how advanced students in the EFL context generally might attempt to complete these assignments successfully. We chose this specific student because class discussions demonstrated that the student is capable of thinking through the problems, has ideas for entrepreneurship, is involved in advanced tech research that has entrepreneurial potential, and can research the problems presented at reasonable speed, and with acceptable efficiency in terms of information access, selection, and content production.

The following case studies are expected to allow us to explore ChatGPT and Google Search as complementary tools in the writing and documentation processes for online businesses.

- Employment profile creation and targeting specific job opportunities with print resume building and LinkedIn profile.
- Customer profile creation and setting up a strategic plan for targeting customer groups on social media.
- Job application with video resume
- Customer journey mapper for a specific type of product from a specific e-commerce website
- Employer profile evaluation for a specific job type and self-judging the candidate fit.
- Industry 4.0 presentation and concept building

These case studies were chosen because it's difficult to generate a suitable response to the above projects with Chat GPT in a way that is customized for a given context. The specificity of the context could invoke iterative development of specific information, and that in turn will help readers see and extend the connection between different ideas generated in the original response. Other applications of ChatGPT in e-commerce include proofreading, generating SEO titles, keyword searches, generating FAQs, writing emails, market, and competitor research, responding to user reviews, etc. However, for this chapter, we are focused on those specific case studies for business communication courses where Chat GPT will act as a good starting point providing a general framework for the response, but will not be as context-driven, and students will need to go back and forth between ChatGPT and Google, generating visuals and creating more specific information for the given situation. Following a design thinking approach, the idea is to create a basic preliminary text paragraph response for the above cases based on Chat GPT and Google search (with citations).

Table 12.1 provides examples of how the prompt types could fit in a business communication context. This example set will allow us to better comprehend a possible optimal use of ChatGPT by both my instructors and students in a business communication course. Table 12.2 shows how to further develop the initial prompt examples.

We do not always need to use ChatGPT for all purposes. For example, there are professional AI-generated resume builders that will walk people through in terms of content choice, formatting, design and writing, and presentation tips. However, for EFL courses, we need conversational interfaces where students are taught to create

Table 12.1 Common prompt types and examples in a business communication context

Prompt type-Initial	Example
Question prompt:	How should I design a resume that best demonstrates my profile as a software engineer? How should I turn these printed resume instructions into a video resume?
Completion prompt	Read my presentation slides content on smart museums. Now help me design a UML diagram based on step-by-step instructions suggesting the typical strategies that I should adopt for people of my small city in Japan to make them feel attracted to the concept of smart museums as tourists
Story prompt	Develop persuasive narratives specific to specific customer types based on the customer segmentation model that we used for this analysis related to the smart museum. Write these narratives in separate paragraphs for better understanding
Dialogue prompts	Set up a typical conversation between a customer and a government official from the city tourism department who oversees promoting the concept of the smart museum
Creative prompts	Develop a creative advertising tagline to get people to visit the smart museum set up in Aizuwakamatsu, Japan Write an email encouraging software engineers in the city to participate with creative ideas for the revitalization of the local economy with relevant tech-oriented projects

Table 12.2 More advanced requests with prompting

Prompt type-Follow-up information	Example
Offering context	Design a resume based on the following information: I have worked as a software engineer for 3 years with background skills that range from Python, R, SPSS, and other related software. I have also done internships at ………; and I graduated from the University of Aizu in 2020
Helpful information upfront	Please design this resume in a way that shows I have enough technical skills, and have experience working on relevant projects, both as part of my courses and during internships
Examples in the prompt	Customize my resume for companies such as Accenture, Toshiba, and Mitsubishi asking for software engineers. Please read the specific resume type and develop something very similar
Length of expected response	The length of this resume should be no more than a single A4 page with a 1-inch margin on all 4 sides
Define expected format	I want a chronological/functional/combo resume format
Ask for help to come up with an appropriate Prompt	Write a prompt where I would like to ask ChatGPT to develop a request to customers asking what they like about the ideas on smart museums. This should be a professional request sent as part of a formal mass email

comprehensive prompts, get a response, and then complete the tasks related to the choice of content and writing styles, design, formatting, etc. based on the suggestions made by ChatGPT as an advisor.

This article not only focused on prompts but also reflected on typical reflections of such use by a non-native graduate student who went through the assignments discussed. This is just one example reflection of ChatGPT use in an EFL context, and shouldn't be overly generalized, but help readers understand how a typical usage of ChatGPT might look like and the impressions formed along the way.

For the following assignments, the details of the assignment instructions and how it fosters a constructivist mindset with the use of ChatGPT and Google search (as applicable) are in focus.

12.5 Content Production and Analysis: Employment Profile Creation and Targeting Specific Job Opportunities

As part of this assignment, the student was provided with ChatGPT-generated guidance on how to proceed with designing a general professional resume, and then two separate customized resumes geared towards two specific types of industry jobs. ChatGPT was asked to guide how to design a resume catering to specific types of industries. The idea was to help the student prepare the resume and customize it accordingly. We left it to the judgment of the student as to how much customization was necessary for specific job types. It's important to understand that the use of ChatGPT is not only for students to complete their assignments but is also a valuable tool for instructors to guide students to complete assignments, based on detailed prompts that could be otherwise time-consuming to generate.

Figure 12.1 demonstrated the assignment provided to the student in a way that would likely help complete the assignment without having to focus on supplementary material for structuring the resumes, and the LinkedIn profiles. These instructions are generated by ChatGPT along with the link to resume types. The idea is for the student to use these instructions to create more intermediate prompts that would help with more customized and focused responses when designing LinkedIn profiles and resumes.

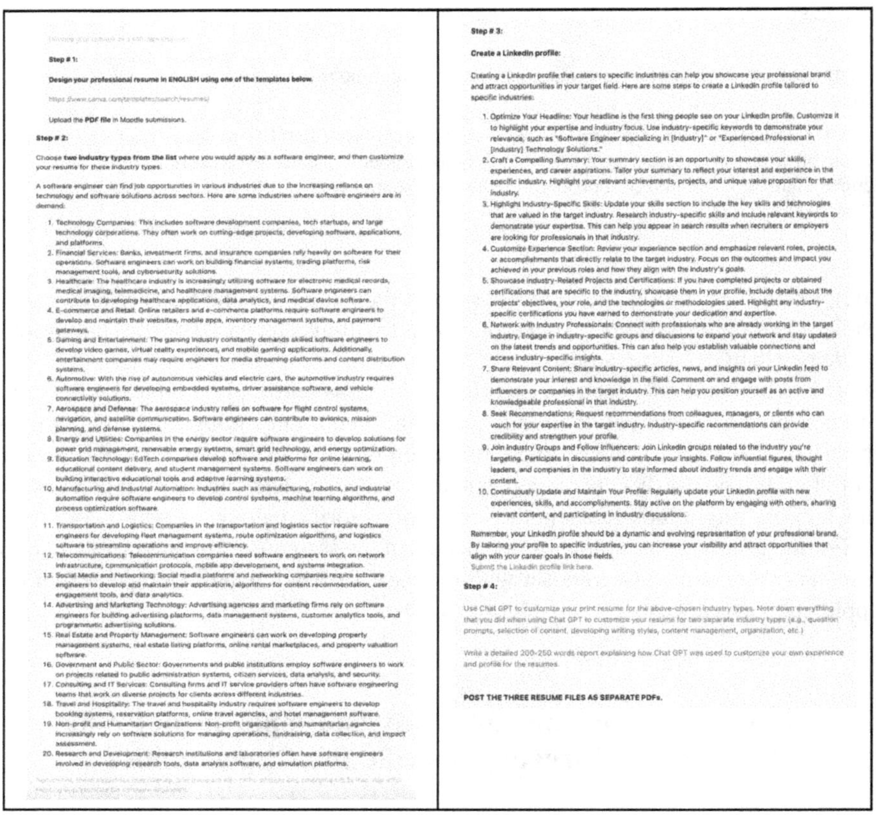

Step # 1:

Design your professional resume in ENGLISH using one of the templates below.

https://www.canva.com/templates/search/resumes/

Upload the **PDF file** in Moodle submissions.

Step # 2:

Choose **two industry types from the list** where you would apply as a software engineer, and then customize your resume for these industry types.

A software engineer can find job opportunities in various industries due to the increasing reliance on technology and software solutions across sectors. Here are some industries where software engineers are in demand:

1. Technology Companies: This includes software development companies, tech startups, and large technology corporations. They often work on cutting-edge projects, developing software, applications, and platforms.
2. Financial Services: Banks, investment firms, and insurance companies rely heavily on software for their operations. Software engineers can work on building financial systems, trading platforms, risk management tools, and cybersecurity solutions.
3. Healthcare: The healthcare industry is increasingly utilizing software for electronic medical records, medical imaging, telemedicine, and healthcare management systems. Software engineers can contribute to developing healthcare applications, data analytics, and medical device software.
4. E-commerce and Retail: Online retailers and e-commerce platforms require software engineers to develop and maintain their websites, mobile apps, inventory management systems, and payment gateways.
5. Gaming and Entertainment: The gaming industry constantly demands skilled software engineers to develop video games, virtual reality experiences, and mobile gaming applications. Additionally, entertainment companies may require engineers for media streaming platforms and content distribution systems.
6. Automotive: With the rise of autonomous vehicles and electric cars, the automotive industry requires software engineers for developing embedded systems, driver assistance software, and vehicle connectivity solutions.
7. Aerospace and Defense: The aerospace industry relies on software for flight control systems, navigation, and satellite communication. Software engineers can contribute to avionics, mission planning, and defense systems.
8. Energy and Utilities: Companies in the energy sector require software engineers to develop solutions for power grid management, renewable energy systems, smart grid technology, and energy optimization.
9. Education Technology: EdTech companies develop software and platforms for online learning, educational content delivery, and student management systems. Software engineers can work on building interactive educational tools and adaptive learning systems.
10. Manufacturing and Industrial Automation: Industries such as manufacturing, robotics, and industrial automation require software engineers to develop control systems, machine learning algorithms, and process optimization software.
11. Transportation and Logistics: Companies in the transportation and logistics sector require software engineers for developing fleet management systems, route optimization algorithms, and logistics software to streamline operations and improve efficiency.
12. Telecommunications: Telecommunication companies need software engineers to work on network infrastructure, communication protocols, mobile app development, and systems integration.
13. Social Media and Networking: Social media platforms and networking companies require software engineers to develop and maintain their applications, algorithms for content recommendation, user engagement tools, and data analytics.
14. Advertising and Marketing Technology: Advertising agencies and marketing firms rely on software engineers for building advertising platforms, data management systems, customer analytics tools, and programmatic advertising solutions.
15. Real Estate and Property Management: Software engineers can work on developing property management systems, real estate listing platforms, online rental marketplaces, and property valuation software.
16. Government and Public Sector: Governments and public institutions employ software engineers to work on projects related to public administration systems, citizen services, data analysis, and security.
17. Consulting and IT Services: Consulting firms and IT service providers often have software engineering teams that work on diverse projects for clients across different industries.
18. Travel and Hospitality: The travel and hospitality industry requires software engineers to develop booking systems, reservation platforms, online travel agencies, and hotel management software.
19. Non-profit and Humanitarian Organizations: Non-profit organizations and humanitarian agencies increasingly rely on software solutions for managing operations, fundraising, data collection, and impact assessment.
20. Research and Development: Research institutions and laboratories often have software engineers involved in developing research tools, data analysis software, and simulation platforms.

Step # 3:

Create a Linkedin profile:

Creating a Linkedin profile that caters to specific industries can help you showcase your professional brand and attract opportunities in your target field. Here are some steps to create a Linkedin profile tailored to specific industries:

1. Optimize Your Headline: Your headline is the first thing people see on your Linkedin profile. Customize it to highlight your expertise and industry focus. Use industry-specific keywords to demonstrate your relevance, such as "Software Engineer specializing in [Industry]" or "Experienced Professional in [Industry] Technology Solutions."
2. Craft a Compelling Summary: Your summary section is an opportunity to showcase your skills, experiences, and career aspirations. Tailor your summary to reflect your interest and experience in the specific industry. Highlight your relevant achievements, projects, and unique value proposition for that industry.
3. Highlight Industry-Specific Skills: Update your skills section to include the key skills and technologies that are valued in the target industry. Research industry-specific skills and include relevant keywords to demonstrate your expertise. This can help you appear in search results when recruiters or employers are looking for professionals in that industry.
4. Customize Experience Section: Review your experience section and emphasize relevant roles, projects, or accomplishments that directly relate to the target industry. Focus on the outcomes and impact you achieved in your previous roles and how they align with the industry's goals.
5. Showcase Industry-Related Projects and Certifications: If you have completed projects or obtained certifications that are specific to the industry, showcase them in your profile. Include details about the projects' objectives, your role, and the technologies or methodologies used. Highlight any industry-specific certifications you have earned to demonstrate your dedication and expertise.
6. Network with Industry Professionals: Connect with professionals who are already working in the target industry. Engage in industry-specific groups and discussions to expand your network and stay updated on the latest trends and opportunities. This can also help you establish valuable connections and access industry-specific insights.
7. Share Relevant Content: Share industry-specific articles, news, and insights on your Linkedin feed to demonstrate your interest and knowledge in the field. Comment on and engage with posts from influencers or companies in the target industry. This can help you position yourself as an active and knowledgeable professional in that industry.
8. Seek Recommendations: Request recommendations from colleagues, managers, or clients who can vouch for your expertise in the target industry. Industry-specific recommendations can provide credibility and strengthen your profile.
9. Join Industry Groups and Follow Influencers: Join Linkedin groups related to the industry you're targeting. Participate in discussions and contribute your insights. Follow influential figures, thought leaders, and companies in the industry to stay informed about industry trends and engage with their content.
10. Continuously Update and Maintain Your Profile: Regularly update your Linkedin profile with new experiences, skills, and accomplishments. Stay active on the platform by engaging with others, sharing relevant content, and participating in industry discussions.

Remember, your Linkedin profile should be a dynamic and evolving representation of your professional brand. By tailoring your profile to specific industries, you can increase your visibility and attract opportunities that align with your career goals in those fields.

Submit the Linkedin profile link here.

Step # 4:

Use Chat GPT to customize your print resume for the above-chosen industry types. Note down everything that you did when using Chat GPT to customize your resume for two separate industry types (e.g., question prompts, selection of content, developing writing styles, content management, organization, etc.)

Write a detailed 200-250 words report explaining how Chat GPT was used to customize your own experience and profile for the resumes.

POST THE THREE RESUME FILES AS SEPARATE PDFs.

Fig. 12.1 ChatGPT-generated instructions for completing customized professional resumes and linkedin profiles

Figure 12.2 provides a detailed self-report on the efficiency with which ChatGPT was used for the process. Further, the results of the text similarity check were provided. A readability score is a number that tells how easy it is likely to be for someone to read a particular piece of text. Generally, a readability score is based on the average length of sentences and words in a document, using a formula known as the Flesch reading-ease test.

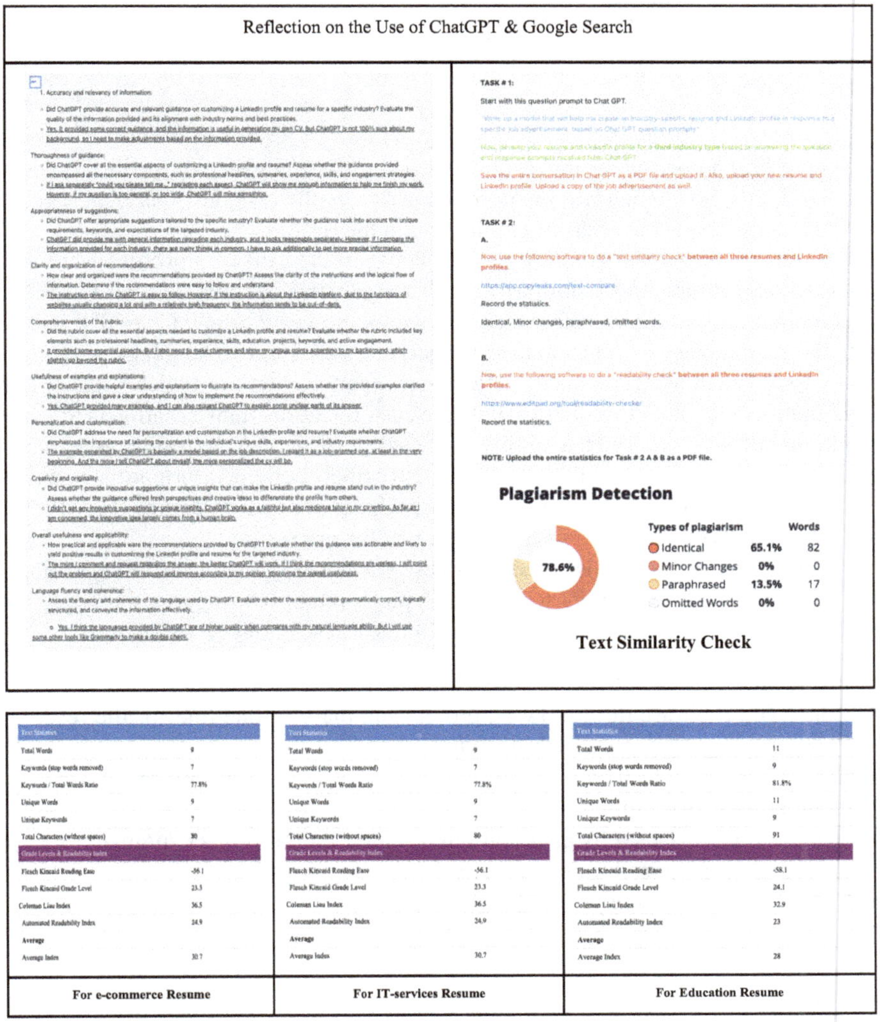

Fig. 12.2 Self-reports on use of ChatGPT, text similarity and; readability check for resume and; linkedin profiles

> When designing my profile in the education technology field, I first asked "What kind of knowledge is necessary for an engineer to work in the education technology field" so that I can understand better about the CV content.
> In the answer, I read and agree that "Data Analysis" is an important skill, so in the education part of my LinkedIn profile, I mentioned my third award in SUST Mathematics Modeling Competition, and R skills in UoA.
> Project Management is also mentioned in the answer, and due to there being a great match between this keyword and my current major, I mentioned my skills in project management, teamwork, and leadership.
> Since "experiences in a traditional IT company can contribute positively to your CV when applying for a role in the education technology field", I didn't delete all my experiences in the traditional IT field, since these can say about my ability in software development.
> As for the experiences not directly related to the target industry, ChatGPT gave me a valuable statement: "If you have experiences from other fields that showcase valuable transferable skills, you may consider including them briefly to demonstrate a well-rounded skill set. " I then keep in mind to make a good statement that what transferable skills I learned, applied and practised in one experience.

Fig. 12.3 A self-report explaining how chat GPT was used to customize students' own experience and resume profile

Figure 12.2 helped us see the results of text similarity between the resume types—in other words, how much was the need to customize the resume, other than specific word choices in specific areas of application in the resume?

The results of the text similarity check between the LinkedIn profiles were also measured and attained 60% similarity. Such assignments not only help use ChatGPT in combination with other references but also help measure how ChatGPT helped with the customization of information in specific cases. English language instructors could assign such assignments to help readers see the extent to which ChatGPT in combination with other tools helped with the customization of information in relevant situations and projects. The readability check assignment is an interesting assignment because it's important to know the extent to which certain information is easily comprehensible and could be successfully used in professional forums (Fig. 12.3).

Using ChatGPT4 to Improve a Customized Resume:

A valuable exercise in this context is to ask students to generate different prompts that could be asked of ChatGPT to make such improvements as suggested in Fig. 12.4. Follow-up prompts should automatically follow. These could be as follows:

- Is this project a good fit for the following resume?
- Could you rephrase the following project summary in one line for the resume? Please highlight my responsibilities for the project.
- Create a 3-point bulleted list to highlight skills gained during the blockchain technology internship.
- Explain my work as a localization intern based on the project summary. Please highlight my responsibilities for the project.
- Please highlight all the projects with action verbs.

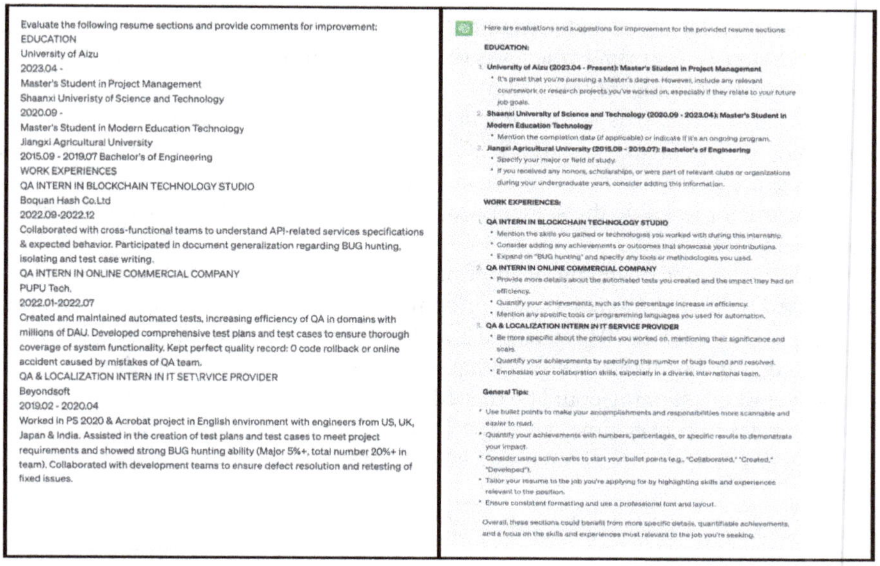

Evaluate the following resume sections and provide comments for improvement:
EDUCATION
University of Aizu
2023.04 -
Master's Student in Project Management
Shaanxi Univeristy of Science and Technology
2020.09 -
Master's Student in Modern Education Technology
Jiangxi Agricultural University
2015.09 - 2019.07 Bachelor's of Engineering
WORK EXPERIENCES
QA INTERN IN BLOCKCHAIN TECHNOLOGY STUDIO
Boquan Hash Co.Ltd
2022.09-2022.12
Collaborated with cross-functional teams to understand API-related services specifications & expected behavior. Participated in document generalization regarding BUG hunting, isolating and test case writing.
QA INTERN IN ONLINE COMMERCIAL COMPANY
PUPU Tech.
2022.01-2022.07
Created and maintained automated tests, increasing efficiency of QA in domains with millions of DAU. Developed comprehensive test plans and test cases to ensure thorough coverage of system functionality. Kept perfect quality record: 0 code rollback or online accident caused by mistakes of QA team.
QA & LOCALIZATION INTERN IN IT SET\RVICE PROVIDER
Beyondsoft
2019.02 - 2020.04
Worked in PS 2020 & Acrobat project in English environment with engineers from US, UK, Japan & India. Assisted in the creation of test plans and test cases to meet project requirements and showed strong BUG hunting ability (Major 5%+, total number 20%+ in team). Collaborated with development teams to ensure defect resolution and retesting of fixed issues.

Here are evaluations and suggestions for improvement for the provided resume sections:

EDUCATION:

1. **University of Aizu (2023.04 - Present): Master's Student in Project Management**
   - It's great that you're pursuing a Master's degree. However, include any relevant coursework or research projects you've worked on, especially if they relate to your future job goals.
2. **Shaanxi University of Science and Technology (2020.09 - 2023.04): Master's Student in Modern Education Technology**
   - Mention the completion date (if applicable) or indicate if it's an ongoing program.
3. **Jiangxi Agricultural University (2015.09 - 2019.07): Bachelor's of Engineering**
   - Specify your major or field of study.
   - If you received any honors, scholarships, or were part of relevant clubs or organizations during your undergraduate years, consider adding this information.

WORK EXPERIENCES:

1. **QA INTERN IN BLOCKCHAIN TECHNOLOGY STUDIO**
   - Mention the skills you gained or technologies you worked with during this internship.
   - Consider adding any achievements or outcomes that showcase your contributions.
   - Expand on "BUG hunting" and specify any tools or methodologies you used.
2. **QA INTERN IN ONLINE COMMERCIAL COMPANY**
   - Provide more details about the automated tests you created and the impact they had on efficiency.
   - Quantify your achievements, such as the percentage increase in efficiency.
   - Mention any specific tools or programming languages you used for automation.
3. **QA & LOCALIZATION INTERN IN IT SERVICE PROVIDER**
   - Be more specific about the projects you worked on, mentioning their significance and scale.
   - Quantify your achievements by specifying the number of bugs found and resolved.
   - Emphasize your collaboration skills, especially in a diverse, international team.

General Tips:

- Use bullet points to make your accomplishments and responsibilities more scannable and easier to read.
- Quantify your achievements with numbers, percentages, or specific results to demonstrate your impact.
- Consider using action verbs to start your bullet points (e.g., "Collaborated," "Created," "Developed").
- Tailor your resume to the job you're applying for by highlighting skills and experiences relevant to the position.
- Ensure consistent formatting and use a professional font and layout.

Overall, these sections could benefit from more specific details, quantifiable achievements, and a focus on the skills and experiences most relevant to the job you're seeking.

Fig. 12.4 1st draft customized resume copy/pasted in ChatGPT3.5 for evaluation & further improvements followed

12.6 Content Production and Analysis: Employment Profile Creation and Targeting Specific Jobs with Video Resumes

The following assignment explored how students use the information generated in the first assignment with print resumes and LinkedIn profiles, to generate video resumes of three different lengths. This assignment required the use of ChatGPT to summarize and paraphrase information based on the content generated in the first assignment, and then use it for speaking for videos of three different lengths, and for customized job positions (task # 3).

The idea behind such customized exercises is to allow students to think in different ways and for specific demands. It allows students to reassess one's qualifications and explore how a specific type of skillset may also be applicable in other areas; how a specific company might have demand for different kinds of candidates with different skills, and how to research and apply with customized projection of one's strengths. Such assignments typically should allow students to engage with ChatGPT in a conversational style and use chain-of-thought prompting, in an attempt to figure out how a specific company might need different combinations of skills with varying degrees of expertise and experience.

Task 4 (see Fig. 12.5) assignment has been completed below based on a self-report. The self-reports above suggested that the student had a definite plan to use ChatGPT and Google search. The above report enables us to think if ChatGPT and

Google search were used indiscriminately, or if the student exercised caution and selective behavior regarding when and the types of information to use and select for completing the assignment (Fig. 12.6).

From the usability perspective, it's important to highlight if there is any common trend and pattern in the use of ChatGPT, in combination with other resources. Following are some examples of chain-of-thought prompting that could be used by the instructor to guide the student toward engaging with ChatGPT for the above assignment (Table 12.3).

Chain-of-thought modeling of information may develop further layers of questions based on the yes/no responses, or opinions shared by ChatGPT. For example, with a question such as "How should I address these points in the video if I have shortcomings in any of these skills? What should I say in the video if that's the case?", a sub-layer of questioning could be related to "How should I portray the skillset that I aspire to develop over time, and how should I portray my motivation to do so?" "How should I explain that I am aware of what the company wants, and the industry trends?".

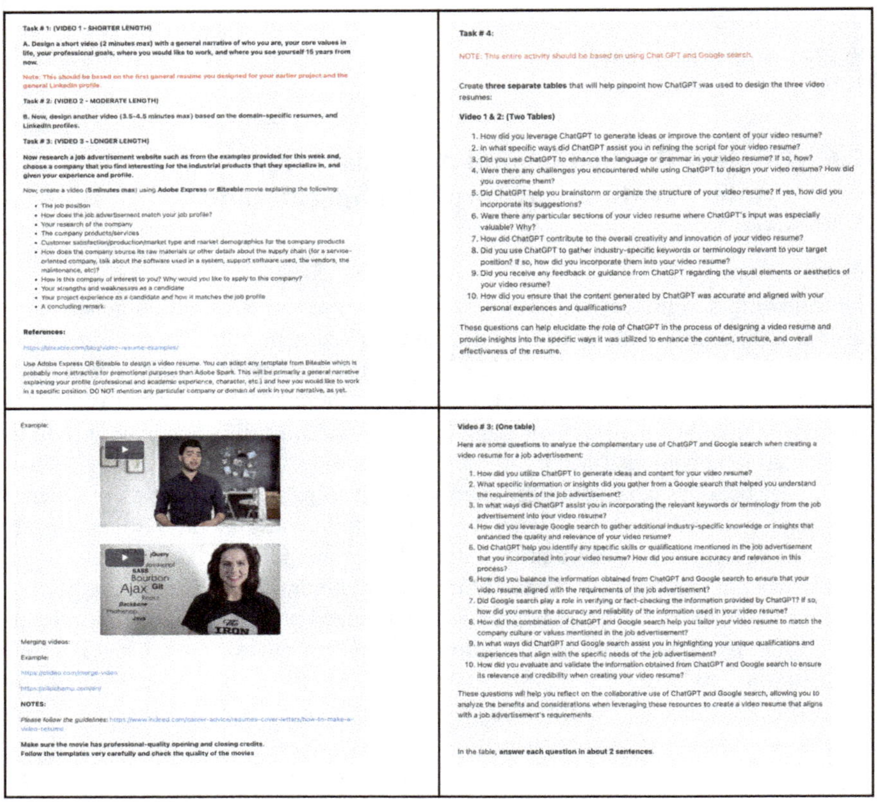
Task # 1: (VIDEO 1 - SHORTER LENGTH)

A. Design a short video (2 minutes max) with a general narrative of who you are, your core values in life, your professional goals, where you would like to work, and where you see yourself 15 years from now.

Note: This should be based on the first general resume you designed for your earlier project and the general LinkedIn profile.

Task # 2: (VIDEO 2 - MODERATE LENGTH)

B. Now, design another video (3.5-4.5 minutes max) based on the domain-specific resumes, and LinkedIn profiles.

Task # 3: (VIDEO 3 - LONGER LENGTH)

Now research a job advertisement website such as from the examples provided for this week and, choose a company that you find interesting for the industrial products that they specialize in, and given your experience and profile.

Now, create a video (**5 minutes max**) using **Adobe Express** or **Biteable** movie explaining the following:

- The job position
- How does the job advertisement match your job profile?
- Your research of the company
- The company products/services
- Customer satisfaction/production/market type and market demographics for the company products
- How does the company source its raw materials or other details about the supply chain (for a service-oriented company, talk about the software used in a system, support software used, the vendors, the maintenance, etc)?
- How is this company of interest to you? Why would you like to apply to this company?
- Your strengths and weaknesses as a candidate
- Your project experience as a candidate and how it matches the job profile
- A concluding remark.

References:

https://biteable.com/blog/video-resume-examples/

Use Adobe Express OR Biteable to design a video resume. You can adapt any template from Biteable which is probably more attractive for promotional purposes than Adobe Spark. This will be primarily a general narrative explaining your profile (professional and academic experience, character, etc.) and how you would like to work in a specific position. DO NOT mention any particular company or domain of work in your narrative, as yet.

Example:

Merging videos:

Example:

https://clideo.com/merge-video

https://clipchamp.com/en/

NOTES:

Please follow the guidelines: https://www.indeed.com/career-advice/resumes-cover-letters/how-to-make-a-video-resume

Make sure the movie has professional-quality opening and closing credits.
Follow the templates very carefully and check the quality of the movies

Task # 4:

NOTE: This entire activity should be based on using Chat GPT and Google search.

Create **three separate tables** that will help pinpoint how ChatGPT was used to design the three video resumes:

Video 1 & 2: (Two Tables)

1. How did you leverage ChatGPT to generate ideas or improve the content of your video resume?
2. In what specific ways did ChatGPT assist you in refining the script for your video resume?
3. Did you use ChatGPT to enhance the language or grammar in your video resume? If so, how?
4. Were there any challenges you encountered while using ChatGPT to design your video resume? How did you overcome them?
5. Did ChatGPT help you brainstorm or organize the structure of your video resume? If yes, how did you incorporate its suggestions?
6. Were there any particular sections of your video resume where ChatGPT's input was especially valuable? Why?
7. How did ChatGPT contribute to the overall creativity and innovation of your video resume?
8. Did you use ChatGPT to gather industry-specific keywords or terminology relevant to your target position? If so, how did you incorporate them into your video resume?
9. Did you receive any feedback or guidance from ChatGPT regarding the visual elements or aesthetics of your video resume?
10. How did you ensure that the content generated by ChatGPT was accurate and aligned with your personal experiences and qualifications?

These questions can help elucidate the role of ChatGPT in the process of designing a video resume and provide insights into the specific ways it was utilized to enhance the content, structure, and overall effectiveness of the resume.

Video # 3: (One table)

Here are some questions to analyze the complementary use of ChatGPT and Google search when creating a video resume for a job advertisement:

1. How did you utilize ChatGPT to generate ideas and content for your video resume?
2. What specific information or insights did you gather from a Google search that helped you understand the requirements of the job advertisement?
3. In what ways did ChatGPT assist you in incorporating the relevant keywords or terminology from the job advertisement into your video resume?
4. How did you leverage Google search to gather additional industry-specific knowledge or insights that enhanced the quality and relevance of your video resume?
5. Did ChatGPT help you identify any specific skills or qualifications mentioned in the job advertisement that you incorporated into your video resume? How did you ensure accuracy and relevance in this process?
6. How did you balance the information obtained from ChatGPT and Google search to ensure that your video resume aligned with the requirements of the job advertisement?
7. Did Google search play a role in verifying or fact-checking the information provided by ChatGPT? If so, how did you ensure the accuracy and reliability of the information used in your video resume?
8. How did the combination of ChatGPT and Google search help you tailor your video resume to match the company culture or values mentioned in the job advertisement?
9. In what ways did ChatGPT and Google search assist you in highlighting your unique qualifications and experiences that align with the specific needs of the job advertisement?
10. How did you evaluate and validate the information obtained from ChatGPT and Google search to ensure its relevance and credibility when creating your video resume?

These questions will help you reflect on the collaborative use of ChatGPT and Google search, allowing you to analyze the benefits and considerations when leveraging these resources to create a video resume that aligns with a job advertisement's requirements

In the table, **answer each question in about 2 sentences**.

Fig. 12.5 The assignment for video resume and use of ChatGPT

How did you leverage ChatGPT to generate ideas or improve the content of your video resume?	I confirmed the basic structure and flow of my speaking logic. I also asked if some topics should be mentioned in video.
In what specific ways did ChatGPT assist you in refining the script for your video resume?	The organization of the structure is important. So I ask ChatGPT to give me advice regarding the structure.
Did you use ChatGPT to enhance the language or grammar in your video resume? If so, how?	No. Instead, I use Grammarly to finish this task.
Were there any challenges you encountered while using ChatGPT to design your video resume? How did you overcome them?	ChatGPT sometimes fail to give correct feedback for the first time. I usually ask yes-no question to double check the answer of ChatGPT.
Did ChatGPT help you brainstorm or organize the structure of your video resume? If yes, how did you incorporate its suggestions?	Yes. This is the most important function provided by ChatGPT. I then adjusted my structure base on the advice.
Were there any particular sections of your video resume where ChatGPT's input was especially valuable? Why?	For Video2, when introducing my working background, I specially confirm the selection of them. For Video 1, as I mentioned in previous answer, I adjusted my structure.
How did ChatGPT contribute to the overall creativity and innovation of your video resume?	Base on the structure, I make the brainstorm by myself, and ask ChatGPT if this content is suitable for this video.
Did you use ChatGPT to gather industry-specific keywords or terminology relevant to your target position? If so, how did you incorporate them into your video resume?	Yes. I got these keywords and then mentioned these keywords in my introduction regarding my working experiences.
Did you receive any feedback or guidance from ChatGPT regarding the visual elements or aesthetics of your video resume?	I can ask questions by typing, but ChatGPT cannot evaluate the video and give me feedback.
How did you ensure that the content generated by ChatGPT was accurate and aligned with your personal experiences and qualifications?	I cannot make sure the content is correct, so I use google and my current knowledge to check if it is correct.

Video # 1: Shorter Length

How did you leverage ChatGPT to generate ideas or improve the content of your video resume?	I confirmed the basic structure and flow of my speaking logic. I also asked if some topics should be mentioned in video.
In what specific ways did ChatGPT assist you in refining the script for your video resume?	The organization of the structure is important. So I ask ChatGPT to give me advice regarding the structure.
Did you use ChatGPT to enhance the language or grammar in your video resume? If so, how?	No. Instead, I use Grammarly to finish this task.
Were there any challenges you encountered while using ChatGPT to design your video resume? How did you overcome them?	ChatGPT sometimes fail to give correct feedback for the first time. I usually ask yes-no question to double check the answer of ChatGPT.
Did ChatGPT help you brainstorm or organize the structure of your video resume? If yes, how did you incorporate its suggestions?	Yes. This is the most important function provided by ChatGPT. I then adjusted my structure base on the advice.
Were there any particular sections of your video resume where ChatGPT's input was especially valuable? Why?	For Video2, when introducing my working background, I specially confirm the selection of them.
How did ChatGPT contribute to the overall creativity and innovation of your video resume?	Base on the structure, I make the brainstorm by myself, and ask ChatGPT if this content is suitable for this video.
Did you use ChatGPT to gather industry-specific keywords or terminology relevant to your target position? If so, how did you incorporate them into your video resume?	Yes. I got these keywords and then mentioned these keywords in my introduction regarding my working experiences.
Did you receive any feedback or guidance from ChatGPT regarding the visual elements or aesthetics of your video resume?	I can ask questions by typing, but ChatGPT cannot evaluate the video and give me feedback.
How did you ensure that the content generated by ChatGPT was accurate and aligned with your personal experiences and qualifications?	I cannot make sure the content is correct, so I use google and my current knowledge to check if it is correct.

Video # 2: Moderate Length

How did you utilize ChatGPT to generate ideas and content for your video resume?	I usually firstly have some ideas ins my mind, and then ask if some topics should be mentioned in video.
What specific information or insights did you gather from a Google search that helped you understand the requirements of the job advertisement?	Google can provide me with multiple links of articles so that I myself can dig the information of these articles.
In what ways did ChatGPT assist you in incorporating the relevant keywords or terminology from the job advertisement into your video resume?	I firstly paste the information, and require ChatGPT to generate a CV for me. And then, I content some key words of CV into my video resume.
How did you leverage Google search to gather additional industry-specific knowledge or insights that enhanced the quality and relevance of your video resume?	I read each related Google search result. Then I adapt reasonable information and adjust my video resume.
Did ChatGPT help you identify any specific skills or qualifications mentioned in the job advertisement that you incorporated into your video resume? How did you ensure accuracy and relevance in this process?	Yes. I usually copy and paste the job ad. to ChatGPT, and observe the CV generated by ChatGPT. Instead of keywords given directly by AI, I think it is better.
How did you balance the information obtained from ChatGPT and Google search to ensure that your video resume aligned with the requirements of the job advertisement?	Yes. Both these two resources, as well as my own idea and opinions, are important.
Did Google search play a role in verifying or fact-checking the information provided by ChatGPT? If so, how did you ensure the accuracy and reliability of the information used in your video resume?	Yes. If Google and ChatGPT told me different stories, I will check more resources and even abandon these contents.
How did the combination of ChatGPT and Google search help you tailor your video resume to match the company culture or values mentioned in the job advertisement?	Yes. I think multiple resources work better.
In what ways did ChatGPT and Google search assist you in highlighting your unique qualifications and experiences that align with the specific needs of the job advertisement?	I usually think my unique point by myself, and then search and ask if it is really good for me to mention that.
How did you evaluate and validate the information obtained from ChatGPT and Google search to ensure its relevance and credibility when creating your video resume?	I think human thing need to have higher priority. I need to make sure I myself make the finial decisions.

Video # 3: Longer Length

Fig. 12.6 Self-reported comments on the use of ChatGPT for video resume production

Table 12.3 Chain-of-thought prompting: An Example for the Video Resume Project

Initial question prompt	Chain-of-thought prompt
• How should I design a video resume for a software developer position? What could be a compact 5-step process?	• Is there only one way to design such a video resume? • Do I need to use all the 5 steps for designing the video? • Can it be done more simply with lesser use of editing skills? Is there a software that can make it easier? • Can I do it in TikTok?
• What types of information should be included in a 3 min video resume?	• How should I balance the narration in the video? • How can I manage the content narration without using a script? • How can I appear more natural in the video?
• Should I relate more to my overall professional aspirations or specific types of skills?	• What should be the balance with such emphasis? • How important is discussing overall professional aspirations and experience? • Should I discuss class projects or work-in-progress? • How deep should I go into explaining the projects?
• How should I customize my narration for the specific company?	• Should I discuss at length my experience with the company? • Where should I get the information about the company? • Should I study LinkedIn profiles for people working in these companies? • Should I first need to enrich my contact list on LinkedIn to get more information and references?
• Could you tell me 5 major things that _______ (won't work without a brand-name company) looks for in a software developer?	• How should I address these points in the video if I have shortcomings in any of these skills? What should I say in the video if that's the case? • If I have a strength that matches what the company is looking for, how should I highlight it?
• Do you think my skill set would be of interest to this company?	• How should I assess and evaluate the current demand in the company? • How should I approach someone on LinkedIn who works in the company in the area where I am applying? • What should be my questions to this person? How could he/she help me with my video resume or content enrichment?

(continued)

Table 12.3 (continued)

Initial question prompt	Chain-of-thought prompt
• What more can I do (or how can I update myself) to fit into this company?	• Input your resume and put ChatGPT in a consulting role asking it how the candidate should update oneself, in general as a software developer, and for a specific position in a company such as ________? • How much should I portray my research about a company as part of my video resume? Should it be over or under-emphasized in the video narration?
• What would be the typical skillset for a software developer working for two years in ________?	• How should I get this information? Should I research any company and try to form a general impression? • Should I try LinkedIn and see if I can get someone working in the same or similar type of company to ask for suggestions and opinions?
• I would like to know more about on-the-job training for software developers in this company. Where should I get such information?	• How should I get this information? Should I research any company and try to form a general impression? • Should I try LinkedIn and see if I can get someone working in the same or similar type of company to ask for suggestions and opinions?

Such prompting skills will most likely enrich the conversational situation with ChatGPT and guide the student towards the information to look for with Google search. Most importantly, ChatGPT has tremendous potential for knowledge building if the prompting is appropriate. Google search can provide specific sources for knowledge building, but ChatGPT has the potential to act as an instructor engaging in a conversational style for knowledge building and switching between different related contexts and responses much faster and comprehensively.

Figure 12.7 is an example of how the transcripts from the video resume were copied/pasted in ChatGPT4 for evaluation, both in terms of content and presentation for the video. The suggestions made by ChatGPT 3.5 in her role as a reviewer could lead to multiple iterations and provide further improvements in the video resume. Some information was misconstrued in the transcript, failing to decipher non-native accents, but that is inconsequential in the given context. The idea behind the following exercise is to show how to improve on the video content, based on specific instructions. The course instructor could verify the suggestions made by ChatGPT 3.5, and if deemed appropriate, could help the student improve on the specific points mentioned in a phased manner.

Video # 1: Shorter Length -
Prompt: Evaluate the video transcript as a reviewer for a general video resume (Instructions for ChatGPT)

Strengths:

1. **Clarity:** The speaker's speech is clear and easy to understand. They articulate their words well, which is important in a video resume.
2. **Educational Background:** The candidate clearly mentions their academic pursuits, indicating that they are pursuing two master's degrees, one at the University of Israel and another at Sanji University of Science and Technology. This demonstrates a commitment to education and a desire for continuous learning.
3. **Professional Experience:** The candidate highlights their three years of experience working in a company as a software quality assurance team member. They provide a brief overview of their responsibilities, including quality assurance, schedule planning, test phase design, and test case implementation. This gives potential employers an idea of their relevant work experience.
4. **Testing Skills:** The candidate mentions their familiarity with various types of testing, including manual testing, functional testing, API testing, and automated testing. This demonstrates a diverse skill set that could be valuable to potential employers.
5. **Teamwork:** The candidate expresses a willingness to work as part of a team, emphasizing cooperation and the development of products to improve the lives of ordinary people. This demonstrates a positive attitude toward collaboration.

Areas for Improvement:

1. **Introduction:** The candidate starts with their name but could have included a brief self-introduction, such as mentioning their career goals or what specifically interests them about the job they're applying for. This would help engage the viewer right from the beginning.
2. **Organization:** The speech could be better organized. It jumps from academic pursuits to professional experience and testing skills rather abruptly. A more structured approach, such as starting with an introduction, followed by education, experience, and skills, would make it easier for the viewer to follow.
3. **Grammar and Clarity:** There are some minor grammatical errors and awkward phrasings in the transcript. For example, "Implement of test case" should be "implementation of test cases," and "knife of Ordinary People" should be "quality of life of ordinary people." Proofreading for clarity and grammar is essential.
4. **Enthusiasm:** While the candidate expresses a willingness to work with a team and develop products, their tone could be more enthusiastic. Adding a bit more passion and energy to their delivery could make the video more engaging.
5. **Conclusion:** The candidate mentions looking forward to talking with the viewer soon, but it might be beneficial to include a call to action, such as encouraging the viewer to contact them for an interview or providing contact information.

What would be your score for this video resume out of 10, if this is used for general job applications?

I would rate this video resume a 6 out of 10 for general job applications.

The candidate has highlighted their educational background, professional experience, and testing skills, which are all important aspects of a resume. However, there is room for improvement in terms of organization, grammar, and enthusiasm in their presentation. With some refinements, this video resume could be more effective in capturing the attention of potential employers and showcasing the candidate's qualifications.

Video # 2: Moderate Length -
Prompt: Evaluate the video transcript as a reviewer for a field-specific video resume and compare it to the earlier one in terms of content and presentation (Instructions for ChatGPT)

Strengths:

1. **Clarity:** The speaker's speech is clear and easy to understand. They articulate their words well, which is important in a video resume.
2. **Educational Background:** The candidate clearly mentions their academic pursuits, indicating that they are pursuing two master's degrees, one at the University of Israel and another at Sanji University of Science and Technology. This demonstrates a commitment to education and a desire for continuous learning.
3. **Professional Experience:** The candidate highlights their three years of experience working in a company as a software quality assurance team member. They provide a brief overview of their responsibilities, including quality assurance, schedule planning, test phase design, and test case implementation. This gives potential employers an idea of their relevant work experience.
4. **Testing Skills:** The candidate mentions their familiarity with various types of testing, including manual testing, functional testing, API testing, and automated testing. This demonstrates a diverse skill set that could be valuable to potential employers.
5. **Teamwork:** The candidate expresses a willingness to work as part of a team, emphasizing cooperation and the development of products to improve the lives of ordinary people. This demonstrates a positive attitude toward collaboration.

Areas for Improvement:

1. **Introduction:** The candidate starts with their name but could have included a brief self-introduction, such as mentioning their career goals or what specifically interests them about the job they're applying for. This would help engage the viewer right from the beginning.
2. **Organization:** The speech could be better organized. It jumps from academic pursuits to professional experience and testing skills rather abruptly. A more structured approach, such as starting with an introduction, followed by education, experience, and skills, would make it easier for the viewer to follow.
3. **Grammar and Clarity:** There are some minor grammatical errors and awkward phrasings in the transcript. For example, "Implement of test case" should be "implementation of test cases," and "knife of Ordinary People" should be "quality of life of ordinary people." Proofreading for clarity and grammar is essential.
4. **Enthusiasm:** While the candidate expresses a willingness to work with a team and develop products, their tone could be more enthusiastic. Adding a bit more passion and energy to their delivery could make the video more engaging.
5. **Conclusion:** The candidate mentions looking forward to talking with the viewer soon, but it might be beneficial to include a call to action, such as encouraging the viewer to contact them for an interview or providing contact information.

In comparison to the previous video resume, this one is more field-specific and tailored to the candidate's background. However, it could benefit from improved clarity, organization, and visual presentation to make a stronger impact on potential employers. I would rate this field-specific video resume a 7 out of 10.

Fig. 12.7 ChatGPT Evaluation based on Video Transcript—Preparation for Subsequent Iterations for Video Improvement

Video # 3:
Prompt: Longer Length - Evaluate the following field-specific video transcript in comparison to the earlier ones above (Instructions for ChatGPT)

Strengths:	**Areas for Improvement:**
1. **Educational Background:** Similar to the earlier field-specific video resume, this candidate also mentions their academic pursuits, indicating that they are pursuing two master's degrees, one at the University of ISO and another at Samsi University of Science and Technology, both related to modern education technology. This demonstrates their commitment to education and relevant expertise. 2. **Professional Experience:** The candidate mentions their three years of experience working in a company as a software quality assurance team member, which is relevant to the field. They provide a brief overview of their responsibilities, including quality assurance, schedule planning, test phase design, and test case implementation. 3. **Technical Skills:** The candidate highlights their familiarity with various types of testing, including manual testing, functional testing, API testing, and automated testing. They also mention their programming skills, including Python and JavaScript, and their experience as a front-end developer. 4. **Multilingual Skills:** Similar to the previous field-specific video resume, this candidate also mentions their ability to speak multiple languages, including Mandarin, English, Japanese, Spanish, Thai, and Vietnamese, which can be an asset in international settings. 5. **Teamwork and Collaboration:** The candidate expresses a strong willingness to work as part of a team and emphasizes cooperation and product development to improve the lives of ordinary people. This collaborative attitude aligns well with many job roles in the field.	1. **Clarity:** While the candidate's speech is generally clear, there are some minor pronunciation issues, such as "person" instead of "Python," which may affect clarity. Ensuring clear pronunciation of technical terms is important. 2. **Structure:** Similar to the earlier field-specific video resume, this one lacks a clear structure. It would be more effective with a well-organized flow, including an introduction, education, skills, work experience, and a strong conclusion. 3. **Grammar and Language:** There are some grammatical issues, such as "implement of test case" instead of "implementation of test cases." Proofreading for grammar and language consistency is important. 4. **Visual Presentation:** As a video resume, the visual aspect is important. The candidate could consider including relevant visuals or text overlays to highlight key points and make the video more engaging. 5. **Length:** Like the previous field-specific video resume, this one is also quite lengthy, which may lead to viewer disengagement. It's important to keep video resumes concise and focused on the most relevant information.
Overall, this field-specific video resume is similar in content and presentation to the earlier one but shares some common areas for improvement, such as clarity, structure, grammar, and visual presentation. I would rate this field-specific video resume a 7 out of 10, just like the previous one.	

Fig. 12.7 (continued)

12.7 Content Production and Analysis: Customer Profile Creation & Strategic Plan Setup for Targeted Customer Groups on Social Media

As part of this project, the student was asked to use ChatGPT and Google search extensively to provide a visual representation of how his pet project, a smart museum could be used towards an entrepreneurial initiative. The student, until that time, focused on the technical dimensions of the project, but never considered the entrepreneurial dimensions. The student was asked to do extensive research with Google search and use ChatGPT to obtain text responses to the prompts generated based on the questions asked (Fig. 12.8) in the assignment.

As seen in Fig. 12.8, we left it to the student on his/her ability to understand the concepts of customer segmentation from the links provided in the assignment, and then design visual models for customer segmentation using Lucid Chart. The idea was to encourage the student to use ChatGPT and Google search in some combination to either understand the concepts beyond the links provided by the instructor, formulate a response, or figure out how to represent the information clusters or metadata visually in Lucid Chart. The idea was not to allow the student to come up with a direct text response on customer segmentation which could be directly generated with ChatGPT in some form, even when customization of the response could be a problem in this case.

The visual responses below offer a template on how the student might have thought through the problem, even when the use of ChatGPT and Google search were encouraged in the way the student desired.

PART A:

Design a 7-8 slide argument explaining your project on the smart museum.

Which of the following aspects of a smart museum are part of your plan?

1. Interactive Exhibits: Smart museums use interactive displays, touchscreens, and sensors to enable visitors to actively engage with the exhibits. They can explore digital content, access additional information, and participate in interactive activities related to the artworks or artifacts.
2. Virtual Reality and Augmented Reality: VR and AR technologies are used to create immersive experiences where visitors can virtually step into historical environments, view 3D reconstructions of ancient artifacts, or interact with virtual objects overlaid on real-world exhibits.
3. Mobile Apps and Audio Guides: Many smart museums provide mobile applications or audio guides that offer detailed information about the exhibits, allowing visitors to customize their experience and delve deeper into the artwork's background, history, and significance.
4. Location-based Services: Using IoT and beacon technology, smart museums can provide location-specific information to visitors. For example, as visitors move through the museum, they can receive notifications about nearby exhibits, upcoming events, or additional multimedia content related to the artworks.
5. Personalized Recommendations: AI algorithms can analyze visitor preferences and behavior to provide personalized recommendations on exhibits, artworks, or related events based on their interests and past interactions.
6. Digital Preservation: Smart museums employ digital preservation techniques to safeguard and preserve artifacts and artworks. They may use high-resolution scanning, 3D modeling, and digital archiving to create virtual replicas, ensuring the long-term accessibility and conservation of cultural heritage.

Your presentation must end with a slide or two explaining why your smart museum idea is saleable to the general public or the city administration.

PART B:

Use https://www.lucidchart.com/pages/ to complete the following projects:

A. Based on the information from the following link - https://www.linkedin.com/pulse/different-customer-segmentation-models-identifying-consumer-agrawal/, design a SINGLE visual representation to explain the different customer segmentation models, and how they are related or unrelated to each other and should be seen as closed segments or interconnected segments. You can create a network diagram or concept map style of visual representation.

B. Visually demonstrate in a diagrammatic format the basic features of the following different software types for effective customer segmentation. Your visual representation should help the management of a company decide which software could be likely used for their own business customer segmentation.
Reference: https://influencermarketinghub.com/ai-software-customer-segmentation/
Make use of Chat GPT along with Google search (for the websites of the specific software types) to further customize your response.

PART C:

1. Based on your question prompts to Chat GPT, and by using Google search to customize your response, set up a model to create two separate customer profiles that will help us understand who could be **two different kinds of potential buyers** of your smart museum idea and applications.

Create a visual representation to explain such a project.
Use the following UML diagram creation software - https://www.lucidchart.com/pages/

2. Create a visual representation to demonstrate how you would create a customer segmentation model for your smart museum project. Who would be the customer types, and how should they be represented and clustered?

Create a visual representation to explain such a project.
Use the following UML diagram creation software - https://www.lucidchart.com/pages/

PART D:

Ask Chat GPT, how you can set up techniques like segmentation, clustering, or machine learning algorithms to identify patterns and trends for customer profile creation.
Now use both Chat GPT and Google search to explain how you would use these techniques for your smart museum project, in your specific geographical location and demographic specifics.

Create a visual representation to explain such a project.
Use the following UML diagram creation software - https://www.lucidchart.com/pages/

Fig. 12.8 Instructions on Designing Customer Profile Creation

The instructors' evaluations suggested that even when ChatGPT and Google search were used for designing the slides as was required in Part A of the assignment, the information was not copied/pasted in any form, but rather adapted in a very summarized form for the slides, and adequately paraphrased. The length of the sentences in the slide and class observation when completing the assignment suggested that although ChatGPT and Google search were used randomly, time was spent on understanding the concepts, and generating own responses. That's probably why the slide explanations were incomplete at best or could have been better explained. A copy-pasted (from ChatGPT and/or Google search) explanation would have yielded a better overall response. In other words, the student did not depend on generating prompts on how to incorporate each of the points mentioned in Part A in the slide response. As for Part B, C & D, where Lucid Chart was used, we see a similar trend where Google search or ChatGPT response couldn't have been directly incorporated. So, here, based on the class observation that showed the student using Google search and ChatGPT, we can safely assume that the responses obtained were customized and incorporated in some capacity in the visualization.

Figures 12.9 and 12.10 demonstrated cases, where ChatGPT could have been used in a much better way, but students need proficiency and training in prompt generation. Figure 12.9 remained a brainstorming mind map which is impossible to use in the context of comprehending either smart museum in the local context, or towards understanding how customer segmentation could happen. There is no explanation for the nodes created in the segmentation model.

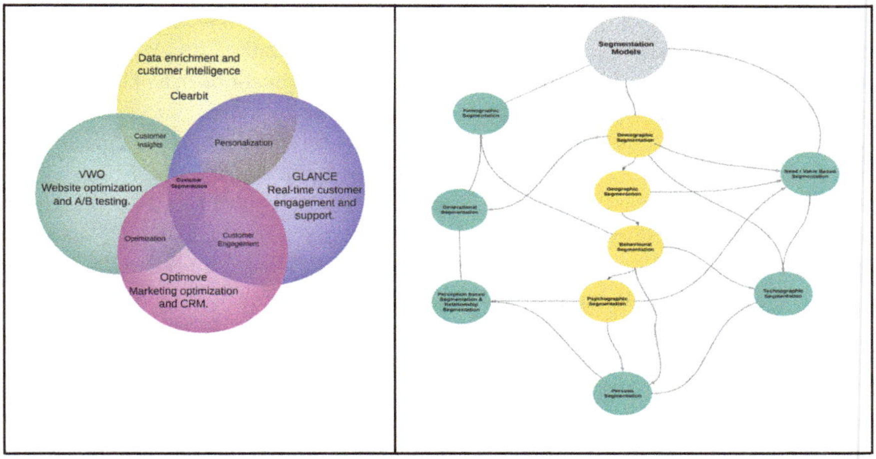

Fig. 12.9 Minimally Customized Response (More use of Google Search/LinkedIn References rather than ChatGPT)

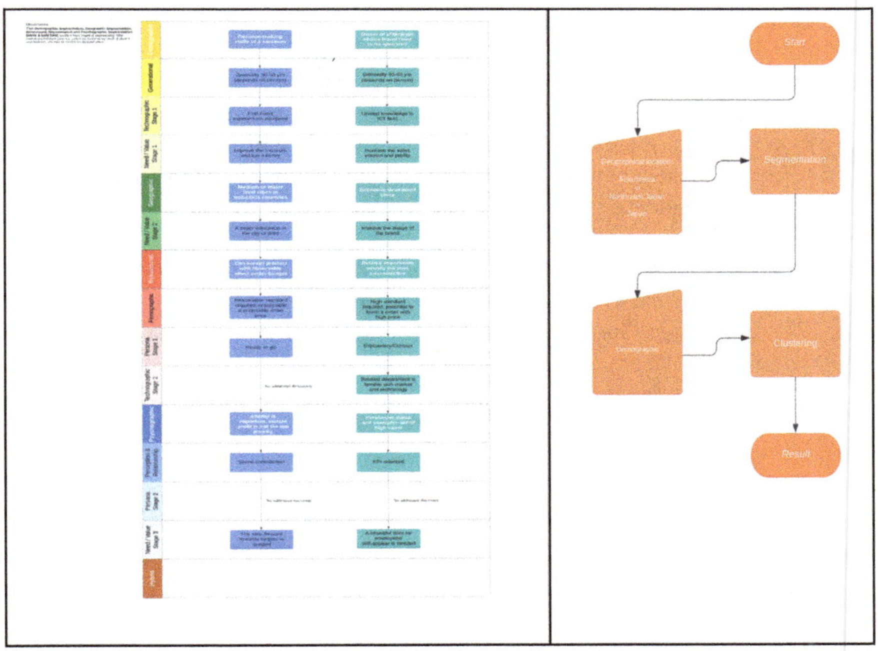

Fig. 12.10 Relatively Customized Response

Knowing the student's proficiency with the English language, instructors concluded that it's not a question of ability to generate general to specific prompts, but more a case of getting trained on how to ask specific questions in a phased manner and closing on the context with an inverted pyramid approach.

The information provided in Fig. 12.10 misinterpreted many of the categories mentioned. For example, the demographic segmentation was largely underrepresented, and/or misunderstood, or minimally mentioned. If the reference articles provided in the assignment do not bring clarity to the concept, ChatGPT could have been asked for a more specific understanding. ChatGPT could provide many general to specific examples, clarifying positions situations, and contexts, without reference. So, ChatGPT is not only about copying/pasting information or generating text for assignments; its productive use could first and foremost lead to concept clarity.

Figure 12.10 provided a somewhat customized model that might help provide some clarity, but the student struggled to situate the model in the context of the smart museum and related customer segmentation. Figure 12.10 could be a good starting point for the students to generate question prompts for ChatGPT to ensure gradual clarity in a phased manner with at least a few-shot prompting. The following prompt layered example in Fig. 12.11 could potentially help us understand the extent to which the Fukushima smart museum case could be customized without providing references but simply based on ideas. The student could then approach Google Search to look for articles (Google Search or Google Scholar) that cater to the following points mentioned in ChatGPT's second-, third-, and fourth-layer prompts. Figure 12.11 suggests the prompts that could have helped with a better and more detailed response.

The above is somewhere in between few-shot prompting and chain-of-thoughts prompting, and students could be further trained toward targeted customization, depending on the context. An interesting point to note here is that depending on the topic being searched in Google, the results may not be enterprising and specific enough, suggesting that a more customized response could be quickly obtained from ChatGPT, rather than Google search. For example, for a Google search with the phrase "demand for smart museums in Japan", we get examples of different museums in Japan, even e-museums. Such examples could be adapted in the response, but there is nothing to directly answer the question asked in terms of demand for smart museums, nor a recommendation easily obtained as to where to look for such information. So, getting a definite response could be problematic. When the same question was asked to ChatGPT, we did not get a definite response, but it guided us toward the following for more specific statistics, that may not be part of public records that are easily accessible.

A. First Layer of Prompt Generation (An Example from the Instructor)

ChatGPT 1st Layer Prompts	Summarized Points
Explain demographic customer segmentation for establishing a smart museum.	• **Visitor Engagement and Experience Enhancement.** • Curate content and collections • Targeted marketing campaigns • Help museums ensure that their exhibits and facilities are accessible to all visitor groups. • Provide personalized experiences. • Visitor Analytics • Long-term Planning
Explain generational customer segmentation for establishing a smart museum.	• **Understanding Technological Aptitude.** • **Content Relevance.** • **Communication Channels.** • **Customized Experiences.** • **Learning Styles.** • **Accessibility and Inclusivity.** • **Social Sharing.** • **Future Planning.** • **Data Collection & Analysis.**

B. Second Layer of Prompt Generation (An Example from the Instructor)

ChatGPT 2nd Layer Prompts	Response
Different generations have varying degrees of familiarity and comfort with technology. How would you evaluate this situation for Japan when it comes to adapting the idea for a smart museum for tourists visiting Japan? In conclusion, adapting the idea of a smart museum for tourists visiting Japan requires a thoughtful approach that takes into account the diverse generational preferences and cultural values of the country. Providing options, respecting traditions, and ensuring accessibility are key factors in creating a smart museum experience that resonates with a wide range of visitors.	When evaluating the situation of adapting the idea of a smart museum for tourists visiting Japan, it's important to consider the technological landscape and generational preferences within the country. Japan has a unique blend of traditional values and advanced technology, which can impact the way different generations interact with smart museum concepts. Here's a breakdown of how this evaluation might look: 1. **Generational Breakdown:** • **Elderly Population (65+):** Japan has a significant elderly population. Many individuals in this age group might not be as comfortable with advanced technology. They might prefer more traditional forms of communication and interaction. For this demographic, offering simple and intuitive interfaces, guided tours led by human guides, and clear signage is crucial. • **Baby Boomers and Generation X:** While some individuals in these generations might be tech-savvy, there will still be a range of comfort levels. Providing options for both traditional and digital experiences would be beneficial. Audio guides, tactile exhibits, and interactive elements with clear instructions can cater to these age groups. • **Millennials and Generation Z:** Younger generations are likely to be more open to technology-driven experiences. They might engage enthusiastically with augmented reality (AR), virtual reality (VR), mobile apps, and interactive displays. Social media sharing opportunities and gamified experiences can also resonate with them. 2. **Cultural Sensitivity:** • Japan's culture places importance on respect, etiquette, and preserving traditions. When introducing technology into cultural experiences like museums, it's important to strike a balance between innovation and maintaining the cultural reverence that tourists and locals expect. 3. **Accessibility:** • Ensuring accessibility for all visitors is crucial. For the elderly population, offering clear instructions, larger text, and comfortable seating areas can enhance their experience. For tech-savvy generations, having multilingual interfaces and information is essential, as many tourists might not speak Japanese. 4. **Technology Integration:** • Consider integrating technology in a way that doesn't overwhelm or alienate those less familiar with it. Providing assistance, tutorials, or staff to help visitors navigate and interact with technology can bridge the gap. 5. **Bridging the Gap:** • To cater to all generations, offer a variety of experiences that allow visitors to choose their preferred level of technology engagement. For instance, a smart museum could have both interactive touchscreens and traditional exhibit panels. 6. **Education and Communication:** • Effective communication is key. Providing clear information about the available technology and its benefits can encourage hesitant visitors to give it a try. Educating museum staff to assist visitors with technology-related queries is also important. 7. **Testing and Iteration:** • It's wise to start with a pilot phase or temporary exhibits to gauge visitor responses and preferences. Based on feedback and data, the museum can iterate and refine its approach to technology integration. 8. **Collaboration with Local Experts:** • Collaborating with experts in Japanese culture, history, and art can help design technology experiences that align with local sensitivities and preferences.

Fig. 12.11 Four Layer of Prompts that Could Have Been Ideally Pursued

Some example recommendations from ChatGPT:

> **Check reports or publications from the Japanese government's cultural or tourism departments.**
> **Organizations like the Japan Museum Association and Japan National Tourism Organization (JNTO) might publish relevant reports or articles on museum trends.**
> **Market research firms specializing in cultural and tourism trends might offer reports that include statistics on the demand for smart museums.**

Even Google Scholar was not helpful with the search phrase "rise of demand for smart museums in Japan".

C. Third Layer of Prompt Generation (An Example from the Instructor)

ChatGPT 3rd Layer Prompts	Response
Based on the above response, how should we bridge the generational gap in a remote prefecture such as Fukushima? In remote areas like Fukushima, building a sense of community and inclusivity is crucial. By combining technology with traditional methods and creating spaces for intergenerational interaction, you can bridge the generational gap and create an inclusive smart museum experience that resonates with all members of the community.	Bridging the generational gap in a remote prefecture like Fukushima involves considering the unique challenges and opportunities presented by the local context. Here's a tailored approach to address this situation: 1. **Community Engagement:** • Begin by engaging with local residents across generations to understand their needs, preferences, and comfort levels with technology. This can be done through surveys, focus groups, and community meetings. 2. **Digital Literacy Programs:** • Develop digital literacy programs specifically targeted at the elderly population. These programs could be held at local community centers or libraries and teach basic technology skills, such as using smartphones, tablets, and computers. 3. **Hybrid Experiences:** • Create hybrid experiences that combine traditional elements with technology. For example, offer guided tours led by knowledgeable locals while integrating technology like QR codes for additional information or translations. 4. **Local Stories and Histories:** • Focus on showcasing the unique stories and histories of Fukushima. Curate exhibits that highlight local traditions, cultural practices, and historical events. This can help capture the interest of older generations who have a deeper connection to these stories. 5. **Interactive Workshops:** • Organize workshops and interactive sessions that cater to different age groups. For instance, hands-on craft workshops, traditional tea ceremonies, and storytelling sessions can attract older visitors. 6. **Multilingual Support:** • Recognize that remote areas might attract international tourists as well. Provide multilingual information and technology interfaces to accommodate visitors who don't speak the local language. 7. **Local Partnerships:** • Collaborate with local schools, colleges, and community organizations to involve younger generations in the museum's activities. This can foster a sense of ownership and engagement among the youth. 8. **Intergenerational Activities:** • Design activities that encourage interactions between different age groups. For example, organizing events where younger generations assist the elderly in using technology can create a sense of unity and shared learning. 9. **Empower Local Guides:** • Train local guides to lead tours and provide explanations using technology. These guides can act as bridges between different generations, offering insights that resonate with both traditional and modern perspectives. 10. **Continuous Feedback Loop:** • Implement a feedback mechanism to gather insights from visitors of all generations. Use this feedback to make iterative improvements and adjustments to the museum experience. 11. **Cultural Sensitivity:** • Ensure that the integration of technology is culturally sensitive and aligned with local values. This can help overcome any resistance to technology among older generations. 12. **Storytelling Platforms:** • Utilize various platforms, including traditional oral storytelling and digital mediums, to convey the history and culture of Fukushima. This approach can engage both old[er] ... younger audiences.

D. Fourth Layer of Prompt Generation (An Example from the Instructor)

ChatGPT 4th Layer Prompts	Response
How should we specifically implement this in the context of smart museums? Provide a summary response. 11. **Respectful Integration:** Ensure that technology doesn't overshadow the cultural and historical significance of exhibits. Maintain a balance that resonates with both traditionalists and technology enthusiasts. 12. **Empowerment of Local Guides:** Train local guides, including students, to assist visitors with technology usage. Their friendly assistance can encourage older visitors to engage with the smart features. By integrating technology thoughtfully and creatively, focusing on community engagement, and offering a variety of tailored experiences, a smart museum in Fukushima can effectively bridge the generational gap. The goal is to create an inclusive, enriching, and enjoyable museum experience that appeals to all members of the community, regardless of age or technological familiarity.	Implementing the bridging of the generational gap in a smart museum context in a remote prefecture like Fukushima involves a balanced and inclusive approach: 1. **Customized Experiences:** Develop smart exhibits that offer various levels of interaction. Provide touchscreens with user-friendly interfaces for tech-savvy visitors and traditional panels with clear explanations for those less familiar with technology. 2. **Guided Technology Tours:** Organize guided tours led by knowledgeable locals who can navigate technology while offering deeper insights into the cultural and historical significance of exhibits. 3. **QR Code Enhancements:** Incorporate QR codes on exhibit labels. These codes can link to detailed digital content, allowing visitors to access additional information at their own pace. 4. **Hybrid Guides:** Design mobile apps that combine GPS-based audio guides with augmented reality features. Visitors can explore the museum, receiving contextual audio explanations while accessing virtual overlays that enhance their understanding. 5. **Intergenerational Workshops:** Organize workshops where younger generations teach older ones how to use smartphones and tablets. This encourages dialogue and shared learning while fostering community cohesion. 6. **Local Stories Podcasts:** Create a podcast series featuring local residents sharing personal stories and memories related to the exhibits. This can engage older generations through a familiar storytelling format. 7. **Interactive Storytelling Stations:** Set up interactive stations where visitors can listen to and record their own stories about the exhibits. This collaborative effort encourages participation across generations. 8. **Multilingual Accessibility:** Provide multilingual interfaces and content to accommodate international tourists, ensuring that both local and foreign visitors can engage with the museum. 9. **Feedback Kiosks:** Install kiosks where visitors can provide feedback on their experience. This helps in continuous improvement and tailoring the museum's offerings to different generations. 10. **Local School Partnerships:** Collaborate with local schools to organize field trips to the museum. This introduces technology experiences to younger generations and promotes cross-generational learning.

Fig. 12.11 (continued)

Here is a self-report from the student as to how ChatGPT and Google Search were used in combination for the above assignment. The comments show relative trust in a conversational versus an open-search pattern for information access.

I used ChatGPT but asked them questions including "Who will be the potential customer of a [project name]? And [project name] means [explanation]." When I got the response from ChatGPT, I asked deeper questions, for example, "Could you please tell me more about [customer type]?" "Do you think all the [customer type] will be potential customers of my project?" so that ChatGPT will tell me in a more detailed way.
There is one single pattern with my questions. Chat GPT told me some basically right things, but I went to extreme intentionally (although I don't really think so) so that ChatGPT will tell me more complicated details to persuade me "things don't go so simple". I repeated this kind of question many times. This is also where I understand that the "government", as a type of customer, can be very different inside. It is very easy to understand now, that the government of a developing country may have different expectations when compared with that of a developed country, but due to the insufficient knowledge of related fields, it is such conversation with ChatGPT that makes me realize this issue.

As for Google search, in this assignment, I used it only after I felt very confident about my conclusion. Google can provide me thousands, or, let's say millions of information points related to my keywords. But if my keywords are incorrect or not suitable, the search result can only slow down my speed and confuse me. That is why I used Google service later than ChatGPT in this assignment.

12.8 Content Production and Analysis: Customer Journey Mapping for [Specific Product] on [E-commerce Website]

This is an assignment where the student was asked to develop a customer journey map based on instructions and an open-ended questionnaire developed by the instructor.

The above instructions helped the student understand the concept clearly and how to approach the task step-by-step. The idea behind using ChatGPT to provide concrete instructions was to guide the student into understanding how such instructions could be generated from ChatGPT if the correct question prompts are generated. The student could then combine such understanding with the references provided as a supplementary knowledge builder (Figs. 12.12 and 12.13).

Figures such as the ones we see above (Fig. 12.14) must be evaluated by the instructor. Although Chat GPT4 could be used for image identification, it can't be used for business intelligence data analytics purposes. We need a business intelligence software such as Power BI from Microsoft for visual text and data analysis. Power BI is a collection of software services, apps, and connectors that work together to turn your unrelated sources of data into coherent, visually immersive, and interactive insights. It has a feature called a "smart narrative for a visual" that textually explains graphs/charts etc. Such a narrative statement could then be copied/pasted into ChatGPT for casual and general analysis and discussion beyond the BI software.

The arrow in Fig. 12.15 shows the summary that could be copied/pasted into ChatGPT for further evaluation.

Further, it's also important to understand how ChatGPT could be used to create visuals such as system design block diagrams, sequence diagrams, etc., in technical and business communication contexts, and then use ChatGPT to evaluate the visuals to evaluate different stages of the process (Tam, 2023) (Fig. 12.16).

The above mind map could be further evaluated by asking students to represent it textually as a text paragraph demonstrating the association between different nodes, and then asking ChatGPT to regenerate the associations between these nodes in terms of codes and generate a different type of diagram to demonstrate it using Mermaid

Assignment: Customer Journey Mapping for [Specific Product] on [E-commerce Website]

Introduction: Customer journey mapping is a valuable technique used by businesses to gain a deeper understanding of their customers' experiences and interactions throughout the buying process. In this assignment, you will act as a customer journey mapper for a specific type of product on a chosen e-commerce website. Your task is to map the customer's journey, identify key touchpoints, and analyze the overall customer experience. This assignment will help you develop insights into customer behavior and provide recommendations for improving the customer journey.

Reference:

https://www.ohgroup.com/articles/analyze-customer-journey-map/

Objective: To create a comprehensive customer journey map for a specific product on a chosen e-commerce website, and analyze the customer's experience at various touchpoints.

Instructions:

1. **Product Selection:** Choose a specific type of product that is available on the e-commerce website. Examples include electronics, clothing, books, home appliances, etc. Be specific about the product category (e.g., smartphones, men's shirts, fiction novels, kitchen appliances).
2. **E-commerce Website Selection:** Select a specific e-commerce website where the chosen product is sold. Examples include Amazon, eBay, Walmart, Best Buy, etc.
3. **Customer Persona:** Create a customer persona who would be interested in purchasing the chosen product. Define the persona's demographic information, interests, preferences, and motivations. This will help you understand the customer's perspective throughout the journey.
4. **Customer Journey Mapping:** Create a visual representation of the customer journey for purchasing the chosen product on the selected e-commerce website. Include the following elements:
   - Pre-purchase phase: Awareness, consideration, and decision-making stages.
   - Purchase phase: Product search, evaluation, selection, and checkout process.
   - Post-purchase phase: Delivery, product usage, customer support, and feedback.
5.

   Identify Touchpoints: Identify the key touchpoints where the customer interacts with the e-commerce website. These touchpoints can include homepage, search results page, product listing page, product details page, shopping cart, customer service, etc.
6. **Analyze the Customer Experience:** Analyze the customer experience at each touchpoint and evaluate the effectiveness of the e-commerce website in providing a seamless journey. Consider factors such as ease of navigation, product information, customer reviews, checkout process, customer support, etc.
7. **Identify Pain Points and Opportunities:** Identify any pain points or areas of improvement in the customer journey. Consider how the e-commerce website could enhance the customer experience, reduce friction, and increase customer satisfaction.
8. **Recommendations:** Provide recommendations for improving the customer journey based on your analysis. Suggest specific actions or strategies that the e-commerce website could implement to enhance the overall customer experience and increase customer loyalty.
9. **Conclusion:** Summarize your findings and key insights from the customer journey mapping exercise. Reflect on the importance of understanding the customer journey and its impact on the success of an e-commerce business.

Deliverables:

- **Create a customer journey mapping report** - https://www.qualtrics.com/m/assets/wp-content/uploads/2019/01/Customer-Journey-Template-Download.pdf?utm_lp=hub-customer-journey-mapping-inline1
- **Design your report based on the framework mentioned here** - https://www.qualtrics.com/experience-management/customer/customer-journey-mapping/
- Customer persona description.
- Customer journey map (visual representation).
- Analysis of the customer experience at each touchpoint.
- Identification of pain points and opportunities.
- Recommendations for improving the customer journey.
- Conclusion summarizing key insights.

Note: Please feel free to adapt the assignment to fit your specific requirements and make it more relevant to your learning objectives.

Fig. 12.12 The Customer Journey Map Assignment

Have you utilized Chat GPT (or similar AI-powered language model) for customer journey mapping? If yes, please describe how you utilized it and elaborate on the specific tasks or aspects of customer journey mapping that it assisted you with.
I ask ChatGPT true/false questions to confirm my idea. For example, I am not sure if activity A belongs to stage B, and personally believe it is. So I will ask ChatGPT. If Chat GPT also says yes, then I will insist on my judgment. If the result is no, then I will reconsider it.
What were the specific benefits or advantages you experienced when using Chat GPT for customer journey mapping? Please provide examples or instances where Chat GPT significantly contributed to your understanding of the customer journey.
The specific thing is that it provides me with a chance to reconsider and reevaluate my understanding regarding a specific issue. It is important to have someone or something to check my work almost instantly, although the finial judgment is in my hand.
Conversely, were there any limitations or challenges you encountered while using Chat GPT for customer journey mapping? If yes, please explain the difficulties you faced or any areas where Chat GPT fell short in meeting your expectations or requirements.
The more I talk with it, the weakness the relationship between us will be. Continuous communication makes me realize that it just generates words based on the information it can get, so is not responsible for what it provided. In other words, it cannot make a judgment if the information it owns is correct or not. It is me who make this decision.
Did you rely on Google Search (or any search engine) during the process of customer journey mapping? If yes, please describe how you used Google Search and specify the types of information or insights you sought through it.
I directly use Google to search for keywords I am concerned about. I enjoy making decisions myself more. For example, I can read a lot in the search result pages, and then check some, maybe all, of them based on my understanding and necessary.
How did the utilization of Google Search contribute to the accuracy or comprehensiveness of your customer journey mapping report? Please provide specific examples where Google Search played a crucial role in gathering relevant data or enhancing your understanding of the customer journey.
The keyword search in Google can provide me more information. The more information I get, the better decision I can make.
In your opinion, how effective was the combination of Chat GPT and Google Search in enhancing the quality of your customer journey mapping report? Please share your thoughts on how the synergy between these tools impacted the overall accuracy, depth, and insights derived from the customer journey mapping exercise.
The combination of both ChatGPT and Google Search can contribute to better quality because one can improve the output of another. For example, I search on Google first and use the Google search result as input of ChatGPT, and then use to output of Chat GPT to improve my use of the result of Google Search.
Besides Chat GPT and Google Search, were there any other tools, resources, or methodologies that you utilized for customer journey mapping? If yes, please provide details on the additional resources and explain how they complemented or supplemented the usage of Chat GPT and Google Search.
I use Lucid to generate the graphic. But I think it is separate from the Google Search and ChatGPT since the function is quite different.
Based on your experience, how do you envision AI-powered language models and search engines continuing to impact the field of customer journey mapping in the future? Please share your insights on how advancements in AI and search technology may shape and transform the way we approach customer journey analysis.
I think it can help us to do that, but cannot make any revolutionary impact. I still firmly believe that the customer journey analysis is a heart-based work. We need to use our heart or let's say brain, to understand our customers. AI cannot do that, they just analyze information, and information is only a part of the customer journey analysis activities.
Is there any additional feedback, insights, or recommendations you would like to share regarding the usage of Chat GPT and Google Search for customer journey mapping? Please feel free to provide any further thoughts, reflections, or suggestions based on your experience.
The attitude is important. In which position I should put Chat GPT on? I will also think about this in my future practices.

Fig. 12.13 Self-Reports based on ChatGPT generated Instructions & Open-ended Questionnaire for Developing a Customer Journey Map

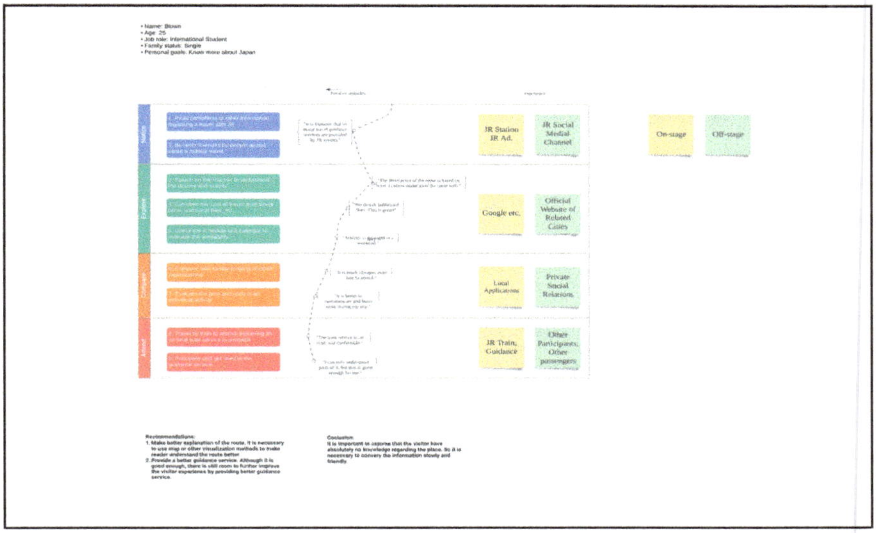

Fig. 12.14 Customer Journey Map for JR East Japan Railway Service

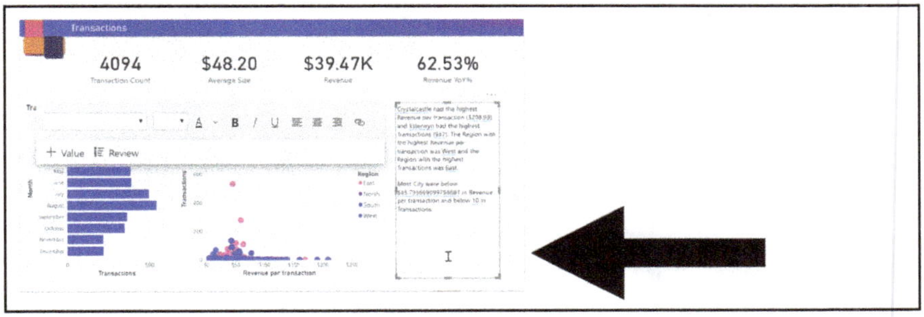

Fig. 12.15 Smart Narrative for a Visual (Representative Visual) (https://learn.microsoft.com/en-us/power-bi/visuals/power-bi-visualization-smart-narrative#smart-narrative-for-a-visual)

Live Editor (https://mermaid.js.org/intro/). In each step, the process could be further enriched with class discussion among students.

Content Production and Analysis: Employer Profile Evaluation for a Specific Job Type and Self-Judging the Candidate Fit

As part of this assignment, the idea was to evaluate an employer profile to judge the candidate's fit for a specific job type. Figure 12.17 helped students approach the assignment in a phased manner.

The text in the customer journey map should be first evaluated by the course instructor in terms of the appropriateness of the points mentioned. The course instructor could then follow up on this customer journey map table format and ask

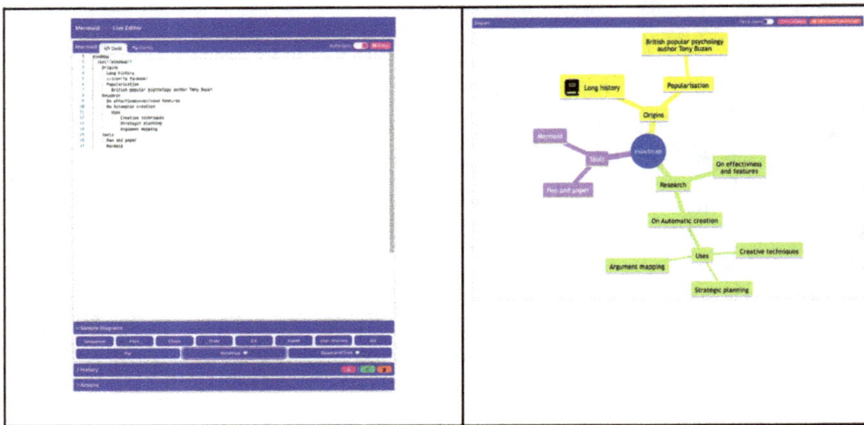

Frame A: Codes Could be Created in ChatGPT based on specific prompts	Frame B: Output Created in Mermaid Live Editor

Fig. 12.16 Coding Generated by ChatGPT and Visual Mind Map Output Created in Mermaid Live Editor

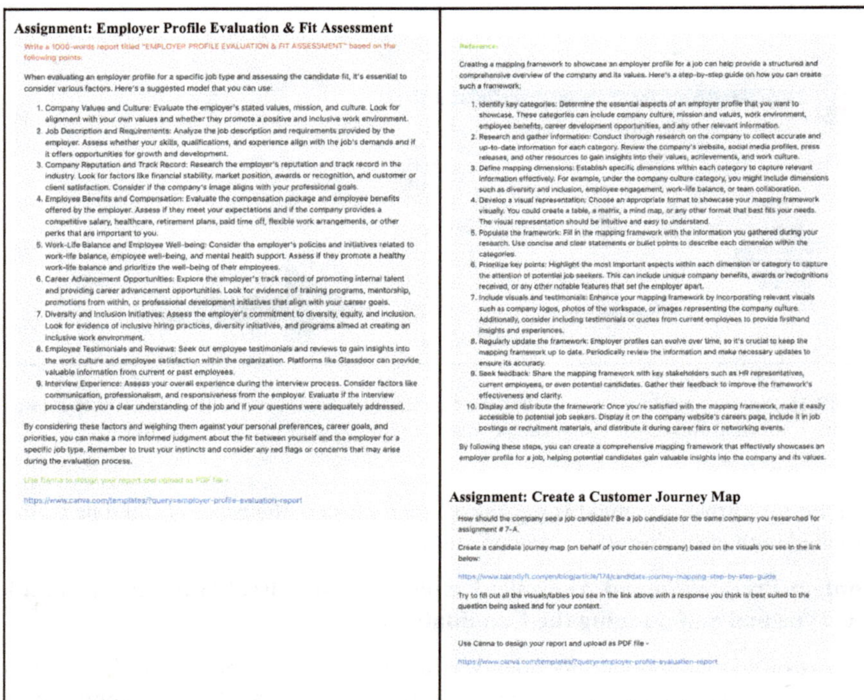

Assignment: Employer Profile Evaluation & Fit Assessment

Write a 1000-words report titled "EMPLOYER PROFILE EVALUATION & FIT ASSESSMENT" based on the following points:

When evaluating an employer profile for a specific job type and assessing the candidate fit, it's essential to consider various factors. Here's a suggested model that you can use:

1. Company Values and Culture: Evaluate the employer's stated values, mission, and culture. Look for alignment with your own values and whether they promote a positive and inclusive work environment.
2. Job Description and Requirements: Analyze the job description and requirements provided by the employer. Assess whether your skills, qualifications, and experience align with the job's demands and if it offers opportunities for growth and development.
3. Company Reputation and Track Record: Research the employer's reputation and track record in the industry. Look for factors like financial stability, market position, awards or recognition, and customer or client satisfaction. Consider if the company's image aligns with your professional goals.
4. Employee Benefits and Compensation: Evaluate the compensation package and employee benefits offered by the employer. Assess if they meet your expectations and if the company provides a competitive salary, healthcare, retirement plans, paid time off, flexible work arrangements, or other perks that are important to you.
5. Work-Life Balance and Employee Well-being: Consider the employer's policies and initiatives related to work-life balance, employee well-being, and mental health support. Assess if they promote a healthy work-life balance and prioritize the well-being of their employees.
6. Career Advancement Opportunities: Explore the employer's track record of promoting internal talent and providing career advancement opportunities. Look for evidence of training programs, mentorship, promotions from within, or professional development initiatives that align with your career goals.
7. Diversity and Inclusion Initiatives: Assess the employer's commitment to diversity, equity, and inclusion. Look for evidence of inclusive hiring practices, diversity initiatives, and programs aimed at creating an inclusive work environment.
8. Employee Testimonials and Reviews: Seek out employee testimonials and reviews to gain insights into the work culture and employee satisfaction within the organization. Platforms like Glassdoor can provide valuable information from current or past employees.
9. Interview Experience: Assess your overall experience during the interview process. Consider factors like communication, professionalism, and responsiveness from the employer. Evaluate if the interview process gave you a clear understanding of the job and if your questions were adequately addressed.

By considering these factors and weighing them against your personal preferences, career goals, and priorities, you can make a more informed judgment about the fit between yourself and the employer for a specific job type. Remember to trust your instincts and consider any red flags or concerns that may arise during the evaluation process.

Use Canva to design your report and upload as PDF file -

https://www.canva.com/templates/?query=employer-profile-evaluation-report

Reference:

Creating a mapping framework to showcase an employer profile for a job can help provide a structured and comprehensive overview of the company and its values. Here's a step-by-step guide on how you can create such a framework:

1. Identify key categories: Determine the essential aspects of an employer profile that you want to showcase. These categories can include company culture, mission and values, work environment, employee benefits, career development opportunities, and any other relevant information.
2. Research and gather information: Conduct thorough research on the company to collect accurate and up-to-date information for each category. Review the company's website, social media profiles, press releases, and other resources to gain insights into their values, achievements, and work culture.
3. Define mapping dimensions: Establish specific dimensions within each category to capture relevant information effectively. For example, under the company culture category, you might include dimensions such as diversity and inclusion, employee engagement, work-life balance, or team collaboration.
4. Develop a visual representation: Choose an appropriate format to showcase your mapping framework visually. You could create a table, a matrix, a mind map, or any other format that best fits your needs. The visual representation should be intuitive and easy to understand.
5. Populate the framework: Fill in the mapping framework with the information you gathered during your research. Use concise and clear statements or bullet points to describe each dimension within the categories.
6. Prioritize key points: Highlight the most important aspects within each dimension or category to capture the attention of potential job seekers. This can include unique company benefits, awards or recognitions received, or any other notable features that set the employer apart.
7. Include visuals and testimonials: Enhance your mapping framework by incorporating relevant visuals such as company logos, photos of the workspace, or images representing the company culture. Additionally, consider including testimonials or quotes from current employees to provide firsthand insights and experiences.
8. Regularly update the framework: Employer profiles can evolve over time, so it's crucial to keep the mapping framework up to date. Periodically review the information and make necessary updates to ensure its accuracy.
9. Seek feedback: Share the mapping framework with key stakeholders such as HR representatives, current employees, or even potential candidates. Gather their feedback to improve the framework's effectiveness and clarity.
10. Display and distribute the framework: Once you're satisfied with the mapping framework, make it easily accessible to potential job seekers. Display it on the company website's careers page, include it in job postings or recruitment materials, and distribute it during career fairs or networking events.

By following these steps, you can create a comprehensive mapping framework that effectively showcases an employer profile for a job, helping potential candidates gain valuable insights into the company and its values.

Assignment: Create a Customer Journey Map

How should the company see a job candidate? Be a job candidate for the same company you researched for assignment # 7-A.

Create a candidate journey map (on behalf of your chosen company) based on the visuals you see in the link below:

https://www.talentlyft.com/en/blog/article/174/candidate-journey-mapping-step-by-step-guide

Try to fill out all the visuals/tables you see in the link above with a response you think is best suited to the question being asked and for your context.

Use Canna to design your report and upload as PDF file -

https://www.canva.com/templates/?query=employer-profile-evaluation-report

Fig. 12.17 ChatGPT Guidance on Employer-Profile Evaluation & Fit Assessment students to complete a text report based on the points highlighted in the matrix format. More detailed research should be highlighted in such a report. This customer journey map table should only be considered the first step toward company evaluation and candidate fit. But before a text report is written, more information is required.

Based on Salazar, the course instructor can follow up on the next iteration by asking students to focus on the following points:

1. Highlight areas where expectations are not met.
2. Highlight unnecessary touchpoints or interactions.
3. Identify points of friction
4. Identify high friction points.
5. Provide time duration for each stage of the journey.
6. Highlight very important issues that cannot be sidelined.
7. Identify areas where expectations are met or exceeded.

The above information should be part of the text report. ChatGPT can then evaluate the complete report and identify areas for improvement. So, Fig. 12.18 is barely the first step in the process. There are many levels of iterations required and content enrichment must happen with and without evaluation from ChatGPT (Fig. 12.19).

[candidate persona]
Age:25
Location: Aizuwakamatsu, Japan
Profession: Engineer, now as master's student
Education: Master's in Engineering
Skills: C, Python, Java, JavaScript, R, QA related tools and operation
Interest: Sports, Transportation vehicles
Salary: Exception salary is 220k+ yen

	Awareness	Consideration	Interest	Application	Selection	Hire
Thinking	There is a IT company in this area.	They provide multiple services in ICT field.	This company has a bright environment. Individual can be accepted and encouraged here.	Did my application received and accepted by the HR department?	1. Is there better company for me? (Shall I try more now? Or just wait for this company?) 2. Is there better candidate for company? (Will they select others due to limited positions?)	1. Will I accept that? 2. How can I reject other companies?
Touch point	Summer Internship Program	Brochure provided to students	1. Talk to related person directly and make judgement. 2. Compare and consider the comments.	1. Click to make application. 2. Communicate in telephone or email with HR.	1. Wait and make telephone call. 2. Send, wait and receive email.	1. Receive phone call about the result. 2. Receive email or snail mail about very detailed issue.
Channel	Career site for students, for example RikuNabi.	Official homepage	1. Communication event. 2. Review on the Internet.	1. Career site for students, for example RikuNabi. 2. Telephone and/or email.	Telephone OR Email	Telephone AND Email, sometimes snail mail as well

Fig. 12.18 First Step Customer Journey Map

1. Did you utilize Chat GPT to seek information or guidance regarding the evaluation of an employer profile for a specific job type? (Yes/No) - *YES.*

2. If yes, how did Chat GPT assist you in understanding the company values and culture? Please provide a brief explanation.
I do try to do that, but cannot get useful information. Chat GPT tend to say "As an AI language model, I don't have access to real-time data, and my knowledge is limited up to September 2021. I am not aware of specific details about ... ". So turn to Google service then.

3. Did you use Google search to research the employer's reputation and track record? (Yes/No) *YES.*

4. If yes, what specific information did you find through Google search regarding the employer's reputation and track record?
Google can guide me to websites which I never realize. When I search "(name of company) comment" or "(name of company) reputation", the search result will guide me to different third-party websites. And the website is different base on country.

5. How did you leverage Chat GPT or Google search to gather insights about the job description and requirements? Please explain the process briefly.
I copy and paste the JD and requirement to Chat GPT, and ask it to "say it in another way, emphasizing the skill needed". Furthermore, I ask questions like "list the skill I need to study base on the following description: ...".

6. Did you employ Chat GPT to gather information about employee benefits and compensation? (Yes/No) - *YES.*

7. If yes, how did Chat GPT assist you in understanding the compensation package and employee benefits offered by the employer?
The problem for Chat GPT is the same as I mentioned regarding question 2. They don't have latest data.

8. How did you utilize Google search to assess the employer's policies and initiatives related to work-life balance and employee well-being?
Google is good, but not so useful in Japan. Actually Google will guide me to Rikunabi or Mynabi every time. In the end, I directly go to these website without using Google.

9. Did you employ Chat GPT or Google search to gather information about career advancement opportunities? (Yes/No) - *YES.*

10. If yes, please describe how Chat GPT or Google search helped you explore the employer's track record of promoting internal talent and providing career growth opportunities.
Chat GPT don't have the latest information, google cannot provide sufficient for typical Japanese companies. I myself will, and sometimes Google will guide me to, visit the official homepage of a Japanese company, and then go to the 'About Us' part of it for this information. In conclusion, the third-party website always works, the Google keeps helping me to find new third-party websites.

11. How did you utilize Chat GPT or Google search to evaluate the employer's commitment to diversity and inclusion?
The method is similar, but here Chat GPT can work in another way. I firstly ask, "What is diversity in an IT company?" to guide Chat GPT about this topic, and then keep copy and paste comments on third-party websites to Char GPT, and ask "give me a mark (1-10, 1 is lowest) to this comment regarding the diversity."

12. Did you refer to employee testimonials and reviews using Chat GPT or Google search? (Yes/No) - *YES.*

13. If yes, please explain how Chat GPT or Google search assisted you in accessing employee testimonials and reviews.
Google guide me to third-party websites, and Chat GPT give me his score regarding these comments based on my standard.

14. How did you leverage Chat GPT to seek guidance on the interview experience with the employ
For this task, I think I should do it by myself. I have my own strengths and weakness, so I would like to make a judgement by myself.

15. Finally, how effectively do you believe Chat GPT and Google search helped you gather the necessary information to evaluate the employer profile and candidate fit? Please rate on a scale of 1 to 5, with 1 being least effective and 5 being highly effective.
I will give 2 in general. 1 to ChatGPT, and 4 to Google.

Fig. 12.19 Self-Reported Use of Chat GPT towards Completing Fig. 12.17

12.9 Content Production and Analysis: Industry 4.0 Presentation

This is a concept-building assignment for entrepreneurship. As part of this assignment, the student was asked to develop the fundamental concepts of Industry 4.0 and 5.0 as much as possible, within the scope of the course. No specific industry was focused, and neither was there any specific expectation about the concepts developed.

Section B (Media Enrichment) allowed the instructor an opportunity to explore the extent to which the student could speak about the concepts with confidence. Figure 12.17 highlights the results based on the ChatGPT review (Fig. 12.20).

Figure 12.21 is a guidance on how prompt engineering could help instructors and students see through the recommendations made for improvement and strengths identified. Based on the points made, further prompts could be generated to understand exactly how to approach a certain recommendation. For example, a prompt based on the areas for improvement could ask, exactly what kind of visuals would

Follow the assignment steps as suggested.

Step # 1: Concept Building
Study the concept of Industry 4.0 and Industry 5.0 as much as possible from the links made available.

Step # 2:

A. Presentation Design
Now, once you think there is enough understanding of the concepts related to Industry 4.0, design a **16-slide** keynote presentation introducing an industry (within Industry 4.0 types) that is of interest to you based on your own area of research. Feel free to use both Google Search and Chat GPT.

B. Media Enrichment
Once you have designed the presentation, save it with added narration and video/picture clippings demonstrating examples of Industry 4.0 as you see it, for your industry of choice.
Save it as a movie file and upload it to YouTube with restricted access.

C. Developing a Business Proposal Report
The idea of the above presentation is for you to initiate a start-up company in Japan that incorporates Industry 4.0 in the design and implementation plan for such an enterprise.
So, based on the above presentation, use Chat GPT to design a **business proposal report** that incorporates the above idea. Your business proposal report should be written to ask and convince the University of Aizu to fund such a start-up idea as part of a venture capital initiative. Contact the university to know more about how such initiative takes shape and how the university supports such initiative.

Reference:
https://u-aizu.ac.jp/en/research/venture/
https://u-aizu.ac.jp/en/stories/matching-service.html

D. Developing Advertising Poster-Infographic
Use Chat GPT to develop a convincing **one-page poster-infographic** that will convince the management about the value of such a start-up venture at the University of Aizu. This should take the shape of an **advertising poster-infographic**.
https://www.canva.com/create/infographics/

Slide Layout Design - A Template:

Slide 1: Title Slide, Name, ID, university affiliation, etc.

Slide 2: The concept of Industry 4.0 in general

Slide 3: A video clipping showing an example of Industry 4.0 - explain the video with your own text and what it shows.

Slides 4-6: Show a few industry types and production and supply chain and automation structures that are consistent with Industry 4.0.

Slides 7-8: How realistic and practical and feasible is the targeted upgrade from Industry 4.0 to Industry 5.0? Provide a narrative discussion in your own words.

Slides 9 -10: An industry 4.0 of your choice where you would like to work. Explain the industry.

Slides 11 - 12: How is your academic and professional experience consistent with the demands of that industry?

Slide 13: Explain a project experience you have that is consistent with Industry 4.0.

Slide 14: Show a job advertisement that is precisely in your domain of education and experience.

Slide 15: How do you think the world is academically and professionally preparing for Industry 4.0 and moving forward to Industry 5.0?

Slide 16: Conclusion as you see the future of Industry 4.0 in your area of interest.

Put narration for your slides (no reading from the slides), export as a movie file, upload on YouTube with restricted access, and upload the URL in Moodle.

Fig. 12.20 Utilizing Chat GPT and Google Search to Design a Presentation and Writing a Business Proposal on Industry 4.0 and Funding a Start-up Company at the University of Aizu

make for an interesting presentation on Industry 4.0; why should such visuals make it more interesting and/or informative for the audience; if the length of the video is too long, how could it be shortened?; What kind of content should be excluded in the revised version to make it shorter or how to prepare a better summarized version without excluding major content areas?

Figure 12.22 provided a self-report on how and to what extent ChatGPT and Google search were used to develop the concepts of Industry 4.0 and complete the assignment in the process. The questionnaire is also aimed at understanding the extent to which these tools helped develop the concepts of Industry 4.0 such that the student could reflect and comment on questions asked about Industry 4.0.

12.9.1 Discussion on Incorporating Prompt Engineering in EFL Courses

The above section was an attempt to provide enough examples for EFL course instructors to develop assignments and for students to develop responses based on the guidelines provided by ChatGPT as part of the assignment. Using prompt engineering in EFL classrooms could be effective in guiding students to develop deeper thinking, creativity, and critical analysis.

We need the following structured approach to implementing prompt engineering in EFL classrooms.

- Develop an Understanding of the Learning Objectives: What specific learning objectives or skills do we want our students to develop? This could include skills such as analyzing a business report, writing persuasively to push a professional agenda, or exploring business themes.

Prompt: Evaluate the following presentation on Industry 4.0

Strengths:

1. **Clarity and Structure:** The presentation is well-structured with clear sections, making it easy to follow. The introduction effectively grabs the audience's attention, and the speaker maintains a logical flow throughout.
2. **Content Knowledge:** The presenter demonstrates a strong understanding of Industry 4.0 and its significance. They explain the concept clearly and provide examples, such as the history of industrial revolutions and the role of emerging technologies in Industry 4.0.
3. **Engagement:** The use of multimedia, such as the video, enhances engagement and supports the presenter's points. The video helps illustrate key concepts and keeps the audience interested.
4. **Multilingual Skills:** The presenter's ability to speak multiple languages, including Mandarin, Japanese, English, Spanish, Vietnamese, and Thai, is highlighted as an asset, demonstrating their readiness for a global workforce.
5. **Relevance to Career Goals:** The presenter clearly articulates their career goals related to Industry 4.0, expressing a desire to work as a software engineer in the manufacturing sector, particularly in Nagoya, Japan. This aligns well with the topic of the presentation.

Areas for Improvement:

1. **Pronunciation:** While the presenter's multilingual abilities are a strength, there are occasional pronunciation issues. Clear pronunciation, especially of technical terms, is essential for effective communication.
2. **Visual Presentation:** While the presentation makes good use of a video, there could be further improvements in visual elements. Adding more visuals, slides, or graphics directly related to the content could enhance understanding and engagement.
3. **Length:** The presentation is quite long, and at times, it feels like there is an information overload. It might benefit from more concise and focused content to maintain audience engagement.
4. **Transitions:** While the presentation is generally well-structured, there could be smoother transitions between sections. Clear transitions help guide the audience and maintain their interest.
5. **Grammar and Language:** There are some grammatical errors and awkward phrasings throughout the presentation. Proofreading for grammar and language consistency is important, especially for a professional context.

Prompt: Which specific content area discussion on Industry 4.0 is missing in the video and something likely to enrich this presentation?

Cybersecurity and Data Privacy in Industry 4.0:

Cybersecurity and data privacy are critical aspects of Industry 4.0, given the increasing connectivity and data exchange between physical and digital systems. Discussing these topics would enhance the presentation in the following ways:

1. **Importance:** Emphasizing the significance of cybersecurity and data privacy in Industry 4.0 is crucial. As more devices and systems become interconnected, the risk of cyber threats and data breaches also grows. This discussion can underscore the need for robust security measures.
2. **Challenges:** Addressing the specific cybersecurity challenges that come with Industry 4.0, such as protecting sensitive data, securing IoT devices, and defending against cyber-attacks, can provide a comprehensive view of the topic.
3. **Solutions:** Exploring cybersecurity solutions and best practices for safeguarding connected systems and data is valuable. This could include discussing encryption, access control, intrusion detection systems, and incident response strategies.
4. **Regulations and Compliance:** Touching upon relevant regulations and compliance requirements, such as GDPR in Europe or industry-specific standards, would demonstrate an awareness of the legal and ethical aspects of data handling.
5. **Case Studies:** Sharing real-world examples of cybersecurity breaches in Industry 4.0 and their consequences can make the topic more relatable and highlight the need for proactive security measures.
6. **Future Trends:** Discussing emerging trends in cybersecurity, such as AI-driven threat detection or blockchain for data integrity, can show that the presenter is forward-thinking and aware of evolving security solutions.

Including a discussion on cybersecurity and data privacy would not only enrich the presentation but also underscore the presenter's awareness of the multifaceted challenges and considerations within the realm of Industry 4.0. It would demonstrate a holistic understanding of the topic and its implications for businesses and industries undergoing digital transformation.

Prompt: Why did you choose cybersecurity and data privacy as an area that has not been mentioned in the video transcript?

I chose cybersecurity and data privacy as an area that has not been mentioned in the video transcript because these topics are highly relevant and crucial within the context of Industry 4.0. Industry 4.0 is characterized by the increased integration of digital technologies, the Internet of Things (IoT), and data-driven processes in manufacturing and industrial settings. As a result, it introduces new challenges and risks related to cybersecurity and data privacy. Here's why I selected this topic:

1. **Relevance:** Cybersecurity and data privacy are paramount concerns in the context of Industry 4.0. The interconnectivity of machines, devices, and systems creates vulnerabilities that need to be addressed to ensure the safety and security of operations.
2. **Emerging Challenges:** With the proliferation of IoT devices and the collection of vast amounts of data, Industry 4.0 introduces new challenges related to protecting sensitive information, ensuring data integrity, and safeguarding against cyber threats.
3. **Impact:** A breach in cybersecurity or a violation of data privacy can have severe consequences, including operational disruptions, financial losses, and reputational damage. It's an area of concern that deserves attention.
4. **Professional Development:** In the context of career and professional development, knowledge and expertise in cybersecurity and data privacy are valuable skills, particularly for individuals seeking roles in Industry 4.0-related fields.
5. **Holistic Understanding:** Discussing cybersecurity and data privacy provides a more comprehensive view of the challenges and considerations within Industry 4.0. It emphasizes the need for a balanced approach that ensures both technological advancement and security.

Prompt: What makes you think the presenter would and should be interested in the topic in this context?

My suggestion to include cybersecurity and data privacy in the presentation was not based on a specific indication that the presenter is interested in these topics in the provided video transcript. Instead, it was a general recommendation to enrich the presentation on Industry 4.0 with an essential aspect that is often relevant to discussions of Industry 4.0, particularly when it comes to technology, manufacturing, and data-driven processes.

In many discussions about Industry 4.0, the importance of cybersecurity and data privacy is acknowledged due to the increased digitization and connectivity of industrial systems. While the presenter did not explicitly express interest in these topics in the video transcript, it is common for individuals involved in Industry 4.0, including project managers, engineers, and technology enthusiasts, to recognize the significance of cybersecurity and data privacy as integral components of the digital transformation.

The recommendation was made in the context of providing a more holistic and well-rounded presentation on Industry 4.0, considering the broader implications and challenges associated with this industrial revolution. It's essential for professionals in the field, including those interested in Industry 4.0, to be aware of these topics, as they play a vital role in ensuring the security, integrity, and ethical use of data and technology in modern industries.

Ultimately, the inclusion of cybersecurity and data privacy in the presentation would align with the overarching goal of promoting a comprehensive understanding of Industry 4.0 and its implications, regardless of the specific interests of the presenter in the provided transcript.

Prompt: Is it possible to know from the video transcript the speaker's specialization?

Based on the video transcript, it is clear that the speaker is a master's student with a background in project management and software engineering. The speaker mentions pursuing two master's degrees, one related to modern education technology and the other in a graduate school of the University of Aisha. They also discuss their experience working as a software quality assurance team member and their proficiency in programming languages like C, Visual Basic, Java, JavaScript, and others.

While the speaker's primary specialization appears to be in project management and software engineering, they also express an interest in Industry 4.0 and its applications, particularly in manufacturing and technology-related fields. The transcript does not provide specific details about their academic major or specialization beyond the general areas of project management and software engineering.

Fig. 12.21 Prompts to Identify and Improve on the Video and Concepts of Industry 4.0

Section 1: Demographics

1. Gender: Male
2. Age: 25
3. Current Academic Status:
   - Undergraduate student
   - Graduate student
   - Faculty/Staff
   - Other (please specify)

Section 2: Familiarity with Industry 4.0 and Start-up Funding

4. How familiar are you with the concept of Industry 4.0?
   - Very familiar
   - *Somewhat familiar*
   - Not familiar
5. Have you had any prior experience or knowledge of start-up funding and writing business proposals?
   - Yes, extensive experience
   - *Yes, some experience*
   - No experience

Section 3: Usage of Chat GPT and Google Search

6. Have you used Chat GPT (e.g., OpenAI's GPT-3.5) for any academic or professional tasks?
   - *Yes*
   - No
7. If you answered "Yes" to the previous question, please describe how you used Chat GPT for your presentation and business proposal on Industry 4.0 and start-up funding at the University of Aizu. (Please provide a brief description) - *I will confirm my structure with the chatgpt and ask it to give me some feedback, for example, if it is reasonable enough, and where need to be improved.*
8. Have you used Google Search or any other search engine as part of your research for the presentation and business proposal?
   - *Yes*
   - No
9. If you answered "Yes" to the previous question, please explain how Google Search or other search engines assisted you in your research. (Please provide a brief description) - *Google will guide me to many interesting website related, so I have more chances to explore that by myself.*

Section 4: Benefits and Challenges

10. In your opinion, what were the main benefits of using Chat GPT for designing the presentation and writing the business proposal? (Open-ended response) - *It will provide the basic idea, especially when I don't have a clear understanding of it. Even if I have one, it will help me to confirm if it is reasonable enough.*
11. Were there any challenges or limitations you encountered while using Chat GPT and Google Search for your academic tasks? If yes, please specify. (Open-ended response) - *Yes. They will give me contradictory answers sometimes. But I understand that they are not human, so I need to confirm the information by myself.*

Section 5: Perception of Industry 4.0 and Start-up Funding

12. Based on your research and understanding, what do you perceive as the key value of Industry 4.0 for start-up companies? (Open-ended response)- *In the beginning stage, survival is an important task. If there is no correct strategy, the business without Industry 4.0 idea may lose from the very beginning.*
13. How do you envision funding a start-up company related to Industry 4.0? What are the potential challenges and opportunities? (Open-ended response) - *I think it is a good idea to do that, especially when compare to a company without any relation to 4.0. I regard this kind of relationship as an advantage, not risk.*

Section 6: General Feedback

14. Overall, how satisfied were you with the support received from Chat GPT and Google Search in completing your academic tasks?
- Very satisfied
- *Satisfied*
- Neutral
- Dissatisfied
- Very dissatisfied

15. Is there anything else you would like to add about your experience using Chat GPT and Google Search for your presentation and business proposal? (Open-ended response) - *Generally, I define Chat GPT as a structure-related tutor and idea-generator. But as for Google, I think it is a possibility-provider.*

Fig. 12.22 Self-Report on the Use of ChatGPT and Google Search for Completing a Basic Assignment on Industry 4.0

- Content Relevance: Choose a case study or theme in business that aligns with the learning objectives. This will serve as the foundation for prompt creation.
- Developing Initial Prompts: Start with the most basic prompts in the context. These should be open-ended, and the responses should make the student think critically.
- Segmentation of Learning Objectives: Analyze each learning objective and identify exactly what kind of skills or concepts do we expect our students to grasp. Then develop prompts that target those skills.

- Diversification of Prompt Types: Teach business communicative situations that will ask for analytical prompts, creative writing prompts, comparative prompts, reflective prompts, and argumentative prompts.
- Consider Scaffolding: The instructor should identify the level and prior knowledge of the students. Develop a scaffolding strategy that will create prompts gradually increasing in complexity from simple to more complex ones.
- Incorporate Bloom's Taxonomy: This will help strategize prompting situations that will encourage different levels of thinking such as remembering and understanding, moving towards applying, analyzing, evaluating, and creating.
- Context and Background Information: Such information will help better understand and engage with the prompts. For example, when setting up a business, provide background information about the situation under which the business setup is being planned. That will help develop a business proposal.
- Peer Review & Feedback: Have class sessions where ChatGPT responses are evaluated and a critical discussion follows related to the appropriateness and completeness of the responses, and if further prompting is required for better representation and/or use of the information.
- Develop Assessment Rubric: Assessment Rubrics should judge the extent to which and how ChatGPT was used towards complete business documentation. This should help with the grading.
- Iterate and Adapt: Prompts should be continuously evaluated and improved to best represent and customize a specific business situation from general to specifics, as required.
- Discussion & Feedback Sessions: In-class sessions should develop a structured approach to discussing the prompts and encourage responses to such prompts, reflecting on what was learned, how the information was used, and what could be follow-up steps. Instructors should provide constructive feedback on student responses, highlighting on strengths and areas for improvement, and encourage revisions.
- Continuous Course Evaluation: Such courses should be continuously evaluated, reflecting on the effectiveness of the prompts used by both instructors and students and the student's responses. This information should be used to improve future prompt engineering efforts.

Figure 12.23 provides an outlook on how appropriate prompts could develop such a scenario (a case study) that will help create a mix of prompts. The case study was created by ChatGPT 3.5. This could be used to create an effective mix of chain-of-thought (CoT) prompting and few-shot prompting.

Some classic follow-up prompts based on the response above from ChatGPT 3.5:

Structure	Develop a structure that will help the research teams to investigate the markets in Brazil and Japan
Highlight	For this kind of project, which factors should be investigated in any given order of importance?

(continued)

(continued)

Similarities & Differences	Are there any similarities between these markets? Any specific difference that's noteworthy and should be further investigated?
Challenges	Identify challenges for each specific market! How could we overcome it? How should we overcome these challenges in the local market?
Assessments	Please carry out a risk assessment based on the points highlighted for each market!

The points and responses above could also be framed in the context of a SWOT analysis. We have the model for a start-up's SWOT analysis (Ali, 2022), and then, based on the points, chain-of-thought prompts could be created for each category. As we continue to get responses from ChatGPT, the responses could be summarized and interconnected with Lucid Chart or mind mapping. Such an approach not only requires students to identify the issues and learn about them from ChatGPT, but also use a specific framework to connect the dots with the large amount of data generated, both at a micro level (e.g., mapping resources and challenges), and at a macro level (e.g., challenges in the local market, and help required).

Prompt engineering is a flexible process and could be integrated into the language teaching pedagogy effectively. A diligent approach could help suit specific teaching styles and the needs of the students. By carefully designing prompts that align with specific learning objectives, one can foster a more engaging and enriching learning experience in the English classroom.

Tailoring Prompt Engineering for EFL Students

Tailoring prompt engineering for English as a Foreign Language (EFL) students involves several key strategies. Firstly, instructors should be mindful of students' language proficiency levels, providing prompts that match their abilities while gradually introducing more complex language elements. Second should be supported with vocabulary lists and grammar tips. Students should have access to model responses to clarify expectations and start with scaffolded prompts that guide critical thinking. Students should be made aware of cultural backgrounds, incorporating visuals for context and real-world relevance for engagement. Assignment activities should promote collaboration through pair and group activities, emphasizing feedback, revision, and critical thinking. There should be regular prompt creation exercises, reflective journals, and diverse prompt types, alongside peer review and individualized support, ensuring regular practice and teacher feedback. This tailored approach enhances language and critical thinking skills in EFL students.

AI-Powered Prompt Evolution in Business Communication

The strategic use of prompt engineering in an English as a Foreign Language (EFL) business communication course is a dynamic and multifaceted process. By customizing prompts to suit language proficiency levels, providing vocabulary and grammar support, and offering model responses, students can navigate the complexities of real-world business communication. Scaffolded prompts encourage critical

Prompt: Create a case study that will help us create a mix of prompts that mimic real-world business communication situations, including email correspondence, business proposal writing, case study analysis, simulated meetings, and intercultural communication scenarios.

Case Study Title: "Expanding a Tech Startup Internationally"

Background:
XYZ Tech Solutions is a successful tech startup based in the United States. They have developed innovative software for data analytics and have achieved significant success in their home market. Now, they are considering expanding their operations internationally to tap into new markets and customers.

Scenario:
XYZ Tech Solutions has identified two potential international markets for expansion: Japan and Brazil. The company's leadership team is preparing to make a decision on which market to enter first. They have assigned teams of employees to conduct research and present proposals for each market. Additionally, they are planning a series of meetings with potential local partners and investors in both countries.

Prompts for Different Business Communication Situations:

1. **Email Correspondence**:
   * Write an email to a potential business partner in Japan introducing XYZ Tech Solutions and expressing your interest in exploring a partnership. Highlight the key strengths of your software.
2. **Business Proposal Writing**:
   * Develop a business proposal for entering the Brazilian market. Include market analysis, competitive analysis, marketing strategy, and financial projections. Present it as a comprehensive report.
3. **Case Study Analysis**:
   * Analyze the opportunities and challenges of entering the Japanese market for XYZ Tech Solutions. Consider cultural, legal, and market-specific factors. Provide recommendations based on your analysis.
4. **Simulated Meeting - Intercultural Communication**:
   * Role-play a meeting between XYZ Tech Solutions' team and potential Japanese partners. Address intercultural communication challenges, such as differences in communication style, hierarchy, and business etiquette.
5. **Simulated Meeting - Investor Pitch**:
   * Prepare a presentation for a meeting with potential Brazilian investors. Focus on articulating the value proposition of XYZ Tech Solutions and its potential impact on the local market.
6. **Follow-up Email**:
   * Write a follow-up email to a Japanese business partner after the initial meeting. Express gratitude for their time and discuss potential next steps in the partnership.
7. **Intercultural Communication Scenario**:
   * Describe a hypothetical scenario where a misunderstanding occurred during a business negotiation in Brazil due to cultural differences. Analyze what went wrong and propose strategies for avoiding similar issues in the future.
8. **Proposal Evaluation**:
   * Evaluate and compare the strengths and weaknesses of the proposals for entering the Japanese and Brazilian markets. Provide feedback on which proposal aligns better with XYZ Tech Solutions' goals and values.
9. **Negotiation Simulation - Joint Venture**:
   * Role-play a negotiation session with a potential Brazilian partner to establish a joint venture. Address negotiation tactics and strategies for reaching a mutually beneficial agreement.
10. **Market Entry Announcement Email**:
    * Write an internal email to all XYZ Tech Solutions employees announcing the company's decision to enter one of the markets. Explain the rationale behind the choice and outline the next steps.

Fig. 12.23 ChatGPT Created a Case Study that will combine chain-of-thought and few-shot prompting effectively in different stages

thinking while considering cultural context and incorporating visual aids ensures relevance and engagement. Collaboration and feedback, both from peers and instructors, play a pivotal role in refining prompt comprehension and response skills. Encouraging students to create their own prompts fosters deeper thinking about course objectives. Through this tailored approach, EFL students not only develop language proficiency but also critical thinking and communication skills that are vital for success in the global business landscape.

The future of research in the field of prompt engineering for English as a Foreign Language (EFL) business communication courses should focus on several key areas to further enhance teaching and learning in this domain:

1. Adaptive AI-driven Prompt Generation: Explore the integration of artificial intelligence and natural language processing to develop adaptive prompt generation systems that can tailor prompts to individual student needs and learning progress.
2. Multimodal Prompts: Investigate the effectiveness of incorporating multimedia elements (e.g., video, audio) into prompts to cater to diverse learning styles and engage students more effectively.
3. Cross-Cultural Communication: Research how prompts can be designed to better prepare EFL students for cross-cultural communication challenges in the global business environment, including intercultural negotiation and collaboration.
4. Prompt Efficacy Assessment: Develop more sophisticated methods for assessing the efficacy of prompts, including their impact on language acquisition, critical thinking, and problem-solving skills.

5. Technology Integration: Explore how emerging technologies such as virtual reality (VR) and augmented reality (AR) can be used to create immersive business communication scenarios and prompts.
6. Prompt Personalization: Investigate strategies for personalizing prompts based on student's interests, career goals, and cultural backgrounds to increase motivation and relevance.
7. Longitudinal Studies: Conduct longitudinal studies to assess the long-term impact of prompt engineering on students' language proficiency and career success in international business contexts.
8. Professional Development for Instructors: Research the training and professional development needs of instructors in the field of EFL business communication, particularly in the context of prompt design and assessment.
9. Ethical Considerations: Examine ethical considerations related to the use of AI in prompt engineering, including issues of bias, privacy, and transparency.
10. Comparative Studies: Compare the effectiveness of different prompt types, such as traditional written prompts versus multimedia-rich prompts, in terms of language skill development and critical thinking.
11. Interdisciplinary Research: Foster collaboration between linguists, educators, technologists, and business professionals to create a holistic approach to prompt engineering that aligns with real-world industry needs.
12. Global Collaboration: Encourage international research collaboration to gather insights and best practices from different cultural and linguistic contexts.
13. Feedback Mechanisms: Investigate effective feedback mechanisms within prompt-based learning systems, emphasizing both automated feedback and peer evaluation.
14. Accessible Learning: Explore how prompt engineering can be adapted to make business communication education more accessible to diverse student populations, including those with disabilities.
15. Hybrid Learning Environments: Investigate the integration of prompt-based learning with hybrid and online learning environments, especially in the post-pandemic educational landscape.

12.10 Conclusion

In the emerging landscape of professional language teaching for business and specific purposes, the future envisions a revolutionary shift towards prompt engineering methodologies driven by advanced technologies. Future researchers are poised to embark on a journey of deeper exploration within this context, harnessing the power of visualization techniques integrated with concept mapping and ontology creation, firmly rooted in the semantic web paradigm. These innovative approaches will facilitate enhanced case study evaluations through meticulous metadata generation, utilizing XML, RDF, and OWL standards, while adhering to adaptable criteria ranging from simplicity to stringency.

Furthermore, the integration of cutting-edge text-mining software and knowledge graphs promises to unlock new dimensions in prompt and response evaluation, particularly in the realm of ChatGPT's data structuring capabilities. This pioneering visualization-centric approach at the user's end holds the key to unraveling ChatGPT's untapped potential for context-driven projects. The resulting visualization and metadata models will pave the way for the development of highly specific question prompts, elevating cognitive comprehension of the context. This, in turn, empowers students to discern intricate relationships among classes, properties, and domains, fostering idea generation and specificity, all as part of an iterative process leading to comprehensive documentation.

The ultimate written output, a testament to this transformative methodology, will undergo scrutiny through AI-driven text detection software, ensuring originality. Additionally, coherence assessment will ascertain how diverse prompts harmoniously converge within reports, forging holistic coherence. ChatGPT will assume a dual role, both as a prompt generator based on student-engineered responses and as a tool to craft quizzes that challenge students to deconstruct the final text, gauging their content comprehension. This deconstruction process will extend to students formulating literal, inferential, and evaluative questions, all evaluated independently without ChatGPT's assistance.

Embracing the tenets of design thinking, this iterative classroom process safeguards against superficial AI-generated content and plagiarism concerns, anchoring content comprehension as the primary objective. This pedagogical innovation extends to other project-based assignments, potentially replicating this dual content creation model by harnessing the synergy of ChatGPT and Google search methodically.

The future of research in EFL business communication prompt engineering promises a paradigm shift with immense potential to elevate pedagogical practices, harness the full spectrum of technology, and equip students for triumph in the global business arena. Its success hinges on its adaptability to evolving educational demands and the dynamic landscape of emerging technologies, all in pursuit of continually enhancing the educational quality within this domain.

References

Ali, N. (2022). https://www.linkedin.com/pulse/startups-swot-analysis-nida-ali/. Retrieved on 3rd September, 2023. A startup's SWOT analysis.

Alqaheri, K. (2023). https://www.linkedin.com/pulse/how-chat-gpt-can-revolutionize-future-khalil-alqaheri/. Retrieved on 7th September, 2023. How Chat GPT can revolutionize the future of entrepreneurship.

Biswas, S. (2023). https://www.opastpublishers.com/open-access-articles/role-of-chat-gpt-in-education.pdf. Retrieved on 5th August, 2023. Role of ChatGPT in Education. Journal of ENT Surgery Research.

Chan, C. K. Y., & Hu, W. (2023). Students' voices on generative AI: Perceptions, benefits, and challenges in higher education. *arXiv preprint* arXiv:2305.00290.

Diab, M., Herrera, J., Chernow, B. (2022-10-28). Stable diffusion prompt book" (PDF). Retrieved 2023–08–07. Prompt engineering is the process of structuring words that can be interpreted and understood by a *text-to-image* model. Think of it as the language you need to speak in order to tell an AI model what to draw.

Gohil, S. (2023). https://meetanshi.com/blog/chatgpt-vs-google/. Retrieved on 15th September, 2023. ChatGPT vs Google: Is ChatGPT Going to Replace Google?

Gouwas-Stewart, N. (2023). https://masterofcode.com/blog/the-ultimate-guide-to-gpt-prompt-engineering. Retrieved on 20th September, 2023. The ultimate guide to prompt engineering your GPT-3.5-Turbo model.

Greenberg, J., Laura N. (2023). How to prime and prompt ChatGPT for more reliable contract drafting support. *contractnerds.com*. Retrieved 24 July 2023. How to Prime and Prompt ChatGPT for More Reliable Contract Drafting Support.

Kapoor, M. (2023). https://www.greataiprompts.com/chat-gpt/best-chat-gpt-prompts-for-seo/. Retrieved on 20th September, 2023. 100 Best ChatGPT Prompts for SEO in 2023.

Kasneci, E., Seßler, K., Küchemann, S., Bannert, M., Dementieva, D., Fischer, F., & Kasneci, G. (2023). ChatGPT for good? On opportunities and challenges of large language models for education. *Learning and Individual Differences, 103*, 102274.

Kosinski, M. (2023). Theory of mind may have spontaneously emerged in large language models. *arXiv preprint* arXiv:2302.02083.

Liu, G., & Ma, C. (2023). Measuring EFL learners' use of ChatGPT in informal digital learning of English based on the technology acceptance model. *Innovation in Language Learning and Teaching*, pp. 1–14.

Lund, B. D., & Wang, T. (2023). Chatting about ChatGPT: How may AI and GPT impact academia and libraries? *Library Hi Tech News, 40*(3), 26–29.

Mia, D. (2023). https://www.whatischatgpt.co.uk/post/chat-gpt-and-business-ideas-how-ai-is-helping-entrepreneurs. Retrieved on 20th September, 2023. Chat GPT and business ideas: How AI is helping entrepreneurs.

Parsons, N. (2023). https://www.liveplan.com/blog/write-business-plan-with-chatgpt/. Retrieved on 10th September, 2023. Can You Use ChatGPT to Write a Business Plan? Yes, Here's How.

Robinson, J. R. (2023). https://bootcamp.uxdesign.cc/starting-up-on-prompt-engineering-481661e67266. Retrieved on 20th September, 2023. Starting Out on Prompt Engineering.

Robinson, R. (2023). How to write an effective GPT-3 or GPT-4 prompt". *Zapier*. Retrieved 2023–08–14. "Basic prompt: 'Write a poem about leaves falling.' Better prompt: 'Write a poem in the style of Edgar Allan Poe about leaves falling.'

Szefer, J., & Deshpande, S. (2023). Analyzing ChatGPT's aptitude in an introductory computer engineering course. *arXiv preprint* arXiv:2304.06122.

Tam, A. (2023). https://machinelearningmastery.com/generating-graphics-with-chatgpt/. Retrieved on 5th September, 2023. Generating Diagrams with ChatGPT.

Timothy, M. (2023). https://www.makeuseof.com/chatgpt-vs-google-translate-which-is-better-at-translation/. Retrieved on 20th September, 2023. ChatGPT vs. Google translate: Which is better at translation?.

Tsai, S. C. (2022). Chinese students' perceptions of using Google Translate as a translingual CALL tool in EFL writing. *Computer-Assisted Language Learning, 35*(5–6), 1250–1272.

Van Rensburg, A., Snyman, C., & Lotz, S. (2012). Applying Google Translate in a higher education environment: Translation products assessed. *Southern African Linguistics and Applied Language Studies, 30*(4), 511–524.

Wei, J., Wang, X., Schuurmans, D., Bosma, M., Xia, F., Chi, E., & Zhou, D. (2022). Chain-of-thought prompting elicits reasoning in large language models. *Advances in Neural Information Processing Systems, 35*, 24824–24837.

Yadava, O. P. (2023). ChatGPT—a foe or an ally? *Indian Journal of Thoracic and Cardiovascular Surgery, 39*(3), 217–221.

Yalalov, D., & Gaszcz, K. (2023). https://mpost.io/prompt-engineering-ultimate-guide/. Retrieved on 17th September, 2023. Prompt Engineering Ultimate Guide 2023: Beginner to Advanced.

Chapter 13
AI in Language Education: The Impact of Machine Translation and ChatGPT

Louise Ohashi

Abstract This chapter explores the impact of artificial intelligence (AI) within the field of language education. It focuses on two key areas: machine translation (MT) and generative AI chatbot technology, the latter exemplified through ChatGPT. The author positions these tools as real-world resources that teachers and institutions cannot and should not ignore. It introduces them in turn, explaining how they have evolved over time and providing a brief overview of their inner workings and capabilities. Next, empirical studies are reviewed, revealing the key strengths and weakness of these tools and casting light on teacher and student perspectives. Issues such as learning gains, perceived/actual accuracy of output, and plagiarism (or AI-based plagiarism, 'AIgiarism') are covered. In addition to learning-based considerations, issues such as online safety and privacy are discussed. Finally, the need for training and guidelines for teachers and students is addressed and practical activities designed to enhance language development and help with guideline formation are offered for consideration.

Keywords Language education · ChatGPT · Machine translation · Generative AI

13.1 Introduction

The educational landscape has changed beyond imagination within the lifetime of anyone who is reading this chapter, and new developments in artificial intelligence (AI) are about to take it in directions that we cannot yet fathom. My own use of technology for language learning may help readers view the massive advances through a relatable lens. As a teen in the early 1990s in rural Australia, before the internet was in our pockets and existed in few homes, I learnt French with textbooks and the sole form of technology I had access to was my school's audio cassettes. At 15, before going trekking in the Himalayas, I picked up some Nepali by listening to the only cassette I

L. Ohashi (✉)
Department of English Language and Cultures, Gakushuin University, Tokyo, Japan
e-mail: ohashigakushuin@gmail.com

P. Ilic et al. (eds.), *Artificial Intelligence in Education: The Intersection of Technology and Pedagogy*, Intelligent Systems Reference Library 261,
https://doi.org/10.1007/978-3-031-71232-6_13

could get my hands on—a homemade one that had been given to me by someone who had visited Nepal. I had no way to access other learning resources for Nepali until I arrived in Kathmandu and found an English-Nepali-Sherpa phrasebook in a market. In my 20s, I moved to Japan and learnt Japanese through face-to-face lessons and language exchanges, used textbooks and CDs, watched TV with captions and subtitles, and learnt kanji by singing songs at karaoke and exchanging text messages. Most of those activities continued into my 30s, but by then the first iPhone had been released. I began studying French and Japanese with a smartphone through social media, blogs, music videos, and podcasts. At 40, in 2017, I began learning Italian while living in Tokyo and most of my gains were made through my access to digital technology. For example, I used vocabulary, grammar and translation apps, accessed online reading materials, listened to music and podcasts, binge-watched Netflix series, and interacted with others through social media and language exchange apps. By 2020, I was making extensive use of in-built translation and transcription features on apps and websites to support my learning. When I began studying German and Spanish, I made full use of these features and the types of tools I'd used to learn Italian. And now, in 2023, machine translation (MT) is more accurate than ever and chatbot technology, such as ChatGPT, has evolved to the point that it can act as a virtual tutor. The affordances of AI for language learners are abundant, but to fully reap the potential rewards we must learn to use it efficiently and responsibly.

This chapter explores the impact of AI on language education, focusing to two commonly used tools: MT and chatbot technology, with ChatGPT exemplifying the latter. To streamline the flow of information, MT and ChatGPT are explored one by one. Each section begins by introducing the technologies, briefly explaining how they work, and summarising their evolution. Next, relevant research is introduced to highlight key issues, explore their potential affordances, and underscore the problems uncritical use can create. Issues related to educators' and students' readiness to adopt these tools and adequately deal with ethical issues are also raised, with suggestions for policy formation and training offered for educators and institutions. The chapter positions MT and chatbots like ChatGPT as beneficial real-world tools that have a place in language education and beyond, while also acknowledging problematic issues that cannot be ignored. These tools are here to stay, so language educators need to unite to ensure they are used in optimal ways.

13.2 Machine Translation

13.2.1 The Evolution of MT

MT has a long history, dating back to at least the 1950s. In 1954, IBM translated Russian into English with what they described as "an electronic 'brain'" (IBM, 1954, January 8), leading to one of the first widely-known uses of MT. In the 1900s, statistical machine translation (SMT) was the norm, with programmers meticulously coding each and every word with prescribed parameters. Different types of statistical coding were conducted for phrase-based translation, syntax-based translation, then later hierarchical phrase-based translation, which combined the previous two methods. An article showing the reasoning behind coding decisions for numerous individual words and phrases that were programmed into a basic German-English MT tool in the 1970s (Lehmann & Stachowitz, 1972) provides a peek into the complexity of this system. Unfortunately, despite valiant coding efforts, natural translation was still far from reach. There was was a shift away from MT, which was seen by some in that period as a "utopian dream" (Hutchins, 1978, p. 119), but by the end of the 1970s interest had been revived, with Hutchins suggesting the revival "may well have surprised many who have tended in recent years to dismiss it as one of the 'great failures' of scientific research" (1978, p. 119). Renewed interest and investment drove MT forward, but SMT never reached great heights even decades later, with Tan et al. (2020) lamenting: "Incapable of modeling long-distance dependencies between words, the translation quality of SMT is far from satisfactory" (p. 5).

However, tech companies did not give up and eventually more noteworthy progress was made. The Google Research Brain Team (Google DeepMind, n.d.) was established in 2011 and made breakthroughs in AI infrastructure, sequence-to-sequence learning, and "automated machine learning for production use". In 2016, the "utopian dream" of fluid, accurate MT was brought a step closer to becoming a reality when Google publicly released its new deep learning MT system—based on neural networks—and the next year DeepL followed suit. IBM (n.d.) explains neural networks as follows:

> Artificial neural networks (ANNs) are comprised of node layers, containing an input layer, one or more hidden layers, and an output layer. Each node, or artificial neuron, connects to another and has an associated weight and threshold. If the output of any individual node is above the specified threshold value, that node is activated, sending data to the next layer of the network. Otherwise, no data is passed along to the next layer of the network. Neural networks rely on training data to learn and improve their accuracy over time. However, once these learning algorithms are fine-tuned for accuracy, they are powerful tools in computer science and artificial intelligence, allowing us to classify and cluster data at a high velocity.

Figure 13.1 depicts a simple neural network comprised of three key layers. In MT, the input layer is the text (written or spoken) that the user enters into the software. The output layer is what the user receives in their target language. Inside the system, the input data passes through a hidden layer (or layers), which is the part of the neural network where deep learning occurs. In this hidden layer the MT system continually works to refine its accuracy, drawing on the mass of data and training it has received to choose the best output. This figure has one hidden layer but there can be more, as depicted in IBM's (n.d.) deep neural network diagram.

With massive amounts of data being entered daily by millions of people around the world, the accuracy of self-learning neural networks in MT tools like Google Translate and DeepL (two of the most commonly-used tools) is increasing at an unprecedented rate. This occurs because neural networks learn well with sufficient data and training. In the initial stages, these networks undergo "supervised learning", in which "the network is shown different examples over and over again. The

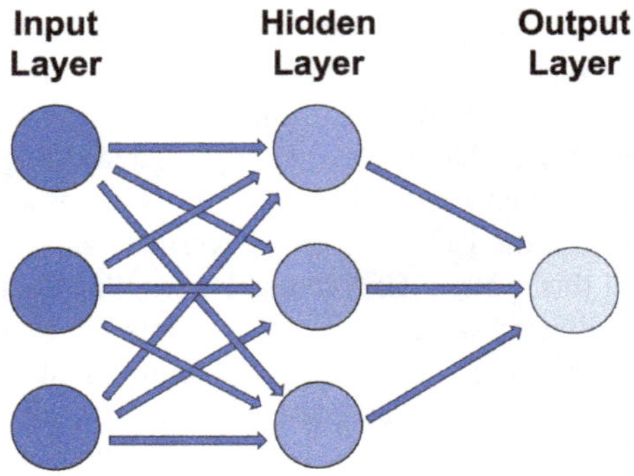

Fig. 13.1 A representation of a basic neural network (author's own work)

network repeatedly compares its own translations with the translations from the training data. If there are discrepancies, the weights of the network are adjusted accordingly" (DeepL, 2021, November 1). Through trial and error, the MT system starts to understand what is acceptable and what is not. DeepL credits the accuracy of its translations to four main features: its network architecture, its targeted training data, its unique training methodology (supervised learning, which is commonly used in machine learning, plus undisclosed methods that give it the competitive edge), and its network size (DeepL, 2021, November 1). We have not yet reached the point where what is input in one language is output as text that offers a target-language mirror of the input in terms of grammatical accuracy and complexity, lexical appropriacy (meaning, formality, pragmatic fit) and cultural context. Despite this, the shift to deep learning has increased accuracy and brought MT into sharper focus within language education.

13.2.2 MT Research in the Field of Language Education

Numerous studies have been conducted on the use of MT in language education, with researchers addressing this issue from a multitude of perspectives. Several systematic literature reviews have been published, including Lee's (2021) 87-article review of work published between 2000 and 2019 and Klimova et al.'s (2023) 13-article review of work published from 2018 to 2021. Furthermore, my own 14-article systematic literature review (Ohashi, 2024) examines open-access articles published between 2020 and 2022. Interested readers are encouraged to seek out these and other systematic literature reviews for a solid overview. The reviews highlight a wide range of affordances and drawbacks. For example, multiple studies have found that MT can help reduce errors and increase writing quality (Kol et al., 2018; Lee, 2020). In an experimental study by Tsai (2020), Chinese learners of English improved their English writing when using Google Translate for revision. Interestingly, students who were not majoring in English showed substantially more favourable attitudes towards it than those majoring in English. However, studies have also suggested that

MT use differs by proficiency level, with more proficient users evaluating output more critically and repairing translation failures more successfully (Chung, 2020), so students' level needs to be considered.

Research suggests that MT may also be useful for assisting with reading comprehension, even when translations are imperfect. A study conducted with English learners in Iran tested whether reading comprehension was affected differently when texts were translated by human translators and MT (Maghsoudi & Mirzaeian, 2020). The results showed no significant difference between reading comprehension test scores in the groups, with the researchers concluding that comprehensible texts could be produced by both humans and machines. Most of the students accepted mistakes related to omission, addition and punctuation, and half of them felt the syntactical errors were acceptable. The vast majority were generally satisfied with MT output, but dissatisfaction was expressed when MT produced errors related to register, idiomatic language and mistranslations. While improvements are constantly being made, concerns over accuracy have been reflected in many studies (Briggs, 2018; Ducar & Schocket, 2018; Kharis et al., 2021; Lee, 2020). Neural networks are driving MT forward, but certain areas, such as the translation of neologisms (newly coined terms), still pose challenges (Awadh & Khan, 2020).

Teachers' Views: Although evidence above suggests that MT can be a useful tool if used appropriately, full integration into language education—in a similar way to a dictionary, for example—is still far from the norm. To understand why this may be the case, it is important to turn our focus to educators. Teacher cognition affects educators' pedagogical decisions, so understanding their experiences and views on MT is essential. Borg has defined teacher cognition as "what teachers know, believe, and think", noting that "teachers are active, thinking decision-makers who make instructional choices by drawing on complex practically-oriented, personalised, and context-sensitive networks of knowledge, thoughts, and beliefs" (2003, p. 81). With this in mind, I conducted a study to understand how teachers use MT themselves, integrate it into their courses, evaluate it as a learning tool, and guide student learning (Ohashi, 2022). The findings from 153 university-level language teachers in Japanese universities showed wide-spread personal use of MT and relatively high levels of support for it as a learning tool, but much lower levels of course integration. Teachers' personal use of MT and course integration tended to follow a similar pattern when considered from the perspective of the four core skills, with reading and writing outranking listening and speaking. Perceptions of its usefulness for different skills followed the same pattern. The majority of teachers suspected that at least some of their students used MT to cheat, but they believed more of their students learned through using it, so very few felt it should be heavily restricted or banned. In terms of guidance, the vast majority agreed it is essential to discuss appropriate and inappropriate use of MT in all language courses and provide guidelines, but only a fifth of the participants actually did this in all courses. In part, this was because teachers themselves needed guidance, with only a third indicating they had enough knowledge to help students use MT effectively to develop their L2 skills, and most indicating they wanted to learn more about this. Uehara's (2023) qualitative study on MT use in English writing courses with four teachers at a Japanese university echoed this need for guidance, with all participants reporting a lack of institutional and departmental MT policies to follow. Thematic analysis of interview data revealed teachers felt it was important to consider proficiency level when using MT and ensure it was used to aid rather than hinder learning, but views on the degree to which MT could be used diverged, further strengthening the need for policies and training.

Students' Views: While teachers' views are important, exploring their perspectives only deals with half of the equation: students' views must also be investigated. A study with 666 teachers and 1,926 students at four Swiss universities addressed this gap, finding that students regularly used MT for academic work and both groups saw it as a relatively useful tool (Delorme Benites et al., 2021). However, under half in the student sample felt students can recognise the potential risks of MT use, suggesting a need for support. Educators were even more sceptical about students' ability to identify risks, but few took action to address this, with almost 80% ignoring MT completely in their courses. Alm and Watanabe (2022) also compared these dual perspectives through a study with 150 students and 12 teachers of five different languages at a university in New Zealand. They found that advanced level learners used MT most critically, made use of multiple tools, and manipulated input to get the best translations. They also found that as proficiency increased, learners were more likely to contextualise word choices through the use of MT and to check the meaning of words while reading.

A study by Ata and Debreli (2021) with 462 low to intermediate level English students at a Turkish university found that the vast majority of students used MT, most commonly for small segments of text (single-word or phrase translation). As the length of translated text increased (word, phrase, sentence, paragraph, entire text) students tended to find MT use less ethical and less reliable. In a Korean study that explored the effect of MT error correction training with low, mid and high proficiency middle school students, successful corrections in Korean-English translations occurred more frequently as proficiency increased (Yoon & Chon, 2022). However, this does not mean that MT is only useful for highly proficient users, as even at lower levels some corrections were possible through the application of multiple error correction strategies. Students at lower levels may see MT as less useful for language development though, due to their limited knowledge of the target language, and teachers may agree. Yoon and Chon sum up the pros and cons for writing assistance well, pointing out that MT has the potential to be a useful aid, but "its effectiveness is manifest only when the users notice what is linguistically inappropriate in the MT output (e.g., linguistic competence), and deploy appropriate correction or revising strategies (i.e., strategic competence)" (2022, p. 168). Some students may discover effective strategies by themselves, but training would help ensure that all learners have the support needed to better deal with MT, so given students' wide-spread use of MT, training should be considered essential within language education.

13.2.3 The Need for Training and Guidance

Research introduced so far suggests that MT is a useful starting point, but that training is desirable. In fact, numerous studies have shown the benefits of training. For example, Chang's (2022) Taiwan-based study documented an increase in critical evaluation of output after training, with students using multiple MT tools to cross-check output and drawing on online dictionaries and search engines (e.g., doing image searches, checking collocations) to make changes when translations were inappropriate. Furthermore, in a study at an Iranian university (Mirzaeian, 2021), 20 English learners were taught MT editing techniques that focused on helping them to improve their accuracy with determiners, collocations and paraphrasing. Post-test gains occurred for all three, with statistical significance reached for determiners.

The positive effects of MT use for overall writing development are evident when appropriate support is provided, with compositions written by students who used MT after training achieving higher grades than those who used it without training (O'Neill, 2019). The group who received MT training also outperformed those who used online dictionaries or had no access to writing assistant tools. The importance of learner training is further emphasised though a study by Olkhovska and Frolova (2020), who found that the translation quality of written work by a group of students who used MT with no post-editing training was actually lower than the group who did not use MT. This was thought to have occurred because students tended to blindly trust MT output when they did not receive training. In other words, there can be benefits to using MT, but not when use is uncritical. When teachers do not take the lead, students may train themselves. Kennedy's (2022) study with English learners in Japan reported that some students provided evidence of strategic MT use when completing written tasks without having received any MT instruction in their course. Their strategies included making L1/L2 draft comparisons to locate differences in content and errors, then seeking to repair them by referring to trusted sources such as dictionaries and textbooks. However, as Olkhovska and Frolova's (2020) study above shows, not all students will take these steps alone, so teacher-led training is recommended.

There is not a simple answer to what is acceptable and unacceptable MT usage and what type of training would most benefit learners, but based on the literature review and my own experiences as a language teacher and learner, the following points warrant consideration:

Considerations for MT Guidelines and Training

- Think about the extent to which it is appropriate to use MT (nothing, single words, phrases, sentence, paragraphs, whole texts). This may vary from skill to skill. For example, are different policies needed for reading, writing, listening and speaking? It may also differ from task to task. For instance, is it acceptable to use it extensively when writing an ungraded lecture summary that aims to checks comprehension, but to restrict it in whole or in part when writing a graded essay?

- Be explicit with students about what is and is not allowed on a whole course and/or task-to-task basis.
- Involve students in discussions about MT (experiences, beliefs, training needs, ethical use, privacy issues, etc.).
- Provide pre-editing training that will lead to more accurate output. For example, if the input language does not explicitly need a subject (like Japanese), advise students to add it so output matches the intended subject.
- Conduct post-editing training to improve the accuracy of translations. For example, teach students to compare the output of multiple MT tools and encourage them to cross-check translated terms with resources like Google Images, a dictionary, or a textbook when accuracy is in doubt.
- Guide students to input content (written or spoken) in their L2 and check output in their L1, scrutinising it for obvious errors or differences in meaning/nuance. Problematic translations in their L1 should be cross-referenced with the L2 input to determine if the cause was the input or a poor translation. This can raise awareness of problems with L2 input that learners can then take steps to address.
- Teach learners to back translate (translate output back to the language that was input) to check for and analyse differences.
- Advise students to use MT to check for useful vocabulary and expressions when preparing to participate in discussions or do oral presentations. As extra steps, learners could use text-to-voice and voice-to-text to learn and practice the pronunciation.
- Explore possibilities that camera translation (image-based translation) offers, such as allowing users to read non-alphabet texts like Chinese and Japanese (and encourage them to listen to the pronunciation if they are still unsure of how to read it). It could also be used to facilitate understanding of texts before a deeper reading in the L2, or to check understanding and resolve misunderstandings after reading in the L2.

13.2.4 Recommendations

MT can support language development, but misuse can be detrimental to learning and contravene ethical standards in formal education contexts. The literature suggests proficiency level has an impact on MT's usefulness to learners so this should be borne in mind. Studies underscore the need for students to receive appropriate training, but a universal set of guidelines did not emerge from the research reviewed. Rather, it is evident that what is suitable in one context will be inappropriate in another, so training and guidelines need to be decided after considering the position of all key stakeholders: institutions, teachers, and students. Institutions would be well advised to be pro-active in contacting teachers and students to gauge their level of understanding and perspectives, then responding by providing MT policies and training that take into account their needs.

13.3 ChatGPT

13.3.1 The Evolution of ChatGPT

ChatGPT (Chat Generative Pre-trained Transformer) has taken the world by storm with its advanced capability to understand and produce responses in human-like ways at a pace that humans cannot match. It is far from the first chatbot to be released on the internet, with models such as ELIZA dating back to the 1960s, but unlike its predecessors it has made remarkable leaps towards acceptance as a mainstream tool. In fact, Reuters estimated that in the first two months of its release, monthly active user rates hit approximately 100 million, "making it the fastest-growing consumer application in history" (Hu, 2023, February 3).

AI in language education, which falls under the term ICALL (Intelligent CALL), has a history of almost 50 years. The initial 30 years have been well documented by Heift and Schulze (2007) who identify an article written in 1978 as the first ICALL paper published. Weischedel et al.'s article, entitled "An artificial intelligence approach to language instruction" outlined a "prototype system for a sophisticated, intelligent tutor for instruction in a foreign language" (1978, p. 225). While significant gains were made in the decades ahead, it was not until ChatGPT's release in November 2022 that use of generative AI chatbots became embedded into public consciousness and started having a far-reaching impact on language education.

ChatGPT's major breakthrough can be attributed to it advanced natural language processing (NLP) capabilities. NLP encompasses multiple skills which have been defined by Schulze as follows:

> Natural language processing deals with both natural language understanding—written or spoken language input is turned into a formal representation which captures phonological/graphological, grammatical, semantic, and pragmatic features of the input—and natural language generation—the reverse process, from a formal representation to natural language output. (2008, p. 2)

ChatGPT is not the first NLP tool available to the general public, with MT already widely used when it was released, but its highly sophisticated way of "interacting" with users captured people's attention and imagination. It outperformed other chatbots because its NLP skills greatly exceeded those of its predecessors, which can be attributed to its exposure to a massive amount of training data. ChatGPT (GPT 3.5) was trained on approximately 600 billion tokens (450 billion words), largely scraped from the internet, and 175 billion parameters (Bell et al., 2023, March 24). In large language models, tokens are units of text and parameters are the settings that control text generation. OpenAI has stated that one token in ChatGPT is approximately 4 English characters, explaining: "Tokens can be thought of as pieces of words. Before the API processes the prompts, the input is broken down into tokens. These tokens are not cut up exactly where the words start or end—tokens can include trailing spaces and even sub-words." (OpenAI, Raf, n.d.)

As with MT, ChatGPT is based on neural network technology, with an input layer, hidden layer and output layer. The main difference is that the input layer in ChatGPT is a "prompt" by the user that instructs it to do something or initiates interaction. For example, if a user asks ChatGPT to write a story in two languages and make a list of 10 idioms or adjectives used in the text, ChatGPT receives the input, and generates output based on what it has previously "learnt". ChatGPT is powered by a large language model developed by OpenAI. Large language model are AI models that are trained to predict what should come next through extensive analysis of language patterns in a massive amount of data. ChatGPT is an autoregressive model, which means it uses the input text to predict the most appropriate word (which becomes phrases, sentences, paragraphs and longer text) that should follow. What has been "learnt" should not be confused with human learning; rather, in the hidden layer this type of AI is highly skilled at making a statistical analysis of the most appropriate words (one word at a time) to respond with. It does this at an incredible speed and draws on a massive database of texts about a vast range of topics. Its output is not always factually correct, but its responses generally reflect typical human use of grammar and vocabulary. It does not have a deep "understanding" of appropriate pragmatic norms so responses may be inappropriate at times, but it can produce relatively natural-sounding language. The majority of its training was conducted with English data so it is especially accurate when using the English language, but it can be used in many other languages too. As of September 13, 2023, OpenAI has listed 57 languages[1] that can be recognised through spoken input (speech-to-text), with a higher but unlisted number available for text input and output (OpenAI, Johanna, n.d.). These capabilities have thrust ChatGPT into the spotlight within foreign language education, sparking a new wave of experimentation and research.

13.3.2 ChatGPT Research in the Field of Language Education

Shortly after ChatGPT was released, there was a media storm that propelled it into the public consciousness like no AI tool before it. Headlines like "Alarmed by A.I.

[1] Afrikaans, Arabic, Armenian, Azerbaijani, Belarusian, Bosnian, Bulgarian, Catalan, Chinese, Croatian, Czech, Danish, Dutch, English, Estonian, Finnish, French, Galician, German, Greek, Hebrew, Hindi, Hungarian, Icelandic, Indonesian, Italian, Japanese, Kannada, Kazakh, Korean, Latvian, Lithuanian, Macedonian, Malay, Marathi, Maori, Nepali, Norwegian, Persian, Polish, Portuguese, Romanian, Russian, Serbian, Slovak, Slovenian, Spanish, Swahili, Swedish, Tagalog, Tamil, Thai, Turkish, Ukrainian, Urdu, Vietnamese, and Welsh.

Chatbots, Universities Start Revamping How They Teach" (Huang, 2023) drew attention to a wave of changes that was rolling in. While journalists were the first ones to widely report on ChatGPT, researchers in the field of language education soon took action. Exploratory work I conducted with Alm, presented at EUROCALL 2023, was one of the first wide-scale exploratory studies to gauge foreign language educators' response to ChatGPT, investigating it through a study with 367 teachers who taught 16 languages at universities in 48 countries/regions (Alm & Ohashi, 2024; Ohashi & Alm, 2023). The survey closed 10 weeks after ChatGPT's release in order to capture an initial response to what was predicted to be a tool that would have a major impact on the field of language education. The findings showed teachers had strong awareness of ChatGPT, but their experience and knowledge varied greatly. The majority of participants were interested in integrating it into their teaching practices, but most had not done so at that point in time. More than half of the participants indicted they were likely to use ChatGPT for creating language learning resources or recommend it to students for self-study, with fewer intending to use it to aid with various automated assessment and feedback tasks. Despite teachers' budding interest in ChatGPT, confidence in their ability to use it to create resources, guide self-study and automate assessment was relatively low ($M = 2.35$ to 2.87 on a 5-point Likert scale). When speculating on the impact ChatGPT would have on the field of education, a mixture of concern and optimism was expressed. Benefits such as an improvement in the accessibility of language education ($M = 3.60$), self-directed learning ($M = 3.53$) and personalised learning ($M = 3.48$) were identified, but these figures were overshadowed by concerns. The potential for an increase in cheating and academic dishonesty ($M = 3.98$) and fears that students would not develop their own language skills and critical thinking abilities due to over-reliance ($M = 3.70$) were of primary concern to teachers, echoing worries voiced by university teachers of other disciplines (Iqbal et al., 2022). The study highlighted the need for guidelines and training, both to increase teachers' ability to use a tool they saw as a potential teaching and learning aid, and to help them deal with predicted negative repercussions.

In 2023, ChatGPT became a trending research topic at language education conferences and published studies started to emerge. Initial studies relied on researchers' own evaluation of ChatGPT. For example, after simulating and evaluating a variety of tasks, Bonner et al. (2023) advocated its use by teachers and students for level-appropriate text summarization, grammar correction, holistic text correction, narrative writing prompt generation, creation of presentation notes, lesson plan generation, and the creation of level-specific texts and reading comprehension questions. However, they stressed that these tasks could not be simply delegated to ChatGPT and accepted without critically evaluating the output. In addition, a report by Barrot (2023) suggested ChatGPT could be used by students to elicit feedback on their written work. After inputting texts with various prompts, he concluded: "In addition to assessing stylistics and mechanics, the tool can evaluate the richness and relevance of content, clarity of audience, clarity of purpose, depth of analysis, focus, and organisation" (2023, p. 4). However, based on his exploration of ChatGPT's writing capabilities, he added that "capturing certain aspects of human writing quality, such as emotional depth, writing voice and identity, and rhetorical flexibility, remains a challenge" (2023, p. 2), and cautioned that this must be considered when employing ChatGPT for feedback purposes.

Learning motivation is another area that has begun to be explored. A study conducted with 80 English learners and teachers found that most participants believed ChatGPT could motivate learners to develop English reading, writing, grammar and vocabulary skills (Ali et al., 2023). Furthermore, the majority felt it could be used by students to manage their learning independently, gain self-confidence, and experience fun and enjoyment while learning. Due to this, while acknowledging that possible negative impacts require further investigation, the researchers concluded that "ChatGPT is recommended to be integrated into English language programs to promote learners' motivation to learn autonomously" (Ali et al., 2023, p. 47).

Studies based on actual practice within language courses are still limited due to ChatGPT's recent release. One study that addresses that gap was conducted with 20 English teachers at a university in Vietnam who had experience using ChatGPT in their writing, reading, grammar, research methodology and/or translation courses (Nguyen, 2023). Survey results showed that they used ChatGPT for most of the lessons in their courses, with 25% using it in every lesson. It was mainly used to help generate learning materials and develop lesson plans and most felt it strongly supported them in doing so. They also saw benefits for students, with strong agreement to the statements "ChatGPT proves useful by suggesting reading resources to students, which can inspire ideas for writing tasks" and "In my view, ChatGPT can serve as an effective tutor in writing classes" (Nguyen, 2023, p. 17). They were less favourable to the notion of it being used to save teachers time when grading and providing feedback on written work, aligning with findings from the worldwide study introduced above that I conducted with Alm (Alm & Ohashi, 2024; Ohashi & Alm, 2023). In follow-up interviews in the Vietnamese study, half of the teachers mentioned that it could help students to check their written work before submission, which was expressed through comments such as, "When finishing a writing assignment, students can get ChatGPT to proofread and give feedback on the language use, organization, and writing styles" (Nguyen, 2023, p. 25). Enthusiasm over benefits was tempered with concerns though, with close to half of the teachers worried about negative impacts on students' academic integrity and almost a third expressing doubts over students' ability to critically judge the output received.

13.3.3 Preparing Teachers for an Educational Shift

When language educators from around the world congregated to share CALL-related research at the University of Iceland for the EUROCALL 2023 conference, Schulze, a long-term specialist in AI research, pointed out that "a disruptive technology such as generative AI is also an opportunity to rethink our pedagogies" (Schulze, 2008). He further advocated the need for research, development and praxis to be "rooted in a good understanding of the technologies and their impact on language education"

(Schulze, 2023). How though, does the average teacher gain this knowledge and skill?

So far, language teaching organisations have played a considerable role in preparing teachers for the challenges and opportunities that ChatGPT poses. For example, in Japan, where I am based, the Mind, Brain and Education SIG (special interest group) of JALT (Japan Association for Language Teaching) co-published a special issue with JALT's CALL SIG on ChatGPT in March 2023 to help educators before the new academic year began in April. In an article titled "Rethinking Education in the Age of AI" (Ohashi, 2023), I prompted educators to consider several key issues that were likely to be of importance when dealing with ChatGPT, including its potential as "someone" to offer guidance at the appropriate level for learners (their "zone of proximal development" in Vygotskian (Vygotsky, 1978) terms), the importance of creativity and critical thinking, and the need to consider ethical issues such as in-built bias and privacy matters. In the same issue, Norton (2023) shared insights gained from asking ChatGPT itself about its potential within language education in his article "ChatGPT's Approach to Teaching: An Interview with the Chatbot". These articles and others in the issue provided teachers with ideas to work with.

Articles often take a great deal of time to write and publish, so they are not usually the first line of support. In fact, webinars and in-person presentations on ChatGPT started proliferating before academic publications began to appear and continue to be a popular mode of information dissemination. The first ChatGPT-related webinar I attended was "GPT and ELT: Productive, Disruptive, or Destructive" (Raine et al., 2023), in January, 2023. Around that time, ChatGPT information and training sessions started being held for language teachers around world, and I became part of this process myself after being invited to share my knowledge, as limited as it was at the time. In the first half of 2023, I conducted sessions for several Japan-based teaching groups within JALT, the Taiwan-based association Pedagogy and Practice in Technology Enhanced Language Learning (PPTELL), the European tech-focused teaching association EUROCALL, the US-based International Language Testing Association (ILTA), and the Moroccan-based teaching organisation Everyone Academy. Efforts by educators were not, however, restricted to formally-organised sessions like these. Teachers banded together on social media, with many of them posting information, questions and responses in groups dedicated to this. For instance, the Facebook group *Online Teaching Japan* quickly became a hotspot for information exchange for Japan-based language teachers.

The examples above show how teachers and teaching organisations responded to the need for teacher support, but where *should* this training come from and who should bear the costs, both financially and in terms of time commitment? From my point of view, foundational training should be provided by institutions, either as in-house training by knowledgeable staff or by inviting experts to conduct it, and ideally it would be counted as work time. If left as an opt-in avenue with only interested teachers seek training, this will surely have an impact upon students who are taught by teachers who have a high level of knowledge and skills and those who do not. If comprehensive training is not provided by educational institutions, it is difficult to imagine how institutional AI policies could be effectively put into place in more than just a token way. After all, even if the "solution" is to ban generative AI, students will surely be exposed to it outside of their educational context and teachers who know little about it will have even more difficulty noticing its use than those who have been trained. Thinking more positively, trained teachers and students can take advantage of potential affordances of this new technology through using it for a wide range of tasks, some of which will be introduced below.

13.3.4 Tasks Teachers and Learners Can Try with ChatGPT

As noted above, journal articles that were published soon after ChatGPT's release generally listed ideas rather than empirically tested methods, providing practical options for teachers to consider and explore within their own contexts. Furthermore, practitioners quickly responded by providing a wealth of ideas and usage reports through blogs, videos and social media posts. Examples on Facebook where such information can be found include very active groups such as *ChatGPT for Teachers* (over 350,000 members as of September 14, 2023) and *Higher Ed Discussion of AI Writing* (4,900 members as of September 14, 2023). Interested readers are encouraged to seek out such information online to enhance their knowledge. Personally, I have explored various uses of ChatGPT by experimenting with it as a language learner and teacher. I will share some of the tasks I have tried here as they may prove useful to others.

Practical Tasks for Teachers and Learners

- Creating reading passages and summaries on topics of interest/course content
- Generating more/less complex versions of texts
- Asking for vocabulary lists to be generated (with definitions, translations, transliterations, synonyms, etc.) from reading passages or about specific topics
- Generating explanations and examples of grammar points
- Chatting (written and spoken conversations) about topics of interest with ChatGPT taking the role of teacher, interviewer, colleague, or another fictional character of choice
- Exploring pragmatics through discussions about tone and suitability of language use
- Co-creating fictional stories of various lengths

As a learner, I have particularly enjoyed co-creating stories in my foreign languages (Japanese, Italian, French, Spanish and German) by prompting ChatGPT with the setting, characters, and other elements. My proficiency in these languages varies considerably, but ChatGPT can produce interesting and appropriate reading materials to suit my level for all of these languages. It struggles most with Japanese, often not knowing which kanji are too difficult for me, but further prompting resolves this. I find it helpful to initially create a story in my first language (English) or one of my more proficient languages, then ask ChatGPT to translate it to other languages, with prompts to simplify it when needed. If the right level has not been reached after asking for a simplified version, a basic prompt like "even more simple" results in a new iteration that is more accessible, and this can be repeated until a comfortable level has been reached. To add interest, I sometimes ask ChatGPT to give me three options to choose from for the next part of the story then select one of them. This additional step has proved popular with students I've co-constructed texts with on ChatGPT. Extensive reading programs emphasise the need for a mass of compelling reading materials that are at the right level for learners (slightly above their current proficiency level) and ChatGPT makes this possible through the generation of tailor-made stories. For extension work, I have prompted ChatGPT for vocabulary lists and summaries (or something more creative, like a Netflix series teaser) and exported the stories to a voice-to-text reader to listen to them. This extends the benefits beyond reading, offering language learners many opportunities for development. For example, once it has become an audio file, learners can simply listen or they can shadow what they hear to work on their pronunciation. Alternatively, they can read the story aloud and capture it through voice-to-text (a function that is generally built into smartphones) then compare the written story with the text captured by their speech. Doing such tasks has helped me identify some areas of weakness in my pronunciation that I have then drilled to improve. For motivated learners like me, ChatGPT supports foreign language development. However, it must be acknowledged that it can be used in disruptive ways within formal education, as explored below.

13.3.5 Graded Work and AIgiarism

According to Sam Altman, CEO of OpenAI, the organisation that created ChatGPT, "We have a new tool in education. Sort of like a calculator for words. […] And the way we teach people is going to have to change and the way we evaluate students is going to have to change." ("Homework", 2023, June 23). One reason for this is the threat of AIgiarism (AI plagiarism), in which students use ChatGPT to complete assignments in lieu of doing the task themselves and do not acknowledge its use. However, despite plagiarism being considered "a heinous crime within the academic community" (Pecorari, 2003, p. 317), what constitutes plagiarism has not always been clear, with studies showing that learners and educators may have different perceptions and students submitting work that they did not intentionally plagiarise (Pecorari, 2003). Defining plagiarism now that ChatGPT and other forms of generative AI are widely available is even more challenging, so discussions between teachers and students are more important than ever.

Teachers may feel they are doing the right thing by issuing strict-sounding policies with harsh penalties in the hope of preventing cheating, and they may feel equally justified calling into question any work that they find suspicious. However, it should be noted that their actions may create a learning environment that positions teachers and students against each other, diminishing student–teacher trust even among honest students, and deflecting attention from learning. A study with 49 students who had sought advice on Reddit after being accused of or suspected of cheating with ChatGPT shone a spotlight on the cheating issue from the perspective of students (Gorichanaz, 2023). Most of the students in the study claimed to have been wrongly accused and were concerned over how to prove it. Legal terms were common in their threads, with students looking for ways to provide "evidence" they had submitted their own work, as well as debate over whether the "burden of proof" lies with students to prove their "innocence" or teachers to prove their "guilt". Many students felt that universities took a "guilty until proven innocent" stance and there were suggestions to document work in the future by enabling version history on software used for assignments (Google Docs and Microsoft Word) or to make screen recordings when doing homework. These strategies were flagged by some as privacy invasions, but students felt they had few other options if they wanted to protect themselves. Analysis of emotional valance showed neutrality for around half of the posts, but strong negative emotions such as hostility, anxiety, fear and defeat for the other half.

Negative feelings among students are warranted, with news reports highlighting serious repercussions when teachers believe they are justified in accusing students of cheating. In a case that went viral, students from a university in the United States failed assignments and had their diplomas withheld after a teacher copied their written work into ChatGPT to check for plagiarism (D'Agostino, 2023, May 19). ChatGPT is not able to detect AI generated text, so the teacher had misused this technology and accused innocent students. The university later provided an update, noting, "Currently, several students have been exonerated and their grades have been issued, while one student has come forward admitting his use of ChatGPT in the course. Several other students have opted to complete a new writing assignment made available to them" (Texas A&M University-Commerce, 2023, May 17). While one student was found to have cheated, this experience was surely traumatic for the innocent students involved.

Concerns over plagiarism/AIgiarism are valid, but they should not lead to knee-jerk reactions to completely ban generative AI. Instead, more thought needs to go into how and when it can be appropriately employed and when its use should be restricted. Educators have already begun exploring ways that teachers can effectively embed ChatGPT into the writing process. For example, in a study by Yan (2023), students received instruction about ChatGPT's basic functions, then practiced using it to write various texts. Next, they participated in peer-learning sessions, sharing tips and techniques for using ChatGPT for L2 writing and discussed how to improve their texts. After writing more texts on various topics, they received peer feedback then learnt how to incorporate AI-based tools to automate writing and improve its quality. While students acknowledged the power of ChatGPT as a writing assistant, they showed concern for its potential to create inequality:

> To many student participants in the practicum, the power of ChatGPT to generate a piece of writing "in the blink of an eye" violated the basic principle of educational equity. Students affirmed that the knowledge and experiences of using ChatGPT brought students "enormous advantages" to outperform their peers. (Yan, 2023)

In addition, students were concerned that the process of "reading-writing-revision" would be replaced with "text-generation and post-editing" which would require a lower degree of language and writing skills. In other words, there are students who share teachers' concerns about the detrimental effects ChatGPT can have when it is used inappropriately. Further teacher-student exploration is likely to strengthen trust and help shape usage guidelines that students do not want to violate, so it is important to open the doors of communication between teachers and students.

Clearly not all students want to cheat, but some will be tempted if the option is there. If teachers really want to minimise cheating, we need to heed the words of ChatGPT's CEO that were introduced at the start of this section, as rethinking how we teach and evaluate is the way forward. ChatGPT and other forms of generative AI have made it necessary for educators to adopt pedagogies that harmonise more with the new societal and educational landscape. Just like the COVID-19 pandemic challenged educators to provide alternatives to high-stakes exams, generative AI is pushing us to question the place the standard take-home essay has in courses and how such tasks should be evaluated. If writing tasks in our courses can easily be done by AI—and AI is already being used in government and business for writing tasks—at what point will we shift our focus to tasks that require more evidence of human thought? Will AI be the starting point to generate ideas in the future? Will it become a leading source of feedback on drafts? Will it be used to paraphrase and put ideas into more or less complex expressions, depending on the writer's target audience? Is there as much value in teaching students how to do these things and post-edit as there is in teaching them how to write without AI? The jury is out for now, but context-based decisions need to be made on how to deal with these issues, so further research that draws together the views of institutions, teachers, and students is needed. We also need to look beyond this to industry; after all, students need to learn real-world skills to secure their position in the job market.

A committee at Cornell University has put together a series of recommendations for their faculty that would be valuable in other contexts. They have recommended that teachers "rethink learning outcomes by integrating GAI [generative AI] into their goals for students; address safety and ethics by educating students about the pitfalls of GAI; and explicitly state their policies to students for the use of GAI in their classes" (Kelley, 2023, September 11). They also recommended the following three policies: "prohibit the use of GAI where it interferes with students developing foundational understanding, skills or knowledge; allow with attribution where GAI could be a useful resource, and require students to take responsibility for accuracy and attribution of GAI content; and encourage and actively integrate GAI into the learning process" (Kelley, 2023, September 11). Educators and institutions in other contexts would be well-advised to consider how these guidelines fit within their programs and to also take into account safety and privacy concerns, which are explored below.

13.3.6 Safety and Privacy Concerns

Reactions to ChatGPT and generative AI have varied greatly in different places and periods in time. For example, in Australia, there have been staunch advocates and opponents within education ministries in different states and territories. Most regions have banned ChatGPT from public education at some point, but this is expected to be overturned in the next academic year, as the Federal Education Minister Jason Clare said regional ministers have reached consensus on a draft framework for educators on how to manage this technology within schools (Belot, 2023, July 9). In one of the regions, South Australia, the Education Minister Blair Boyer has argued that schools need to include instruction on how to use ChatGPT safely to better equip students for their future careers once they graduate (Henebery, 2023, July 13). South Australia is the only state in Australia that has not previously banned generative AI, and has instead focused on finding solutions such as "an eight-week trial in high schools to pilot an Australian-first 'education-safe' version of ChatGPT that has been created in conjunction with Microsoft" (Henebery, 2023, July 13).

Privacy and safety concerns must be taken into account when considering the impact of any new technology and ChatGPT has received a lot of attention over its data collection and usage policies. In April 2023, the Italian data-protection authority instigated a blanket ban throughout Italy, claiming that ChatGPT couldn't legally justify "the mass collection and storage of personal data for the purpose of 'training' the algorithms underlying the operation of the platform" and arguing that the lack of age verification of users "exposes minors to absolutely unsuitable answers compared to their degree of development and awareness" (McCallum, 2023, April 1). Several weeks later the ban was lifted after OpenAI agreed to implement age verification to all new ChatGPT users in Italy and provide the option for European Union users to object to their personal data being used for training (McCallum, 2023, April 28). This is evidence of how external regulation can result in practical changes that reduce intrusions on privacy and lead to more age-appropriate use, but it is important to note that these changes were not enforced globally. People in the European Union are protected by the General Data Protection Regulation, a legally-binding document that was created in 2016 (Publications Office of the European Union, 2016) and came into force in 2018, but such stringent laws are not enforced worldwide.

External pressure has led to other noteworthy changes. For instance, currently (September 2023) when users access ChatGPT's settings, there is a section on chat history and training that says, "Save new chats on this browser to your history and allow them to be used to improve our models. Unsaved chats will be deleted from our systems within 30 days." This option allows users to easily control whether or not their data will be used for training. It was rolled out by OpenAI in April 2023, almost a year and a half after ChatGPT was released. Their statement, "We hope this provides an easier way to manage your data" (OpenAI, 2023, April 25) could be interpreted as tacit acknowledgment of complaints over previous policies which did not make opt-out options readily visible and required users to submit a form to request it. Although this is a step in the right direction, the new policy highlights the transactional value of data to freeware (free software) companies like OpenAI. After all, anyone who does not opt in loses access to their chats after 30 days, whereas those who share their data maintain access to them. It should be noted that even with opt-out options in place, OpenAI has access to the data for viewing purposes for 30 days and maintains "we are not able to delete specific prompts from your history. Please don't share any sensitive information in your conversations" (OpenAI, Natalie, n.d.).

Notwithstanding these safety and privacy issues, ChatGPT has already begun to be integrated into the workforce, even in traditionally cautious domains. For example, on August 23, 2023, the use of generative AI, including ChatGPT, was approved for use by approximately 50,000 employees in Tokyo's Metropolitan Government (NHK World News, 2023, August 23). Its use is authorised for use for tasks such as idea generation and summary creation, with the stipulation that staff do not input personal information or confidential data and check output to ensure copyright violations do not occur. With key organisations in society integrating generative AI into their workflow, students will be left behind if they do not do the same. Therefore, it is advisable to consider safe, responsible ways for ChatGPT and other forms of AI to be used in educational settings.

13.3.7 Recommendations

As with MT management, institutions and teachers should be leading the way for fair and useful integration of ChatGPT (and other generative AI) into language education. Policies and training should be context-based and draw on the needs of the key stakeholders (institutions, teachers, and students). Recommendations such as those introduced by Cornell University serve as a useful starting point, but specific details need to be negotiated at the institutional level to find options that are contextually appropriate. Privacy and safety issues also need to be taken seriously, with training not only covering logistical aspects of use, but also focusing on online security and data protection.

13.4 Conclusion

As this chapter has shown, generative AI has made dramatic advances in recent years and it is now starting to have a far-reaching role on language education and in society at large. Adapting to change does not happen easily, and there will be much disruption before these technologies are successfully integrated. Teachers and students have new challenges ahead that will require novel approaches and steep learning curves.

While MT, ChatGPT and other forms of generative AI present valuable opportunities within language education, adjusting to them will take time and bring forth negative emotions such as frustration and fear along the way. As educators, we need to band together and help one another in whatever ways we can, and do our best to support our students through this process. As researchers, we need to empirically explore pathways towards pedagogies that deter misuse of AI and help users make the most of its affordances. MT, ChatGPT, and other forms of generative AI are real-world tools that we cannot ignore. The path towards effective adoption is still being paved and as language education specialists, we need to unite to pave it well.

References

Ali, J., Shamsan, M. A., Hezam, T., & Mohammed, A. A. Q. (2023). Impact of ChatGPT on learning motivation: Teachers' and students' voices. *Journal of English Studies in Arabia Felix, 2*(1), 41–49. https://doi.org/10.56540/jesaf.v2i1.51

Alm, A., & Ohashi, L. (2024). A worldwide study on language educators' initial response to ChatGPT. *Technology in Language Teaching & Learning, 6*(1), 1–23. https://doi.org/10.29140/tltl.v6n1.1141

Alm, A., & Watanabe, Y. (2022). Online machine translation for L2 writing across languages and proficiency levels. *Australian Journal of Applied Linguistics, 5*(3), 135–157. https://doi.org/10.29140/ajal.v5n3.53si3

Ata, M., Debreli, M. (2021). Machine translation in the language classroom: Turkish EFL learners' and instructors' perceptions and use. *IAFOR Journal of Education*, 9(4), 103–122. https://files.eric.ed.gov/fulltext/EJ1318690.pdf

Awadh, A. N., & Khan, A. S. (2020). Challenges of translating neologisms comparative study: Human and machine translation. *Journal of Language and Linguistic Studies, 16*(4), 1987–2002. https://doi.org/10.17263/jlls.851030

Barrot, J. S. (2023). Using ChatGPT for second language writing: Pitfalls and potentials. *Assessing Writing, 57*, 1–6. https://doi.org/10.1016/j.asw.2023.100745

Bell, G., Burgess, J., Thomas, J., & Sadiq, S. (2023, March 24). Rapid response information report: Generative AI—language models (LLMs) and multimodal foundation models (MFMs). *Australian Council of Learned Academies.* https://www.chiefscientist.gov.au/sites/default/files/2023-06/Rapid%20Response%20Information%20Report%20-%20Generative%20AI%20v1_1.pdf

Belot, H. (2023, July 9). ChatGPT ban in Australia's public schools likely to be overturned. *The Guardian.* https://www.theguardian.com/technology/2023/jul/09/chatgpt-ban-in-australias-public-schools-likely-to-be-overturned.

Bonner, E., Lege, R., & Frazier, E. (2023). Large language model-based artificial intelligence in the language classroom: Practical ideas for teaching. *Teaching English with Technology, 23*(1), 23–41. https://doi.org/10.56297/BKAM1691/WIEO1749

Borg, S. (2003). Teacher cognition in language teaching: A review of research on what language teachers think, know, believe, and do. *Language Teaching, 36*(2), 81–109. https://doi.org/10.1017/S0261444803001903

Briggs, N. (2018). Neural machine translation tools in the language learning classroom: Students' use, perceptions, and analyses. *The JALT CALL Journal, 14*(1), 2–24. https://doi.org/10.29140/jaltcall.v14n1.221

Chang, L.-C. (2022). Chinese language learners evaluating machine translation accuracy. *The JALT CALL Journal, 18*(1), 110–136. https://doi.org/10.29140/jaltcall.v18n1.592

Chung, E. S. (2020). The effect of L2 proficiency on post-editing machine translated texts. *Journal of Asia TEFL, 17*(1), 182–193. https://doi.org/10.18823/asiatefl.2020.17.1.11.182

D'Agostino, S. (2023, May 19). Professor to students: ChatGPT told me to fail you. *Inside Higher Ed*. https://www.insidehighered.com/news/quick-takes/2023/05/19/professor-students-chatgpt-told-me-fail-you

DeepL. (2021, November 1). How does DeepL work? https://www.deepl.com/en/blog/how-does-deepl-work

Delorme Benites, A., Cotelli Kureth, S., Lehr, C., & Steele, E. (2021). Machine translation literacy: A panorama of practices at Swiss universities and implications for language teaching. In N. Zoghlami, C. Brudermann, C. Sarré, M. Grosbois, M., L. Bradley, & S. Thouësny (Eds.) *CALL and professionalisation: Short papers from EUROCALL 2021* (pp. 80–87). Research-publishing.net. https://doi.org/10.14705/rpnet.2021.54.1313

Ducar, C., & Schocket, D. H. (2018). Machine translation and the L2 classroom: Pedagogical solutions for making peace with Google translate. *Foreign Language Annals, 51*(4), 779–795. https://doi.org/10.1111/flan.12366

Google DeepMind. (n.d.). Build AI responsibly to benefit humanity. *Google*. https://deepmind.google/about/.

Gorichanaz, T. (2023). Accused: How students respond to allegations of using ChatGPT on assessments. *Learning: Research and Practice, 9*(2), 183–196. https://doi.org/10.1080/23735082.2023.2254787

Heift, T., & Schulze, M. (2007). *Errors and intelligence in computer assisted language learning: Parsers and pedagogues*. Routledge.

Henebery, B. (2023, July 13). Why the ChatGPT ban in public schools is being reversed. *The Educator Australia*. https://www.theeducatoronline.com/k12/news/why-the-chatgpt-ban-in-public-schools-is-being-reversed/282834

Homework will 'never be the same,' says ChatGPT founder. (2023, June 13). *The Japan Times*. https://www.japantimes.co.jp/news/2023/06/13/business/corporate-business/chatgpt-homework-revolution/

Hu, K. (2023, February 3). ChatGPT sets record for fastest-growing user base—analyst note. *Reuters*. https://www.reuters.com/technology/chatgpt-sets-record-fastest-growing-user-base-analyst-note-2023-02-01/

Huang, K. (2023). Alarmed by A.I. chatbots, universities start revamping how they teach. *The New York Times*. https://www.nytimes.com/2023/01/16/technology/chatgpt-artificial-intelligence-universities.html

Hutchins, W. J. (1978). Machine translation and machine-aided translation. *Journal of Documentation, 34*(2), 119–159. https://doi.org/10.1108/eb026657

IBM. (1954, January 8). 701 Translator. https://mt-archive.net/IBM-1954.pdf

IBM. (n.d.). What are neural networks? Retrieved September 15, 2023 from https://www.ibm.com/topics/neural-networks

Iqbal, N., Ahmed, H., & Azhar, K. A. (2022). Exploring teachers' attitudes towards using ChatGPT. *Global Journal for Management and Administrative Sciences, 3*(4), 97–111. https://doi.org/10.46568/gjmas.v3i4.163

Kelley, S. (2023, September 11). Faculty offered guidance for teaching in the age of ChatGPT. *Cornell Chronicle*. https://news.cornell.edu/stories/2023/09/faculty-offered-guidance-teaching-age-chatgpt?fbclid=IwAR3awklx1DXtAh6__umwioNVzfmGcKiH0I5lNc5pgI8uY4Ltx5QrnC1gxA

Kennedy, O. (2022). The negative impacts of student use of online tools during emergency remote teaching and learning on teacher–student relationships. In T. D. Cooper & J. York (Eds), *Remote teaching & beyond* (pp. 40-53). JALTCALL. https://doi.org/10.37546/JALTSIG.CALL.PCP2021-04

Kharis, M., Kisyani, K., Suhartono, S., & Yuniseffendri, Y. (2021). Takarir: A new simultaneous translator voice to text to promote bi/multilinguality. *Journal of Language and Linguistic Studies, 17*(3), 1175–1183. https://www.jlls.org/index.php/jlls/article/view/2629/869

Klimova, B., Pikhart, M., & Benites, A. D. (2023). Neural machine translation in foreign language teaching and learning: A systematic review. *Education and Information Technologies, 28*, 663–682. https://doi.org/10.1007/s10639-022-11194-2

Kol, S., Schcolnik, M., & Spector-Cohen, E. (2018). Google Translate in academic writing courses. *The EUROCALL Review, 26*(2), 50–57. https://doi.org/10.4995/eurocall.2018.10140

Lee, S.-M. (2021). The effectiveness of machine translation in foreign language education: A systematic review and meta-analysis. *Computer Assisted Language Learning, 36(1-2), 103–125.* https://doi.org/10.1080/09588221.2021.1901745

Lee, S.-M. (2020). The impact of using machine translation on EFL students' writing. *Computer Assisted Language Learning, 33*(3), 157–175. https://doi.org/10.1080/09588221.2018.1553186

Lehmann, W. P., & Stachowitz, R. A. (1972). Development of German-English machine translation system. *University of Texas technical report.* https://eric.ed.gov/?id=ED065008

Maghsoudi, M., & Mirzaeian, V. (2020). Machine versus human translation outputs: Which one results in better reading comprehension among EFL learners? *The JALT CALL Journal, 16*(2), 69–84. https://doi.org/10.29140/jaltcall.v16n2.342

McCallum, S. (2023, April 1). ChatGPT banned in Italy over privacy concerns. *BBC News.* https://www.bbc.com/news/technology-65139406

McCallum, S. (2023, April 28). ChatGPT accessible again in Italy. *BBC News.* https://www.bbc.com/news/technology-65431914

Mirzaeian, V. (2021). The effect of editing techniques on machine translation-informed academic foreign language writing. *The EuroCALL Review, 29*(2), 33–43. https://doi.org/10.4995/eurocall.2021.13120

Nguyen, T. T. H. (2023). EFL teachers' perspectives toward the use of ChatGPT in writing classes: A case study at Van Lang University. *International Journal of Language Instruction, 2*(3), 1–47. https://doi.org/10.54855/ijli.23231

NHK World News. (2023, August 23). Tokyo government now using generative AI at all bureaus. NHK. https://www3.nhk.or.jp/nhkworld/en/news/20230823_33/

Norton, P. (2023). ChatGPT's approach to teaching: An interview with the chatbot. *Mind Brain Ed Think Tank, 9*(3), 46–56. https://www.mindbrained.org/march-2023-chat-gpt/

O'Neill, E. (2019). Training students to use online translators and dictionaries: The impact on second language writing scores. *International Journal of Research Studies in Language Learning, 8*(2), 47–65. https://doi.org/10.5861/ijrsll.2019.4002

Ohashi, L. (2022). The use of machine translation in L2 education: Japanese university teachers' views and practices. In B. Arnbjörnsdóttir, B., B. Bédi, L. Bradley, L., K. Friðriksdóttir, H. Garðarsdóttir, S. Thouësny, & M. J. Whelpton (Eds.) *Intelligent CALL, granular systems, and learner data: Short papers from EUROCALL 2022* (pp. 308–314). Research-publishing.net. https://doi.org/10.14705/rpnet.2022.61.1476

Ohashi, L., & Alm, A. (2023). ChatGPT and language learning: University educators' initial response. In B. Bédi, Y. Choubsaz, K. Friðriksdóttir, A. Gimeno-Sanz, S. Björg Vilhjálmsdóttir, & S. Zahova (Eds.), *CALL for all Languages – EUROCALL 2023 Short Papers* (pp. 31–36). University of Iceland, Reykjavik. https://doi.org/10.4995/EuroCALL2023.2023.16917https://doi.org/10.4995/EuroCALL2023.2023.16917

Ohashi, L. (2023). *Rethinking education in the age of AI. Mind Brain Ed Think Tank, 9*(3), 9–14. https://www.mindbrained.org/march-2023-chat-gpt/

Ohashi, L. (2024). Machine translation in language education: A systematic review of open access articles. *Kenkyu Nenpou: The Annual Collection of Essays and Studies, 70*, 105–125. https://www.gakushuin.ac.jp/univ/let/top/publication/KE_70/KE_70_008.pdf

Olkhovska, A., & Frolova, I. (2020). Using machine translation engines in the classroom: A survey of translation students' performance. *Advanced Education, 15*, 47–55. https://eric.ed.gov/?id=EJ1287521

OpenAI, Johanna. (n.d.). Whisper API FAQ. *OpenAI.* Retrieved September 15, 2023 from https://help.openai.com/en/articles/7031512-whisper-api-faq

OpenAI, Raf. (n.d.) What are tokens and how to count them? *OpenAI.* Retrieved September 15, 2023 from https://help.openai.com/en/articles/4936856-what-are-tokens-and-how-to-count-them

OpenAI, Natalie. (n.d.) What is ChatGPT? Commonly asked questions about ChatGPT. *OpenAI.* Retrieved September 15, 2023 from https://help.openai.com/en/articles/6783457-what-is-chatgpt

OpenAI. (2023, April 25). New ways to manage your data in ChatGPT. *OpenAI.* https://openai.com/blog/new-ways-to-manage-your-data-in-chatgpt

Pecorari, D. (2003). Good and original: Plagiarism and patchwriting in academic second-language writing. *Journal of Second Language Writing, 12*(4), 317–345. https://doi.org/10.1016/j.jslw.2003.08.004

Publications Office of the European Union. (2016). Legislative acts: Regulations. *Official Journal of the European Union, 119*, 1–88. https://eur-lex.europa.eu/legal-content/EN/TXT/PDF/?uri=CELEX:32016R0679

Raine, P., Kerr, P., Ryan, K., & Tomei, J. (2023, January 26). *GPT and ELT: Productive, disruptive, or destructive?* [Webinar]. Technology in Language Teaching and Learning. https://www.eventbrite.com/e/gpt-and-elt-productive-disruptive-or-destructive-tickets-515526602947

Schulze, M. (2008). AI in CALL—Artificially inflated or almost imminent? *CALICO Journal, 25*(3), 510–527. http://www.jstor.org/stable/calicojournal.25.3.510

Schulze, M. (2023, August 17). *What's AI got to do with it? Pedagogic perspectives* [Paper presentation]. EUROCALL2023, University of Iceland, Reykjavik, Iceland.

Tan, Z., Wang, S., Yang, Z., Chen, G., Huang, X., Sun, M., & Liu, Y. (2020). Neural machine translation: A review of methods, resources, and tools. *AI Open, 1*, 5–21. https://doi.org/10.1016/j.aiopen.2020.11.001

Texas A&M University-Commerce (2023, May 17). Texas A&M University-Commerce addresses concerns about ChatGPT in ag classroom. https://www.tamuc.edu/news/texas-am-university-commerce-addresses-concerns-about-chatgpt-in-ag-classroom/

Tsai, S. C. (2020). Chinese students' perceptions of using Google Translate as a translingual CALL tool in EFL writing. *Computer Assisted Language Learning, 3*(5-6), 1250–1272. https://doi.org/10.1080/09588221.2020.1799412

Uehara, S. (2023). Teacher perspectives of machine translation in the EFL writing classroom. In P. Ferguson, B. Lacy, & R. Derrah (Eds.) *Learning from students, educating teachers: Research and practice, pp. 270–279.* JALT. https://doi.org/10.37546/JALTPCP2022-31

Vygotsky, L. S. (1978). *Mind in society: The development of higher psychological processes.* Harvard University Press.

Weischedel, R. M., Voge, W. M., & James, M. (1978). An artificial intelligence approach to language instruction. *Artificial Intelligence, 10*, 225–240.

Yan, D. (2023). Impact of ChatGPT on learners in a L2 writing practicum: An exploratory investigation. *Education and Information Technologies.* https://doi.org/10.1007/s10639-023-11742-4

Yoon, C. W., & Chon, Y. V. (2022). Machine translation errors and L2 learners' correction strategies by error type and English proficiency. *English Teaching, 77*(3), 153–175. https://doi.org/10.15858/engtea.77.3.202209.153

Index

P. Ilic et al. (eds.), *Artificial Intelligence in Education: The Intersection of Technology and Pedagogy*, Intelligent Systems Reference Library 261,
https://doi.org/10.1007/978-3-031-71232-6

www.ingramcontent.com/pod-product-compliance
Lightning Source LLC
Chambersburg PA
CBHW050812231224
19414CB00011B/11

9 783031 712319